MATH/STAT
LIBRARY

Techniques of Semigroup Theory

Techniques of Semigroup Theory

PETER M. HIGGINS

Department of Mathematics,
University of Essex

Oxford   New York   Tokyo
OXFORD UNIVERSITY PRESS
1992

Oxford University Press, Walton Street, Oxford OX2 6DP

Oxford New York Toronto
Delhi Bombay Calcutta Madras Karachi
Petaling Jaya Singapore Hong Kong Tokyo
Nairobi Dar es Salaam Cape Town
Melbourne Auckland
and associated companies in
Berlin Ibadan

Oxford is a trade mark of Oxford University Press

Published in the United States
by Oxford University Press, New York

© *Peter M. Higgins, 1992*

All rights reserved. No part of this publication may be reproduced, stored in a retrieval system, or transmitted, in any form or by any means, electronic, mechanical, photocopying, recording, or otherwise, without the prior permission of Oxford University Press

A catalogue record for this book is available from the British Library

Library of Congress Cataloging in Publication Data
Higgins, Peter M., 1956–
Techniques of semigroup theory / Peter M. Higgins.
p. cm.
1. Semigroups. I. Title.
QA171.H56 1991 512'.2—dc20 91-25928
ISBN 0-19-853577-5

Typeset by Macmillan India Ltd, Bangalore 25
Printed and bound in Great Britain by
Biddles Ltd, Guildford and King's Lynn

This book is dedicated to my wife Theresa, and to our children, Alexander, Genevieve, Vanessa, and Caroline.

Foreword

G. B. Preston

It is some years since we saw a new book giving a broad introduction to semigroup theory, and the appearance of Peter Higgins' book with its vital and fresh approach is a welcome event. The book starts from scratch, assuming no knowledge of semigroup theory, and so the first part, say about a quarter of the book, covers well-worn ground. However, this familiar material is viewed from a fresh perspective, with new emphases and new connexions, that makes the account sparkle with interest. The remainder of the book contains much that has not previously been given the rounded account that a book affords, and in selected areas leads the reader to the frontiers of present research. The author has a light touch and an enviable clarity. The argument moves fast but is always easy to read. The author's enthusiasm for his subject is infectious. The book is equipped with an extensive bibliography and with a collection of illuminating and challenging exercises for the reader.

A major feature of the book is the use of graphical arguments. There is a growing use of graphs in semigroup theory, in particular for replacing or illustrating algebraic manipulations. Such graphs, or diagrams, provide a powerful intuitive guide, as well as often affording elegant proofs. The basic approach to free inverse semigroups, leading to McAlister's P-semigroup theorems, starts from the Munn birooted trees. Chapter 5 exploits the use of diagrams for dealing with word problems, a highlight being the proof that finitely presented semigroups whose defining relations satisfy a 'small overlap condition' have a soluble word problem. In the final chapter the power of graphs for solving equations in the semigroup of transformations of a finite set is again illustrated, for example for finding square roots. Chapter 3 is devoted to the Nambooripad–Easdown theory of biordered sets and exhibits, *inter alia*, the clarifying power and usefulness of the Easdown diagrams.

Finally let me comment on the general remarks, partly historical, partly elucidatory, partly establishing connexions, occasionally pointing to important unsolved problems, that occur regularly throughout the text. These are always illuminating and stimulating and are one of the delights of the book.

The book is an important addition to semigroup literature.

Preface

The first chapter provides sufficient background for any of the subsequent chapters, so the reader should feel no inhibitions in moving from the opening chapter to any later one of particular interest.

Function compositon will, unless otherwise specified, be written from left to right so that the composition fg will mean first f then g.

There are 170 exercises which serve the dual purpose of illustrating the richness of the subject while giving the reader the opportunity to come to grips with the material. Those carrying an asterisk are referred to at some other point in the text, often in some later exercise. It is not expected however that the reader should have to develop substantial amounts of the theory in order to understand the text fully.

I would like to acknowledge the assistance and criticism of Mark Lawson who carefully scrutinized a large portion of the book, together with David Jackson, James Renshaw, and Boris Schein. The time spent at the University of St Andrews visiting John Howie is also appreciated. Finally I would like to thank the editorial staff of the OUP for their approachability and valuable service.

Essex P. M. H.
May 1991

Contents

1. **Fundamentals** — 1
   1.1 Introduction — 1
   1.2 Green's relations — 15
   1.3 Completely 0-simple and completely regular semigroups — 33
   1.4 Properties of regularity — 44
   1.5 Biordered sets — 60
   1.6 Graph theory preliminaries — 67
   1.7 Semigroup diagrams — 73

2. **Free inverse semigroups and the theorems of McAlister** — 81
   2.1 Free inverse semigroups — 81
   2.2 The Scheiblich representation and P-semigroups — 92

3. **Biordered sets** — 110
   3.1 Introduction — 110
   3.2 An axiomatic system for biordered sets — 111
   3.3 Reconstructing some semibands from their biordered sets — 121
   3.4 Biordered sets of certain classes of semigroup — 126

4. **Zigzags and their applications** — 143
   4.1 The Zigzag Theorem — 143
   4.2 Zigzags and amalgams — 158

5. **Semigroup diagrams and word problems** — 167
   5.1 Introduction — 167
   5.2 The word problem for small overlap semigroups — 168
   5.3 Group diagrams and a theorem of Adjan and Remmers — 177

6. **Combinatorial aspects of transformation semigroups** — 187
   6.1 Probabilistic features of T_n — 187
   6.2 The solution of equations in finite full transformation semigroups — 204
   6.3 Products of idempotents in T_n — 223

References 237

Glossary of notation 250

Index 253

1 Fundamentals

1.1 Introduction

The purpose of this chapter is to introduce the basic concepts and theorems of algebraic semigroups, together with the additional ideas that are to be developed in this monograph. Its content forms a concise course in algebraic semigroup theory suitable for the graduate or research student, or as a supplement to an advanced undergraduate course in this subject. Proofs will be provided for all facts relevant to our purposes, although some simple results will be set as exercises or left to the reader to supply the details. There are several other well-known texts containing broadly similar material, the most suitable for undergraduates being Howie (1976), especially the first five chapters. The texts of Clifford and Preston (1961, 1967) provide a particularly careful and thorough account of the work done up until their time of publication, while an early Russian text available in English is that of Lyapin (1974). Other well-known books on the subject include those of Petrich (1973, 1977, 1984) and Lallement (1979), which concentrates on applications to coding and language theory. Also along these lines, Pin (1986) is noteworthy in giving an introducton to the theory of automata and formal languages, together with many worked examples and interesting glimpses of current research topics. Howie (1991) treats similar material, but with the undergraduate audience in mind.

A *semigroup* $(S, \circ)$ is a non-empty set S together with an associative binary operation $\circ$. In accord with the usual practice we often speak simply of a semigroup S when the operation $\circ$ is understood, and abbreviate the product $a \circ b$ (a, b in S) to ab. The symbol S will denote a semigroup without further comment. Associativity is important as it allows products of any length to be written unambiguously without the need for brackets (a fact which can be proved by induction on the length of products). A semigroup with an identity element is called a *monoid*. If a semigroup S is not a monoid we may adjoin a new symbol, 1, to S, with the property that $1a = a1 = a$ for all a in $S \cup \{1\}$. The resultant monoid is denoted by S^1. If S is a monoid we define $S^1 = S$. In the same way we denote by S^0 the semigroup with an adjoined zero element 0, if S does not possess a zero; otherwise $S^0 = S$. By convention, S must have at least two elements to possess a zero.

An element e of S is an *idempotent* if $e = e^2$, and the set of idempotents of S is denoted by $E(S)$, or simply E. For any subset T of S we shall abbreviate

$E(S) \cap T$ by $E(T)$. We define the *natural partial order* on $E(S)$ by $e \leq f$ if and only if $ef = fe = e$ (that $\leq$ does define a partial order is easily checked: the 'lower' idempotent e acts as a zero relative to the 'upper' idempotent f). We say that an idempotent e is a *right identity* [*right zero*] of S if $ae = a$ [$ae = e$] for all a in S. (Corresponding left-sided notions, in this case those of left identity and left zero, are defined dually, and we shall often take them as implicitly defined once their right-handed counterparts have been introduced.) On any non-empty set S we may define an associative product by insisting that all members of S act as right zeros, thus forming a *right zero semigroup*. One easily shows that if a semigroup S has a left identity a and a right identity b then $a = b$ and S is a monoid.

A *homomorphism* $\alpha: S \to T$ is a mapping from a semigroup S into a semigroup T such that $(ab)\alpha = a\alpha b\alpha$ for all a, b in S. We say that α is an *epimorphism* if $\alpha\beta = \alpha\gamma$ implies $\beta = \gamma$, where β and γ are homomorphisms from, say, T to U. In other words α is left (or pre-) cancellative. Any surmorphism (surjective homomorphism) is an epimorphism, but the converse is false in the category of semigroups. However, the dual concept of a right (or post-) cancellative homomorphism, known as a *monomorphism*, coincides with that of injective homomorphisms (Exercise 1.1.7). The concepts of *isomorphism*, *automorphism*, and *endomorphism* are defined as in any universal algebra, an isomorphism being a bijective homomorphism between semigroups, an automorphism being an isomorphism of a semigroup onto itself, while endomorphism is the word that describes any homomorphism from a semigroup to itself. In the semigroup context, a subalgebra is, of course, called a *subsemigroup*. The word 'homomorphism' will on occasion be shortened to simply *morphism*. The *direct product* of semigroups S and T is the semigroup with underlying set the cartesian product $S \times T$, in which multiplication is carried out by components. For example, if L is a left zero semigroup and R is a right zero semigroup, their direct product is known as a *rectangular band*, $L \times R$, and has multiplication

$$(a, b)(c, d) = (a, d) \quad (a, c \text{ in } L, b, d \text{ in } R).$$

The name derives from the observation that if the members of $L \times R$ are pictured in a rectangular grid in the obvious fashion, then the product of two elements lies at the intersection of the row of the first member and the column of the second. In general, a semigroup consisting entirely of idempotents is known as a *band*.

A *commutative semigroup* S is one for which $ab = ba$ for all a and b in S. Examples of these are abelian groups, the natural numbers under addition, and the *null semigroup* on a set S, in which all products equal a distinguished member of S, the zero. A (*lower*) *semilattice* is a poset S in which each pair of elements, a and b, has a greatest lower bound, $a \wedge b$. Clearly, any semilattice is a commutative band with respect to this meet operation and the partial order is the natural partial order on the band. Conversely, it is easy to show that any commutative band B is a semilattice in which $ab = a \wedge b$, where the meet is with respect to the natural partial order on B. We thus identify the classes of semilattices and commutative bands. The existence of a surmorphism of a semigroup S onto a semilattice often yields important structural information about S. In particular, a band admits a surmorphism onto a semilattice such that the inverse image of each member of the semilattice is a rectangular band. This is a special case of a result of Clifford that any 'union of groups' is a 'semilattice of completely simple semigroups' (Theorem 1.3.10).

An important example of a semigroup is the *full transformation semigroup*, $\mathcal{T}_X$, on a non-empty set X, which consists of all self-maps on X under composition. When X is a finite set we shall write T_n in place of $\mathcal{T}_X$, where $n = |X|$, the order of the set X. We shall usually consider the base set of T_n to be $X_n = \{1, 2, \ldots, n\}$, and a typical member $\alpha \in T_n$ can then be specified by listing the images of the members of X_n in order: $(1\alpha, 2\alpha, \ldots, n\alpha)$. The *rank* r of $\alpha \in T_n$ is $|\text{ran } \alpha|$, while the *defect of* α is $n - r$. The significance of $\mathcal{T}_X$ lies in the 'Cayley Theorem' that any semigroup S may be faithfully embedded in $\mathcal{T}_{S^1}$. The proof is similar to the corresponding result for groups: the monomorphism mapping S to $\mathcal{T}_{S^1}$ is defined by $a \mapsto \rho_a$ $(a \in S)$, where ρ_a is the *right inner translation of* S^1 *by* a, which is defined by $x \mapsto xa$ $(x \in S^1)$. The presence of the identity in S^1 ensures that the representation is faithful. This monomorphism is called the *extended right regular representation of* S. If S is used instead of S^1, we have the *right regular representation* which is not necessarily faithful. In general, any morphism $\alpha: S \to \mathcal{T}_X$ is known as a *representation of* S.

A *right ideal* A of a semigroup S is a non-empty subset A of S such that ax is in A for all a in A and x in S. We call A an *ideal* of S if it is both a right and a left ideal of S. Since S is an ideal of itself, and the intersection of [right] ideals is a [right] ideal, we may speak of the [right] ideal generated by a non-empty subset A of S as the smallest [right] ideal of S containing A. This is easily seen to be given by $[AS^1] S^1A S^1$. If A consists of a single element $\{a\}$ we speak of the *principal* [right] *ideal* generated by a. A semigroup S is [right] *simple* if $[aS = S] SaS = S$ for all a in S. A semigroup S is a group if S is both left and right simple (Exercise 1.1.2). A semigroup S is *right cancellative* if $ax = bx$ implies $a = b$ $(a, b, x$ in $S)$, while S is *cancellative* if it is both left and right cancellative.

Unlike the theory of rings, it is not generally true that a homomorphic image of a semigroup S is determined by some ideal of S, as will be explained below.

The subsemigroup generated by a non-empty subset A of S, denoted by $\langle A \rangle$, is the smallest subsemigroup of S containing A, and consists of all finite products of members of A. If A has only one member, a, we speak of the *monogenic semigroup* generated by a. Either $\langle a \rangle$ is isomorphic to the natural numbers under addition, or there exist positive integers r, m such that $\langle a \rangle = \{a, a^2, \ldots, a^r, \ldots, a^{r+m-1}\}$ with $K_a = \{a^r, a^{r+1}, \ldots, a^{r+m-1}\}$ a cyclic group of order m. The integers r and m are called the *index* and *period* of $\langle a \rangle$ respectively. A semigroup is *periodic* if all its monogenic subsemigroups are finite. A still wider class of semigroups is that of *group-bound semigroups*: S is *group-bound* if for each a in S there is some power a^n of a such that a^n is a member of some subgroup of S (that is, a subsemigroup which is a group with respect to the semigroup operation of S).

An element x in S is an *inverse* of a in S if $a = axa$ and $x = xax$. The set of inverses of a is denoted by $V(a)$, and a is *regular* if $V(a)$ is not empty. The (symmetric) relation $V(S)$ is defined as

$$V(S) = \{(a, b) \in S \times S : b \in V(a)\}.$$

The set of regular elements of S is denoted by $\text{Reg}(S)$, and we say that S is *regular* if $\text{Reg}(S) = S$. The equation $a = axa$ implies that $xax \in V(a)$ and that both ax and xa are idempotents. Note that for any $a \in \text{Reg}(S)$ we have $a = aa'a$ ($a' \in V(a)$) so that $aS = aS^1$ and $Sa = S^1 a$. A regular semigroup in which every element a has a unique inverse a^{-1} is called an *inverse semigroup*. It is easy to check that $\mathcal{T}_X$ is a regular monoid (although for the case of an infinite base set X, the axiom of choice is required), while any semilattice and any group is an inverse semigroup.

The definition of inverse is akin to that of a Penrose generalized inverse of a matrix. The link is best realized through the semigroup of all linear maps on a finite-dimensional vector space, which is a regular semigroup under composition. It is possible to show that for any such map α, the set of its inverses, $V(\alpha)$, contains a unique map satisfying certain additional conditions called the Moore–Penrose inverse of α (see, for example, Robinson 1962).

An element a of S is *completely regular* if there exists $x \in V(a)$ such that $ax = xa$, and S is called *completely regular* if all its elements enjoy this property. Obviously, both groups and bands are among the classes of completely regular semigroups.

The *partial transformation semigroup* on a non-empty set X, denoted by $\mathcal{PT}_X$, consists of all functions $\alpha : \text{dom}\, \alpha \to \text{ran}\, \alpha$, where $\text{dom}\, \alpha$ and $\text{ran}\, \alpha$ are subsets of X, and once again the semigroup operation is composition. The composition is carried out to the extent possible, so that for α, β in $\mathcal{PT}_X$, $\text{dom}\, \alpha\beta = (\text{ran}\, \alpha \cap \text{dom}\, \beta)\alpha^{-1}$, and $\text{ran}\, \alpha\beta = (\text{ran}\, \alpha \cap \text{dom}\, \beta)\beta$. If $\text{ran}\, \alpha \cap \text{dom}\, \beta$ is empty, then $\alpha\beta$ is the empty map. Like the full transformation semigroup, $\mathcal{PT}_X$ is also regular.

The partial transformation semigroup on X has $\mathcal{T}_X$ as a subsemigroup, which in turn contains the symmetric group, $\mathcal{G}_X$. We obtain another notable subsemigroup, $\mathcal{I}_X$, by taking the collection of all injective members of $\mathcal{PT}_X$. An injection $\alpha : \text{dom}\, \alpha \to \text{ran}\, \alpha$ has a unique injective inverse in the usual inverse function $\alpha^{-1} : \text{ran}\, \alpha \to \text{dom}\, \alpha$, giving us an inverse semigroup known as the *symmetric inverse semigroup* on X. A map ε is in $E(\mathcal{I}_X)$ if and only if $\varepsilon = 1|\text{dom}\, \varepsilon$, the restriction of the identity map 1 on X to the domain of ε, whence $E(\mathcal{I}_X)$ is isomorphic to the semilattice of the power set of X under intersection.

The symmetric inverse semigroup is regarded as the prototype of the class of inverse semigroups because any inverse semigroup can be embedded in some $\mathscr{I}_X$ (see Theorem 1.1.6 below). However, inverse semigroups also arise through a natural generalization of the concept of automorphism group of an algebraic structure. Let A denote a universal algebra (typically A could be a semigroup, a group, a ring, or a lattice). By a *partial automorphism* of A we mean an isomorphism between two subalgebras of A. The set of all partial automorphisms forms an inverse monoid under composition, denoted by $PA(A)$. (In fact, $PA(A)$ is a submonoid of $\mathscr{I}_A$.) The semilattice of idempotents of $PA(A)$ is, in effect, $\mathrm{Sub}(A)$, the lattice of subalgebras of A, while the 'group of units' of $PA(A)$ (the maximal subgroup containing the identity) is $\mathrm{Aut}(A)$, the group of automorphisms of A. In general, $PA(A)$ contains more information about A than do $\mathrm{Sub}(A)$ and $\mathrm{Aut}(A)$ combined (see, for example, Goberstein 1985) and so one may expect that $PA(A)$ can clarify the structure of A better than $\mathrm{Sub}(A)$ and $\mathrm{Aut}(A)$.

The set of binary relations on X (subsets of $X \times X$) forms a semigroup, $\mathscr{B}_X$ under the relational product

$$\alpha \circ \beta = \{(a, c) \in X \times X \mid \exists b \in X \text{ such that } (a, b) \in \alpha, \ (b, c) \in \beta\}.$$

If these relations are functions, this product reduces to function composition and thus $\mathscr{PT}_X$ is a subsemigroup of $\mathscr{B}_X$. Indeed, partial transformations are just relations α such that, given $a \in X$ there exists at most one $b \in X$ such that $(a, b) \in \alpha$. General binary relations can be regarded as 'many-valued transformations': for each $a \in X$ there exists a (possibly empty) set $a\alpha = \{b \in X : (a, b) \in \alpha\}$ of images of a under α. If Y is a subset of X and $Y\alpha$ denotes the set of all images of Y under α, then the product of binary relations α and β may be expressed in the familiar fashion $\{a\}(\alpha\beta) = (\{a\}\alpha)\beta$ for all a in X or, if we omit the braces around a, simply as

$$a(\alpha\beta) = (a\alpha)\beta.$$

This simple unifying observation, which is nevertheless an important psychological hurdle, is attributed by Schein (1986) to V. V. Wagner.

Let E be an equivalence relation on a set X. Denote the set of equivalence classes of E by S/E, the *factor set* of E. Let $\alpha: S \to T$ be a surjection of sets. The relation $\alpha \circ \alpha^{-1}$ is an equivalence ρ on S ($a \rho b$ if and only if $a\alpha = b\alpha$) known as the *kernel* of α. The *natural mapping* associated with ρ is $\rho^\natural: S \to S/\ker \alpha$, where $a\rho^\natural = a\rho$. The mapping $\psi: S/\rho \to T$, where $(a\rho)\psi = a\alpha$ is then the unique bijection that makes the following diagram commute:

$$\begin{array}{ccc} S & \xrightarrow{\rho^\natural} & S/\rho \\ {\scriptstyle\alpha}\downarrow & \swarrow{\scriptstyle\psi} & \\ T & & \end{array} \quad (1)$$

Furthermore, if $\alpha_1: S \to T_1$, $\alpha_2: S \to T_2$ are surjections, and $\alpha_1 \circ \alpha_1^{-1} \subseteq \alpha_2 \circ \alpha_2^{-1}$, then there exists a unique map $\psi: T_1 \to T_2$ making the diagram commute, defined by $(a\alpha_1)\psi = a\alpha_2$:

 (2)

Now suppose that the surjection α in diagram (1) is a semigroup homomorphism: the set S/ρ can then be regarded as a semigroup because multiplication of the classes by representatives is well-defined, and the bijection ψ becomes an isomorphism. A similar comment applies if the mappings α_1 and α_2 of diagram (2) are semigroup homomorphisms: the mapping ψ is then a surmorphism.

An equivalence ρ on a semigroup is a *right congruence* if $a\rho b$ implies $ac\,\rho\,bc$ for all $c \in S$. If $a\rho b$, $c\rho d$ implies $ac\,\rho\,bd$ (a, b, c, d in S) then ρ is a *congruence* on S, which occurs if and only if ρ is both a right and a left congruence. In summary, we have:

Theorem 1.1.1 (first isomorphism theorem) *Let $\alpha\colon S \to T$ be a surmorphism of semigroups. Then $\rho = \ker \alpha$ is a congruence and there exists a unique isomorphism $\psi\colon S/\rho \to T$ such that $\rho^\natural \psi = \alpha$. Conversely, if ρ is any congruence on S then $\rho^\natural\colon S \to S/\rho$ is a surmorphism of semigroups with kernel ρ.*

Suppose that we have two congruences $\sigma \subseteq \rho$ on a semigroup S. Diagram (2) now translates as

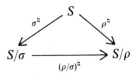

where we denote $\psi \circ \psi^{-1}$ by ρ/σ and $(a\sigma, b\sigma) \in \rho/\sigma$ if and only if $a\rho b$.

Theorem 1.1.2 (second isomorphism theorem) *Let σ, ρ be congruences on a semigroup S such that $\sigma \subseteq \rho$. Then*

$$\rho/\sigma = \{(x\sigma, y\sigma) \in S/\sigma \times S/\sigma \colon (x, y) \in \rho\}$$

is a congruence on S/σ and $(S/\sigma)/(\rho/\sigma) \simeq S/\rho$.

Let σ, ρ be two congruences [right congruences, resp. equivalences] on a semigroup S. Then $\sigma \cap \rho$ is a congruence [right congruence, resp equivalence] on S. Since ω, the universal relation, is a congruence on S containing all other equivalences, it follows that any relation R on S generates a smallest congruence [right congruence, resp. equivalence] containing R (if $R = \varnothing$, it generates the identity congruence ι). Thus the collection of all congruences [right congruences, resp. equivalences] on S forms a lattice under inclusion. Let R^{-1} denote the inverse relation of R ($xR^{-1}y$ if and only if yRx), let R^* denote the congruence generated by R, and denote the least symmetric and reflexive relation containing R by $R^s = R \cup R^{-1} \cup \iota$. For $a, b \in S$ suppose that $a = xcy$, $b = xdy$, and $cR^s d$ for some $c, d \in S$ and $x, y \in S^1$. We say that the passage from a to b or the reverse is an *elementary R-transition*.

Proposition 1.1.3 *Let R be a relation on a semigroup S. Then aR^*b ($a, b \in S$) if and only if b can be obtained from a by some finite sequence of elementary R-transitions.*

Proposition 1.1.4 *Let E be an equivalence relation on a semigroup S. Then there is a largest congruence contained in E, denoted by $E^\flat$:*

$$E^\flat = \{(a, b) \in S \times S : (\forall x, y \in S^1)(xay, xby) \in E\}.$$

Proofs of Propositions 1.1.3 and 1.1.4 are left as exercises.

Let I be an ideal of S. Then $\rho_I = \iota \cup (I \times I)$ is a congruence on S known as a *Rees congruence*, and S/ρ_I is simply denoted by S/I, in recognition of the fact that a Rees congruence is determined by the single congruence class I. (This is the case with a congruence on a group, which is determined by the normal subgroup which forms the class of the identity element, but this is not usually the case with semigroup congruences: for example, any equivalence relation at all on a null semigroup is also a congruence.) We say that S is an *ideal extension* of I by S/I.

We next characterize the class of inverse semigroups within the class of regular semigroups.

Theorem 1.1.5 *The following are equivalent for a semigroup S:*

(i) *S is regular and $E(S)$ is a semilattice;*
(ii) *every principal right ideal and every principal left ideal of S has a unique idempotent generator;*
(iii) *S is an inverse semigroup.*

Proof (i) implies (ii). Let $a \in S$, $a' \in V(a)$; then $aS = aa'S$, $Sa = Sa'a$, so that every principal right ideal and principal left ideal has an idempotent generator. Suppose that $e, f \in E(S)$ with $eS = fS$. Take x, y in S such that $ex = f$ and $fy = e$. Since idempotents commute with one another we obtain $e = fy = f^2y = fe = ef = e^2x = ex = f$.

(ii) implies (iii). Let $a', a'' \in V(a)$ ($a \in S$). Then $aa'S = aS = aa''S$, which implies that $aa' = aa''$, and, dually, $a'a = a''a$. Then

$$a' = a'aa' = a'aa'' = a''aa'' = a''.$$

8 | Techniques of semigroup theory

(iii) implies (i). Let $e, f \in E(S)$ and $x = (ef)^{-1}$. One checks that xe and fx are both inverse to ef, and thus $x = xe = fx$, whence $x^2 = x$. But then $x = x^{-1} = ef$, whence $ef \in E(S)$. One can then verify that fe is inverse to ef and so, by uniqueness of inverses, $ef = fe$, whereupon $E(S)$ is a semilattice. ∎

We now give a 'Cayley' theorem for inverse semigroups. The proof requires the following facts, which are readily checked. For members a, b of some inverse semigroup S the usual laws of inverses hold:

$$a = (a^{-1})^{-1} \quad \text{and} \quad (ab)^{-1} = b^{-1}a^{-1}.$$

Also, for $e, f \in E(S)$, $Se \cap Sf = Sef$ and $Sa = Sa^{-1}a$, $Sa^{-1} = Saa^{-1}$.

Theorem 1.1.6 (Preston–Wagner Theorem) *Any inverse semigroup S can be faithfully embedded in $\mathscr{I}_S$.*

Proof The embedding $\alpha: S \to \mathscr{I}_S$ is given by $a\alpha = \rho_a$ ($a \in S$), where we define $\rho_a: Sa^{-1} \to Sa^{-1}a$ by the rule $x \mapsto xa$ ($x \in Sa^{-1}$). One readily verifies that ρ_a and $\rho_{a^{-1}}$ are mutually inverse mappings of Saa^{-1} and $Sa^{-1}a$ onto each other. Thus $\rho_a \in \mathscr{I}_S$ and $\rho_{a^{-1}} = \rho_a^{-1}$.

Next, if $\rho_a = \rho_b$ then $Saa^{-1} = Sbb^{-1}$, which implies that $aa^{-1} = bb^{-1}$ as both are idempotent. Since $a^{-1} \in Saa^{-1}$ we obtain $a^{-1}a = a^{-1}b$, and thus

$$a = aa^{-1}a = aa^{-1}b = bb^{-1}b = b.$$

Therefore α is injective.

To show that α is a homomorphism it is enough to check that $\text{dom } \rho_a \rho_b = \text{dom } \rho_{ab}$. Now,

$$\text{dom } \rho_{ab} = Sab(ab)^{-1},$$

while

$$\text{dom } \rho_a \rho_b = (\text{ran } \rho_a \cap \text{dom } \rho_b)\rho_a^{-1} = (Sa^{-1}a \cap Sbb^{-1})\rho_a^{-1}$$

$$= Sa^{-1}abb^{-1}\rho_a^{-1} = Sabb^{-1}\rho_a^{-1} \quad (\text{as } Sa^{-1}a = Sa),$$

$$= Sabb^{-1}a^{-1} = S(ab)(ab)^{-1} = \text{dom } \rho_{ab}. \quad \blacksquare$$

The class of regular semigroups is clearly closed under the taking of homomorphisms. That the same is true for inverse semigroups is a consequence of the following particularly useful fact (Lallement 1966).

Lemma 1.1.7 (Lallement's Lemma) *Let S be a regular semigroup and ρ a congruence on S. If $a \in E(S/\rho)$ then $a\rho e$ for some $e \in E(S)$.*

Proof Since $a \in E(S/\rho)$ then $a\rho a^2$. Take $x \in V(a^2)$. Put $e = axa$, then $e^2 = axa^2xa = axa = e$, and so $e \in E(S)$. Hence we obtain

$$e\rho = (axa)\rho = a\rho x\rho a\rho = a^2\rho x\rho a^2\rho = (a^2xa^2)\rho = a^2\rho = a\rho.$$ ∎

Corollary 1.1.8 *The homomorphic image of an inverse semigroup is an inverse semigroup.*

Proof Let $\alpha: S \to T$ be a surmorphism of an inverse semigroup S. Certainly T is regular. Let $a, b \in E(T)$. Then, by Lemma 1.1.7, there are idempotents e, f in S such that $a\alpha = e\alpha$, $f\alpha = b\alpha$. But then

$$a\alpha b\alpha = e\alpha f\alpha = (ef)\alpha = (fe)\alpha = f\alpha e\alpha = b\alpha a\alpha,$$

as required. ∎

An important class of semigroups that lies between the classes of inverse and regular semigroups is the class of *orthodox semigroups*, which are the regular semigroups the idempotents of which form a band. We shall use them mainly as a source of exercises. Chapter VI of Howie (1976) is devoted entirely to their study.

Theorem 1.1.9 *The following are equivalent for a regular semigroup S:*

(i) *S is orthodox:*
(ii) *if $a, b \in S$, $a' \in V(a)$, $b' \in V(b)$ then $b'a' \in V(ab)$;*
(iii) *if $e \in E(S)$, $x \in V(e)$ then $x \in E(S)$.*

Furthermore, in any orthodox semigroup, aea', $a'ea \in E(S)$ ($a \in S$, $a' \in V(a)$, $e \in E(S)$).

Proof The final statement is demonstrated as follows:

$$aea' = aea'aa' = a(ea'a)^2 a' = aea'aea'aa' = aea' \cdot aea',$$

with a similar line of proof to show that $a'ea$ is idempotent.

(i) $\Rightarrow$ (ii). Since S is orthodox we have $abb'a'ab = aa'abb'a'abb'b = a(a'abb')^2b = aa'abb'b = ab$ and, similarly, $b'a'abb'a' = b'a'$.

(ii) $\Rightarrow$ (iii). Since $xe, ex \in E(S)$ it follows from the given property that $ex^2e \in V(xe^2x)$. But $x = xe^2x$, which is inverse to ex^2e, and thus

$$x = x(ex^2e)x = (xex)(xex) = (xex)^2 = x^2.$$

(iii) $\Rightarrow$ (i). Let $e, f \in E(S)$ and take $x \in V(ef)$. Routinely, we check that $efe \in V(fxe)$ and that $fxe \in E(S)$, whence $ef \in E(S)$ by the property given. ∎

We close this section with an introduction to free semigroups and presentations. If X is a non-empty set, denote by F_X the set of all non-empty finite words $x_1 x_2 \ldots x_n$ in the alphabet X under the operation of concatenation.

The semigroup F_X, together with the usual embedding of X into F_X where each $x \in X$ is identified with the corresponding one-letter word x in F_X, form the *free semigroup* on X in the usual sense that if $\varphi: X \to S$ is an arbitrary map into a semigroup S, then there is a unique homomorphism $\psi: F_X \to S$ such that the diagram below commutes. The uniqueness property guarantees that any two free semigroups on base sets of equal order are isomorphic, which is the justification for speaking of '*the* free semigroup on X' as opposed to 'a free semigroup on X':

The map ψ is necessarily defined by $(x_1 x_2 \ldots x_n)\psi = x_1\varphi x_2\varphi \ldots x_n\varphi$. We shall on occasion write $F(X)$ instead of F_X. If we take X to be some generating set for S then $S \simeq F_X/\rho$, where $\rho = \psi \circ \psi^{-1}$. If X is finite and ρ can be generated by a finite set $R = \{(w_1, z_1), \ldots, (w_m, z_m)\}$ of elements $(w_i, z_i) \in F_X \times F_X$, we say that F_X/ρ is *finitely presented* and write $F_X/\rho = \langle x_1, x_2, \ldots, x_n | w_1 = z_1, \ldots, w_m = z_m \rangle$.

We say that F_X/ρ has *generators* $x_1, x_2, \ldots, x_n$ and *relations* $w_1 = z_1, \ldots, w_m = z_m$. In general, a semigroup presentation with generating set X and relation set R will be denoted by $(X; R)$, while the corresponding semigroup will be denoted by $\langle X | R \rangle$.

If we adjoin the empty string Λ to F_X the resulting semigroup is F_X^1, the free monoid on X. Monoid presentations differ only in that defining relations of the form $w = 1$ are permitted.

If X is a countable set and $p, q \in F_X$ we say that S satisfies the identity $p = q$ if $p\varphi = q\varphi$ for every homomorphism $\varphi: F_X \to S$. Informally, we regard the members of X as variables, and we require that the words p and q give rise to equal members of S for every possible assignment of values of S to the variables. A class of semigroups is called a *variety* or an *equational class* if it consists exactly of all those semigroups satisfying some (possibly infinite) collection of identities $p_1 = q_1, p_2 = q_2, \ldots$. We denote the variety defined by this collection by $[p_1 = q_1, p_2 = q_2, \ldots]$. For example, the classes of all commutative semigroups and all bands are the varieties of semigroups $[xy = yx]$ and $[x = x^2]$ respectively. Less obviously, $[x = x^2, xyz = xz] = [x = xyx]$, and both equal the variety of rectangular bands. For a survey on the theory of semigroup varieties, see Evans (1971).

The class of inverse semigroups can be regarded as a class of algebras with one binary and one unary (single argument) operation of inversion. Since this class is closed under the taking of morphic images, subalgebras, and arbitrary direct products it forms, by a theorem of Birkhoff, a variety, and thus a free inverse semigroup on a given base set exists. Defining systems of identities for this variety are not difficult to find (see Ex. 1.1.20). We shall however give

Fundamentals | 11

a direct construction of the free inverse semigroup on a given base set X without reference to varietal considerations.

We define an *anti-homomorphism* α from a semigroup S to another semigroup S' as a mapping which reverses multiplication in that $(ab)\alpha = b\alpha a\alpha$ $(a, b \in S)$. For example, consider the left regular representation of a semigroup S into $\mathcal{T}_{S^1}$, by which $a \mapsto \lambda_a$, where λ_a is the left translation of S by a. Since $\lambda_{ab} = \lambda_b \lambda_a$, this affords an instance of an anti-homomorphism or *anti-representation* of S. If α is also a bijection, then it is an *anti-isomorphism*; for example, any bijection of a right zero semigroup onto a left zero semigroup is an anti-isomorphism. More generally, any semigroup S is anti-isomorphic to its dual S^*, the semigroup defined on the set S with product $*$ whereby $a * b = ba$ $(a, b \in S)$. If $S = S'$, then α is an *anti-automorphism*; if in addition $\alpha = \alpha^2$, then α is called an *involution*.

Let X be a non-empty set and let $'$ be a bijection of a non-empty set X to a disjoint set X', and put $Y = X \cup X'$. Define a unary operation $(\cdot^{-1})$ on the free semigroup $F = F_Y$ by

$$y^{-1} = \begin{cases} x' & \text{if } y = x \in X, \\ x & \text{if } y = x' \in X', \end{cases}$$

and

$$(x_1 x_2 \ldots x_n)' = x_n' x_{n-1}' \ldots x_1', \qquad x_i \in Y.$$

Clearly $(\cdot^{-1})$ is an involution on F.

The free inverse semigroup on X, denoted by FI_X, can now be realized as the quotient of the free semigroup F_Y by the least inverse congruence.

Theorem 1.1.10 *Let ρ be the congruence on F generated by*

$$\{(yy'y, y) | y \in F\} \cup \{(yy'zz', zz'yy') | y, z \in F\}.$$

Then $(F/\rho, i)$ is the free inverse semigroup on X, where $i = \rho^\natural | X$.

Proof Since F/ρ is clearly regular we need to show that the idempotents of F/ρ commute with each other in order to prove that F/ρ is inverse. This will follow from the definition of ρ once we have verified that each member of $E(F/\rho)$ is a product of members of F of the form yy'. We shall denote the ρ-class containing $w \in F$ by $[w]$. Let $[e]$ be a member of $E(F/\rho)$. Then

$$[e] = [e]^2 = [ee'e][ee'e] = [e][e'e][ee'][e] = [e][ee'][e'e][e]$$
$$= [e]^2[e']^2[e]^2 = [ee'][e'e].$$

Now let $\varphi : X \to S$ be any map into an inverse semigroup S. Since F is the free semigroup on Y, there is a unique homomorphism $\alpha : F \to S$ such that $x\alpha = x\varphi$ and $x'\alpha = (x\varphi)^{-1}$ for all $x \in X$, whereupon $F\alpha$ is an inverse subsemigroup of S. By the evident minimality of ρ as an inverse semigroup congruence on F, it follows that $\rho \subseteq \alpha \circ \alpha^{-1}$, whence by Theorem 1.1.2

there exists a homomorphism $\psi: F/\rho \to S$ such that the following diagram commutes:

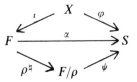

and in particular we have:

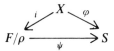

where $i = \iota\rho^{\natural}$. Furthermore, the uniqueness of any homomorphism ψ making the second diagram commute is clear, as its action on the members of the generating set Y is determined. This completes the proof. ∎

Exercise 1.1

1. Each idempotent in a left cancellative semigroup is a left identity element. A cancellative semigroup can contain at most one idempotent which is the identity.

2. A semigroup is a group if and only if it is left and right simple.

3. (a) A group-bound cancellative semigroup is a group.
   (b) Let X be an infinite set. Show that $S = \{\alpha \in \mathcal{T}_X : |\operatorname{ran}\alpha| < \infty\}$ is an infinite periodic semigroup.

4.* Let S denote a full transformation semigroup.

   (a) The constant maps of S form a right zero subsemigroup and S has no left zeros.
   (b) The idempotents of S are the maps which are the identity when restricted to their range.
   (c) Show there is a one-to-one correspondence between the set of principal ideals of the full transformation semigroup and the set of all cardinal numbers $r \leq |X|$ by which the principal ideal corresponding to r consists of all elements α of $\mathcal{T}_X$ of rank $\leq r$ (meaning that $|\operatorname{ran}\alpha| \leq r$). Show that every ideal of T_n is principal, and that the only non-principal ideal of $\mathcal{T}_X$ with denumerable X is the set of all transformations of finite rank.

5. (a) Let $X = \{0, 1, \ldots, r + m - 1\}$ and consider the monogenic subsemigroup $S_{r,m} = \langle a \rangle$ of T_X, where a is the map $(1, 2, 3, \ldots, r + m - 1, r)$. Show that $\langle a \rangle$ has index r and period m.

(b) Any finite monogenic semigroup is isomorphic to some $S_{r,m}$.
(c) Find all monogenic semigroups of order 11 generated by a symbol a such that a^8 is idempotent.
(d) Determine the subsemigroup of T_{12} generated by the map $a = (3, 3, 4, 5, 6, 7, 8, 6, 10, 11, 12, 12)$ by finding its order and period. What are its idempotent and subgroups?

6. (a) Let $\alpha: S \to T$ be a semigroup surmorphism. Let $\mathscr{A}$ denote the set of all subsemigroups of S and let $\mathscr{B}$ denote the set of all subsemigroups of T. The mapping $\alpha: A \mapsto A\alpha$ $(A \in \mathscr{A})$ is an inclusion-preserving map from $\mathscr{A}$ onto $\mathscr{B}$.
(b) Repeat part (a), but let $\mathscr{A}$ and $\mathscr{B}$ represent the set of all ideals of S and T respectively. Draw the corresponding conclusion.
(c) Let I be an ideal of a semigroup S. Let $\mathscr{A}$ be the set of ideals of S containing I and let $\mathscr{B}$ be the set of ideals of S/I. Then the map $\alpha: J \mapsto J/I$ $(J \in \mathscr{A})$ is an inclusion-preserving bijection of $\mathscr{A}$ onto $\mathscr{B}$.

7.* (a) Show that in the category of semigroups and semigroup morphisms a morphism is a monomorphism if and only if it is injective, and that any surmorphism is an epimorphism.
(b) Let $S = \mathscr{T}_X$ with X infinite. Let U be the subsemigroup of S consisting of 1 together with $\{u \in S: |X \backslash \operatorname{ran} u| = \infty\}$. Then U is a proper subsemigroup of S. However, the inclusion $i: U \to S$ is an epimorphism. To see this take $d \in S \backslash U$ and $u \in U$ any injection except 1; define $y \in S$ by

$$ny = \begin{cases} nu^{-1} & \text{if } n \in \operatorname{ran} u, \\ p & \text{otherwise, } p \text{ a fixed member of } S. \end{cases}$$

Observe that $uy = 1$. Now let $\alpha, \beta: S \to T$ be a pair of homomorphisms from S to a semigroup T which agree on U. Show that $du \in U$, and hence prove that $d\alpha = d\beta$.

8. (a) A semigroup that is right simple and left cancellative is a *right group*. If G is a group and E is a right zero semigroup, then $G \times E$ is a right group. Establish the converse by proving that:

  (i) there are idempotents in S;
  (ii) $E = E(S)$ is a right zero subsemigroup of S;
  (iii) that $eb = b$ for every $b \in S$ and $e \in E(S)$;
  (iv) Se is a subgroup of S for every idempotent e;
  (v) if f is a fixed element of E and G is the group Sf, then $S \simeq G \times E$.

(b) Also deduce that a semigroup S is a right group if and only if:

  (i) S is right simple and contains at least one idempotent;
  (ii) the equation $ax = b$ has a unique solution in S $(a, b \in S)$.

9. Let S be the set of non-zero complex numbers with product $a \circ b = |a|b$. Show that S is a right group and express it in the form $G \times E$. Show that

the subsemigroup of non-zero reals is also a right group and express it as a direct product.

10. (a) A semigroup S is a group if and only if S is regular and has a unique idempotent.
    (b) A semigroup S is a right group if and only if S is regular and left cancellative.

11.* Let X be a countably infinite set and let S be the set of injections $\alpha: X \to X$, with the property that $|X \setminus X\alpha| = \infty$. Show that S is a subsemigroup of $\mathcal{T}_X$, the so-called *Baer–Levi semigroup* on X. By finding a one-to-one correspondence between $X \setminus X\alpha$ and $X\alpha \setminus X\alpha^2$ deduce that S is idempotent-free. Hence prove that S is right simple and right cancellative, but is neither a right group nor a left group.

12. Any cancellative commutative semigroup S can be embedded in an abelian group as follows. Let $F = S^1 \times S^1$ and define ρ on F by $(a, b)\rho(c, d)$ if and only if $ad = bc$ $(a, b, c, d \in S^1)$. Show that ρ is a congruence on F, and that F/ρ is an abelian group. Show that S^1 is embedded in F/ρ by the mapping whereby $a \mapsto (a, 1)\rho (a \in S^1)$. Carry out this process on the natural numbers under addition, and on the natural numbers under multiplication.

13.* We write $T_n, PT_n, I_n,$ and B_n to denote $\mathcal{T}_X, \mathcal{PT}_X, \mathcal{I}_X,$ and $\mathcal{B}_X$ when X is a finite set of order n.

   (a) Show that $|T_n| = n^n$.
   (b) Show that $\mathcal{PT}_X$ is isomorphic to the subsemigroup of $\mathcal{T}_{X \cup \{\infty\}}$ of all mappings which fix ∞ $(\infty \notin X)$.
   (c) Hence, or otherwise, prove that $|PT_n| = (n + 1)^n$.
   (d) Show that $|I_n| = \sum_{r=0}^{n} \binom{n}{r}^2 r!$
   (e) Show that $|B_n| = 2^{n^2}$.

14. A semigroup S is *right reductive* if $xa = xb$ for all x in S implies $a = b$. Let S be any semigroup and $\alpha: S \to \mathcal{T}_S$ the right regular representation of S in $\mathcal{T}_S$ by right inner translations.

   (a) Show that α is faithful if and only if S is right reductive.
   (b) Prove that any inverse semigroup is right reductive.

   Call a semigroup *reductive* if $xa = xb$ and $ax = bx$ for all x in S implies that $a = b$.

   (c) Show that a regular semigroup is reductive.
   (d) Find an example of a regular semigroup which is not right reductive.

15. Show that if ρ is a congruence on a periodic semigroup S and $a\rho \in E(S/\rho)$, then there exists an idempotent e such that $a\rho e$. Find a semigroup S and

a congruence ρ for which the conclusion of Lallement's Lemma does not hold.
16. Let $\alpha: S \to T$ be a surmorphism from an inverse semigroup S.
    (a) Show that α is a homomorphism of semigroups with involution; that is, α necessarily preserves inverses.
    (b) Let U be an inverse subsemigroup of T. Then $U\alpha^{-1}$ is an inverse subsemigroup of S.
17. (a) Show that a subsemigroup of a free semigroup F_X is not necessarily free.
    (b) Show that the free semigroup on $\{a, b\}$ contains copies of F_X for any set X of countable cardinality.
18. The monoid $\langle x, y | xyx = 1 \rangle$ is a group isomorphic to the integers under addition.
19. A semigroup S is called an *inflation* of one of its subsemigroups T if $S^2 \subseteq T$ and there is a surjection $\alpha: S \to T$ such that $\alpha^2 = \alpha$ and $(a\alpha)(b\alpha) = ab$ for all $a, b \in S$. Show that the semigroup variety $[xy = y^2x]$ is precisely the class of all inflations of semilattices.
20. Considered as algebras with the two operations of multiplication and inversion, show that inverse semigroups form an equational class determined by associativity and by either of the following systems of identities:
    (a) $x = xx^{-1}x$, $\quad (x^{-1})^{-1} = x$, $\quad x^{-1}xy^{-1}y = y^{-1}yx^{-1}x$;
    (b) $x = xx^{-1}x$, $\quad (x^{-1})^{-1} = x$, $\quad (xy)^{-1} = y^{-1}x^{-1}$,
    $xx^{-1}x^{-1}x = x^{-1}xxx^{-1}$.

A theorem of universal algebra asserts the existence of free objects in any equational class. Thus these realizations of inverse semigroups as a variety of semigroups with involution guarantee the existence of the free inverse semigroup on a given generating set X (Schein 1963).

1.2 Green's relations

Green's relations are five equivalences on a semigroup based on the idea of mutual divisibility of elements. They play no role in group theory since they all coincide with the universal equivalence, but they are important tools in the description and decomposition of semigroups.

Let S be any semigroup. We define $a \mathcal{R} b$ if $aS^1 = bS^1$, and $a \mathcal{L} b$ if $S^1 a = S^1 b$ ($a, b \in S$). The equivalence $\mathcal{H} = \mathcal{L} \cap \mathcal{R}$, while the equivalence $\mathcal{D}$ is $\mathcal{L} \vee \mathcal{R}$, where the join is in the lattice of all equivalences on S; that is, $\mathcal{D}$ is the least equivalence containing both $\mathcal{L}$ and $\mathcal{R}$. Finally, we say that $a \mathcal{J} b$ if $S^1 a S^1 = S^1 b S^1$. Note that $a \mathcal{R} b$ if and only if there exist x, y in S^1 such that

$ax = b$ and $by = a$. Similar remarks apply to $\mathscr{L}$ and $\mathscr{J}$. What is more, $\mathscr{R}$ is a left congruence and $\mathscr{L}$ is a right congruence on S (which does not imply that $\mathscr{H}$ is a congruence). Since $\mathscr{L} \cup \mathscr{R} \subseteq \mathscr{J}$, it follows from the definition of $\mathscr{D}$ that $\mathscr{D} \subseteq \mathscr{J}$: we shall see that equality holds in a number of significant cases. A better description of the $\mathscr{D}$-relation is desirable, however, as it is the only one of the five not given in terms of principal ideals.

Lemma 1.2.1 *Every left congruence $\rho \subseteq \mathscr{R}$ commutes with every right congruence $\lambda \subseteq \mathscr{L}$.*

Proof Take $(a, b) \in \lambda \circ \rho$ so there exists $c \in S$ with $a \lambda c \rho b$, whence there exist $u, v \in S^1$ such that $a = uc$ and $b = cv$. It follows that $av = ucv = ub = d$, say. But $a \lambda c$ implies $av \lambda cv$; in other words, $d \lambda b$. Similarly, $c \rho b$ implies $uc \rho ub$, which is $a \rho d$. Hence $a(\rho \circ \lambda)b$ and so $\lambda \circ \rho \subseteq \rho \circ \lambda$. Dually, we obtain the reverse inclusion. ∎

Corollary 1.2.2 $\mathscr{D} = \mathscr{R} \circ \mathscr{L} = \mathscr{L} \circ \mathscr{R}$.

Proof Since $\mathscr{R} \circ \mathscr{L} = \mathscr{L} \circ \mathscr{R}$ by Lemma 1.2.1, this relation is reflexive and symmetric. Since

$$(\mathscr{R} \circ \mathscr{L}) \circ (\mathscr{R} \circ \mathscr{L}) = \mathscr{R} \circ (\mathscr{L} \circ \mathscr{L}) \circ \mathscr{R} = \mathscr{R} \circ (\mathscr{L} \circ \mathscr{R}) = (\mathscr{R} \circ \mathscr{R}) \circ \mathscr{L} = \mathscr{R} \circ \mathscr{L},$$

it is also transitive, whence $\mathscr{R} \circ \mathscr{L} = \mathscr{D}$. ∎

Notation For any a in S we denote the $\mathscr{L}$-class of a in S by L_a, and in the same way we define R_a, H_a, D_a and J_a. The corresponding principal ideals are written $L(a) = S^1 a$, $R(a) = aS^1$ and $J(a) = S^1 a S^1$ respectively. We define a partial order on the $\mathscr{R}$-classes of S by saying that $R_a \leq R_b$ if $aS^1 \subseteq bS^1$, with a dual partial order defined on $S/\mathscr{L}$. Likewise, for $S/\mathscr{J}$ we say that $J_a \leq J_b$ if $S^1 a S^1 \subseteq S^1 b S^1$, and for two members H_a, H_b of $S/\mathscr{H}$ we write $H_a \leq H_b$ if $L_a \leq L_b$ and $R_a \leq R_b$. Note that for each $a \in S$ and $x, y \in S^1$, $R_{ax} \leq R_a$, $L_{xa} \leq L_a$ and $J_{xay} \leq J_a$. We call S a *simple semigroup* if it has just one $\mathscr{J}$-class, and S is called *bisimple* if it has only one $\mathscr{D}$-class.

The partial order on (say) $\mathscr{R}$-classes of a semigroup S establishes a reflexive and transitive relation on S itself whereby $a \prec b$ if $R_a \leq R_b$. In general, a reflexive and transitive relation on a set S is known as a *pre-order* or *quasi-order*. Any pre-order $\prec$ on S induces an equivalence relation $\sim$ on S whereby $x \sim y$ if and only if $x \prec y$ and $y \prec x$. The resulting factor set $S/\sim$ then inherits a well-defined partial order $\leq$, whereby $[x] \leq [y]$ if and only if $x \prec y$ (where $[x]$ denotes the $\sim$-class of $x \in S$).

Corollary 1.2.2 allows us to represent a $\mathscr{D}$-class by means of an 'egg-box diagram': a rectangular array of squares in which the rows correspond to $\mathscr{R}$-classes, the columns to $\mathscr{L}$-classes, and the square forming the intersection of a row and a column corresponds to an $\mathcal{H}$-class, as Corollary 1.2.2 guarantees that this is not empty. Indeed, as will be shown, all $\mathcal{H}$-classes within the one $\mathcal{D}$-class have the same cardinality, further justifying the egg-box picture.

Example 1.2.3 Let B be the subsemigroup of T_7 generated by the maps $a = (2, 2, 5, 6, 5, 6, 5)$ and $b = (3, 4, 3, 3, 7, 4, 7)$. B is a six-element band: $B = \{a, b, ab, ba, aba, bab\}$. The band has three $\mathcal{J}$-classes: $J_a = \{a\}$, $J_b = \{b\}$, and the remaining $\mathcal{J}$-class, which also is a $\mathcal{D}$-class, consists of two $\mathcal{R}$- and two $\mathcal{L}$-classes in which each $\mathcal{H}$-class is a singleton:

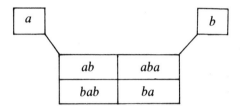

The equality of $\mathcal{D}$ and $\mathcal{J}$ in this example is not by chance: $\mathcal{D} = \mathcal{J}$ in any periodic, indeed, in any group-bound semigroup (Theorem 1.2.20).

The usefulness of Green's relations stems from the next result.

Lemma 1.2.4 (Green's Lemma, right-hand version) *Let $a\mathcal{R}b$ (a, b in S) and take s, s' in S^1 such that $as = b$ and $bs' = a$. Then the mappings $\rho_s|L_a$ and $\rho_{s'}|L_b$ are mutually inverse, $\mathcal{R}$-class preserving bijections of L_a onto L_b, and of L_b onto L_a respectively.*

Proof The map $\rho_s|L_a$ maps into L_b as $\mathcal{L}$ is a right congruence. Let $x \in L_a$; then there exists t in S^1 such that $x = ta$, whence

$$x\rho_s\rho_{s'} = xss' = tass' = tbs' = ta = x.$$

By applying the same argument to L_b and $\rho_{s'}$ we see that $\rho_s|L_a$ and $\rho_{s'}|L_b$ are mutually inverse bijections. To prove the last assertion of the lemma note that if $x \in L_a$ and $y = x\rho_s = xs$, then $ys' = x$, so that $y\mathcal{R}x$. Similarly, $\rho_{s'}|L_b$ is $\mathcal{R}$-class preserving. ∎

The dual of Green's Lemma asserts the existence of mutually inverse, $\mathcal{L}$-class preserving bijections, the left inner translations $\lambda_t, \lambda_{t'}$ say, between any two $\mathcal{R}$-classes in a $\mathcal{D}$-class. The composition $\rho_s|L_a \circ \lambda_t|R_b$ is then a bijection between two given $\mathcal{H}$-classes, thus establishing that all $\mathcal{H}$-classes within the one $\mathcal{D}$-class have equal cardinalities.

If $a\mathcal{L}e$, e in $E(S)$, then $a \in Se$, whence $ae = e$, and so e is a right identity for L_e, and dually e is a left identity of R_e. It follows immediately that no $\mathcal{H}$-class can contain more than one idempotent.

The following seven standard results involving $\mathscr{D}$-classes can all be well visualized in terms of egg-box diagrams. Where not provided, the reader will find that sketching a relevant diagram will aid understanding both of the statements and the proof of a result of this kind.

A crucial 'location' theorem is due to Miller and Clifford (1956).

Theorem 1.2.5 *For any two elements a, b in S, $ab \in R_a \cap L_b$ if and only if $R_b \cap L_a$ contains an idempotent.*

Proof Suppose that $ab \in R_a \cap L_b$. Then ρ_b define a bijection from L_a onto L_b. There exists c in $R_b \cap L_a$ such that $c\rho_b = cb = b$. Since $c\mathscr{R}b$, there exists u in S such that $c = bu$. It follows that $bub = cb = b$, and thus $c^2 = bubu = bu = c$. Conversely, if $e = e^2 \in R_b \cap L_a$ then $eb = b$ and $ae = a$. From $e\mathscr{R}b$ we deduce $a = ae\mathscr{R}ab$, and from $e\mathscr{L}a$ we deduce that $b = eb\mathscr{L}ab$, which shows that $ab \in R_a \cap L_b$. ∎

Corollary 1.2.6 *Let H be an $\mathscr{H}$-class of S. The following are equivalent:*

(i) *H contains an idempotent;*
(ii) *there exists a, b in H such that $ab \in H$;*
(iii) *H is a maximal subgroup of S.*

Proof It is clear that (i) implies (ii). That (ii) implies (i) follows from Theorem 1.2.5, since $H = R_a \cap L_a = R_b \cap L_b$, whence it also follows from Theorem 1.2.5 that H is also a submonoid of S. Finally, the equivalence of (i) and (iii) is a direct consequence of the definition of Green's $\mathscr{H}$ relation and the axioms of a group. Indeed, a subgroup G of a semigroup is maximal if and only if $G = H_e$ for some e in $E(S)$. ∎

Another remarkable consequence of Green's Lemma is the isomorphism of all the maximal subgroups contained within the one $\mathscr{D}$-class.

Theorem 1.2.7 *Any two group $\mathscr{H}$ classes H_e, H_f $(e, f \in E)$ within the same $\mathscr{D}$-class of a semigroup S are isomorphic.*

Proof Take a in $R_e \cap L_f$. Then $ea = a$ and $a'a = f$ for some a' in S. As before, the mapping $\rho_a \circ \lambda_{a'}$ defines a bijection of H_e onto H_f in which e is mapped to $a'ea = a'a = f$. Note that $aa'a = af = a$, whence $aa' \in E(S)$ and $aa'\mathscr{R}a$. Hence for any $z \in R_a$ we have $aa'\mathscr{R}z$ and thus $aa'z = z$. In particular, for $y \in H_e$, $aa'y = y$. In order to complete the proof we verify that the bijection of H_e onto H_f whereby $x \mapsto a'xa$ is a homomorphism. To see this take any x, y in H_e. We obtain

$$a'xya = a'x(aa'y)a = (a'xa)(a'ya),$$

as required. ∎

Fundamentals | 19

Next let us turn our attention to $\mathscr{D}$-classes which contain a regular element a. Take $a' \in V(a)$. In this case $a'a \mathscr{L} a \mathscr{R} aa'$ and so both R_a and L_a contain an idempotent. Conversely, if $a \mathscr{R} e$ for some $e = e^2$, so that $e = ay$, say (y in S^1), we obtain $a = ea = aya$ and thus $yay \in V(a)$. It follows that if an $\mathscr{R}$-class R or $\mathscr{L}$-class L contains a regular element, then it contains an idempotent, whence every element of R or L is regular. Since every $\mathscr{L}$-class of S within its $\mathscr{D}$-class D meets every $\mathscr{R}$-class of S contained in D, it follows that all elements of D are regular, and D is thereby called a *regular $\mathscr{D}$-class*, in which case every $\mathscr{R}$- and $\mathscr{L}$-class of D contains at least one idempotent. Furthermore, for $a \in \mathrm{Reg}(S)$ with $a' \in V(a)$, the fact that $a \mathscr{R} aa' \mathscr{L} a'$ proves that $V(a) \subseteq D_a$, giving us the picture shown in Fig. 1.1.

a			aa'
	·		
	·		
	·		
$a'a$			a'

Fig. 1.1.

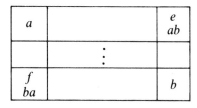

Fig. 1.2.

Conversely, suppose that $b \in S$ such that $R_a \cap L_b$ and $L_a \cap R_b$ contain the idempotents e and f respectively as in Fig. 2. By Theorem 1.2.5, $ab \in H_e$ and $ba \in H_f$. By Green's Lemma, ρ_a defines a bijection of L_b onto L_a, and thus there exists a unique $x \in H_b$ such that $xa = f$. Since $af = a$ it follows that $axa = a$. Furthermore, $ax \in E(S)$ and because λ_a defines an $\mathscr{L}$-class-preserving bijection from R_b onto R_a it follows that $ax \in H_e$, and so $ax = e$. But then $x = xe = xax$. We conclude that x is the unique inverse of a in H_b. This all serves to prove our next result.

Theorem 1.2.8 *The $\mathscr{H}$-class H contains an inverse x of a if and only if $R_a \cap L_b$ and $R_b \cap L_a$ each contain an idempotent. In this case x is the only inverse of a in H_b.*

This yields the following description of regular $\mathscr{D}$-classes.

Theorem 1.2.9 *Let a, b be regular elements of a semigroup S. Then:*

(i) *$a \mathscr{L} b$ if and only if for some [for all] $a' \in V(a)$ there exists $b' \in V(b)$ such that $a'a = b'b$;*
(ii) *$a \mathscr{R} b$ if and only if for some [for all] $a' \in V(a)$ there exists $b' \in V(b)$ such that $aa' = bb'$;*
(iii) *$a \mathscr{H} b$ if and only if for some [for all] $a' \in V(a)$ there exists $b' \in V(b)$ such that $a'a = b'b$ and $aa' = bb'$.*

20 | Techniques of semigroup theory

Proof (i) Let $a\mathscr{L}b$ and take $a' \in V(a)$, whence $a'a\mathscr{L}a$. Take an idempotent e in R_b whereupon by Theorem 1.2.8 there exists b' in $V(b)$ also a member of $R_{a'a} \cap L_e$ which, by Theorem 1.2.5, has the required property.

Conversely, if there exists $a' \in V(a)$, $b' \in V(b)$ with $a'a = b'b$ then $a\mathscr{L}a'a = b'b\mathscr{L}b$, whence $a\mathscr{L}b$. Statement (ii) is the dual of (i), while (iii) follows from Theorem 1.2.8 and the definition of Green's $\mathscr{H}$ relation. ∎

A fact that we shall use below and in Section 1.4 is taken from Hall (1973).

Lemma 1.2.10 *Let $a, b \in \text{Reg}(S)$ with $L_a \geqslant L_b$. Then for each $e \in E(L_a)$ there exists $f \in E(L_b)$ such that $e \geqslant f$.*

Proof Take $e \in E(L_a)$ and $b' \in V(b)$. Using $be = b$, one checks that $eb' \in V(b)$, whence $eb'b \in E(L_b)$ and $eb'b \leqslant e$. ∎

For each regular $\mathscr{D}$-class D of S there is an associated group: the group of any of the (isomorphic) group $\mathscr{H}$-classes in D. Schützenberger (1957) showed that there is a group naturally associated with any $\mathscr{D}$-class of a semigroup which, in the case of a regular $\mathscr{D}$-class, is isomorphic to a group $\mathscr{H}$-class of D. This so-called Schützenberger group of the $\mathscr{D}$-class D is constructed as follows. Let $H \in S/\mathscr{H}$, $h \in H$, $x \in S^1$. It follows from the proof of Green's Lemma that if $hx \in H$ then ρ_x defines a permutation on H, and so $T_r(H) = \{x \in S^1 : Hx = H\}$ is a submonoid of S^1. Define an equivalence $x \sim y$ on $T_r(H)$ whereby $x \sim y$ if $hx = hy$ for some (and thus for all) h in H, which one sees is a congruence on $T_r(H)$: denote $T_r(H)/\sim$ by $\Gamma_r(H)$. The construction ensures that for $h_1, h_2 \in H$ there is a unique $\bar{x} \in \Gamma_r(H)$ such that $h_1 x = h_2 (x \in \bar{x})$ and so $\Gamma_r(H)$ is isomorphic to a regular group of permutations via $\bar{x} \mapsto \rho_x | H$; $\Gamma_r(H)$ turns out to depend only on the $\mathscr{D}$-class D containing H and not on the chosen $\mathscr{H}$-class; it is the so-called *Schützenberger group of D*, and the construction shows it to be a group of cardinality equal to that of H. We outline its characteristic properties: details can be found in, for example, Lallement (1979).

Let H_1, H_2 be two $\mathscr{H}$-classes in D. Then, by the proof of Green's Lemma, there are two mutually inverse, $\mathscr{R}$-class preserving inner translations, ρ_s, ρ_t say, between the two corresponding $\mathscr{L}$-classes, L_1 and L_2. One readily checks that $\Gamma_r(H_1) \simeq \Gamma_r(H_2)$ via the mapping $\bar{x} \mapsto \overline{txs}$. Furthermore, if H is itself a group, each class $\bar{x} \in \Gamma_r(H)$ has a unique representative x from H itself, whence $\Gamma_r(H) \simeq H$. Finally, the dual group $\Gamma_l(H)$ is isomorphic to $\Gamma_r(H)$. To see this let h be a fixed member of H and $\bar{x} \in \Gamma_l(H)$. There exists a unique $\bar{y} \in \Gamma_r(H)$ such that $xh = hy$. Again it is routine to verify that the associated bijection defined by $\bar{x} \mapsto \bar{y}$ is a homomorphism between $\Gamma_l(H)$ and $\Gamma_r(H)$; this follows because any pair of left and right inner translations λ_x, ρ_y of a semigroup are 'linked', meaning that $(\lambda_x(h))\rho_y = \lambda_x(h\rho_y)$, a fact which is but

Fundamentals | 21

a statement of associativity. We are therefore entitled to suppress subscripts, r and l, and speak of the *Schützenberger group*, $\Gamma(D)$.

Theorem 1.2.11 *Let H_1, H_2 be $\mathcal{H}$-classes within a $\mathcal{D}$-class D. Then $\Gamma_r(H_1) \simeq \Gamma_r(H_2) \simeq \Gamma_l(H_2) \simeq \Gamma_l(H_1) = \Gamma(D)$, the group of D; and $|\Gamma(D)| = |H|$ for each $\mathcal{H}$-class H within D. Furthermore, if D is regular, $\Gamma(D)$ is isomorphic to each of its maximal subgroups.*

As an application we prove a result of Hall and Munn (1979). Recall that a semigroup is group-bound if some power of each element lies in a subgroup.

Theorem 1.2.12 *Let S be a group-bound semigroup in which every subgroup is trivial. Then $\mathcal{H}$ is trivial on S.*

Proof Let ρ_t be a right translation, where t is a member of $T_r(H)$ for some $\mathcal{H}$-class H, and write γ_t for the corresponding member of $\Gamma_r(H)$. Since S^1 is also a group-bound semigroup in which every subgroup is trivial, there exists a positive integer n such that t^n is idempotent. Hence $t^n . t^{n+1} = t^{n+1}$ and $t^{n+1} . t^{2n-1} = t^n$, from which it follows that $t^{n+1} \in H_{t^n}$. But H_{t^n} is a group, and so $t^{n+1} = t^n$. Thus $\gamma_{t^n} = \gamma_{t^n} . \gamma_t$, and so since $\Gamma(H)$ is a group, γ_t is the identity of $\Gamma(H)$. Consequently, $|\Gamma(H)| = 1$. But $|\Gamma(H)| = |H|$, and therefore $|H| = 1$. ∎

Let U be a subsemigroup of S, and let $\mathcal{G}$ denote any one of Green's relations. We write $a\mathcal{G}^U b$ to mean that $a\mathcal{G}b$ in the semigroup U. It is obvious that $\mathcal{G}^U \subseteq \mathcal{G}^S \cap (U \times U)$. A partial converse is true for regular subsemigroups (Hall, 1972).

Lemma 1.2.13 *If U is a regular subsemigroup of a semigroup S then $\mathcal{R}^U = \mathcal{R}^S \cap (U \times U)$, $\mathcal{L}^U = \mathcal{L}^S \cap (U \times U)$, $\mathcal{H}^U = \mathcal{H}^S \cap (U \times U)$.*

Proof From the previous comment together with duality and the definition of H, we need only show that $\mathcal{R}^S \cap (U \times U) \subseteq \mathcal{R}^U$. To this end take a, b in U with $a\mathcal{R}^S b$ and $a' \in V(a), b' \in V(b)$. Then $(a, aa') \in \mathcal{R}^U \subseteq \mathcal{R}^S$, $(b, bb') \in \mathcal{R}^U \subseteq \mathcal{R}^S$, whence $(aa', bb') \in \mathcal{R}^S$, and thus $bb'aa' = aa'$, $aa'bb' = bb'$, which tells us that $(aa', bb') \in \mathcal{R}^U$, whence $a\mathcal{R}^U aa' \mathcal{R}^U bb' \mathcal{R}^U b$. ∎

The previous result does not extend to $\mathcal{D}$- and $\mathcal{J}$-classes (see Exercise 1.2.8 below) but the next corollary is along these lines.

Corollary 1.2.14 *If a regular $\mathcal{D}$-class D of S forms a subsemigroup of S then D is a bisimple semigroup.*

Proof Let $a, b \in D$. Then there exists $c \in D$ such that $a\mathscr{L}^S c\mathscr{R}^S b$. Since D is a regular $\mathscr{D}$-class, it is a regular subsemigroup of S, and invoking Lemma 1.2.13 we conclude that $a\mathscr{L}^D c\mathscr{R}^D b$, whence $a\mathscr{D}^D b$, as required. ∎

As an example of an embedding theorem involving Green's relations we give a short proof, taken from Higgins (1990a), of the theorem of Preston (1959) that any semigroup can be embedded in a (regular) bisimple monoid. First a class of bisimple monoids is constructed. Consider the subset of $\mathscr{T}_X$,

$$S = \{\alpha \in \mathscr{T}_X : \text{whenever } Y \subseteq X \text{ and } |Y| = |X| \text{ then } |Y\alpha| = |X|\}.$$

Clearly, S is a submonoid of $\mathscr{T}_X$ (if X is finite, S is the symmetric group). Given $\alpha \in S$ construct a map β as follows. For each $a \in \operatorname{ran}\alpha$ choose a member of $a\alpha^{-1}$ for $a\beta$. Thus $\beta|\operatorname{ran}\alpha$ is injective, and take $\beta|(X\setminus\operatorname{ran}\alpha)$ to be any injection into X. By construction $\alpha\beta\alpha = \alpha$ and $\beta \in S$: to see this let $Y \subseteq X$ with $|Y| = |X|$. Either $|Y \cap \operatorname{ran}\alpha| = |X|$ or $|Y \cap (X\setminus\operatorname{ran}\alpha)| = |X|$. In the first case

$$|Y\beta| \geq |(Y \cap \operatorname{ran}\alpha)\beta| = |Y \cap \operatorname{ran}\alpha| = |X|,$$

and likewise $|Y\beta| = |X|$ in the latter case. Therefore S is a regular monoid. By Lemma 1.2.13, Green's relations $\mathscr{L}$ and $\mathscr{R}$ in S are the restrictions of the corresponding relations in $\mathscr{T}_X$ to $S \times S$ (see Exercise 1.2.4). Hence, to show that S is bisimple it is sufficient to prove that given $\alpha, \beta \in S$ there exists $\gamma \in S$ such that $\ker\alpha = \ker\gamma$ and $\operatorname{ran}\gamma = \operatorname{ran}\beta$. However, by definition of S, $|\operatorname{ran}\beta| = |X|$, which also equals the number of equivalence classes of $\ker\alpha$. A suitable γ is then defined by any bijection from the classes of $\ker\alpha$ onto $\operatorname{ran}\beta$. Such a map γ will inherit the defining property of S from the map α.

Corollary 1.2.15 *Any semigroup may be embedded in a bisimple monoid.*

Proof It suffices to show that this is true for $\mathscr{T}_X$ for any set X. Let Y be the disjoint union of X and Z, where Z is an infinite set of cardinality greater than that of X. Embed $\mathscr{T}_X$ into $\mathscr{T}_Y$ via the map ($\hat{}$), whereby

$$x\hat{\alpha} = \begin{cases} x\alpha & \text{for all } x \in X \ (\alpha \in \mathscr{T}_X) \\ x & \text{for all } x \in Z. \end{cases}$$

To complete the proof we show that ($\hat{}$) embeds $\mathscr{T}_X$ in $S \subseteq T_Y$, where S is the bisimple monoid constructed in the fashion above. Take $U \subseteq Y$ with $|U| = |Y|$. Then $|U \cap Z| = |Y|$, as $|U| = |U \cap X| + |U \cap Z|$ and $|U \cap X| < |Y|$, which is infinite. Thus we obtain $|Y| \geq |U\hat{\alpha}| \geq |(U \cap Z)\hat{\alpha}| = |U \cap Z| = |Y|$, giving equality throughout. ∎

Remark If S is a monoid, the right regular representation of S into $\mathscr{T}_S$, followed by the embedding of the above proof, provide a monoid embedding (identity-preserving semigroup embedding) of S into a bisimple monoid.

These ideas can be adapted to give the corresponding theorem for inverse semigroups, first proved by Reilly (1965): see Exercise 1.2.11. That any orthodox semigroup can be embedded in a bisimple orthodox semigroup has been proved by Ash (1980) using amalgamation techniques. More results along these lines can be found in Byleen (1984, 1988), and in Hall (1980, 1986). Pastijn (1977) first proved that any semigroup can be embedded in a bisimple idempotent-generated semigroup, a result also demonstrated explicitly in Howie (1981a). Laffey (1983) considered the question of embedding finite semigroups in finite idempotent-generated semigroups. A fundamental example, the so-called four-spiral semigroup of Byleen et al. (1978), is bisimple, idempotent-generated but not completely simple. It forms the basis of their subsequent paper (1980) on building bisimple idempotent-generated semigroups.

Chain conditions

Just as chain conditions are important in the theory of rings, they also play a significant role in the theory of semigroups. The chain conditions which have proved to be the most fruitful are descending chain conditions similar to the artinian conditions of ring theory, but defined on principal, as opposed to arbitrary, ideals. They are thus conveniently formulated in terms of the partially ordered sets that arise from Green's relations.

We shall consider seven related conditions; M_L, M_R, M_H, M_J, M_L^*, M_R^*, and GB. A semigroup S satisfies M_L if every non-empty set of $\mathscr{L}$-classes from S has a minimal member or, equivalently, every strictly descending chain of $\mathscr{L}$-classes in the poset $S/\mathscr{L}$ is finite. The conditions M_R, M_H, and M_J are the corresponding chain conditions on the posets $S/\mathscr{R}$, $S/\mathscr{H}$, and $S/\mathscr{J}$ respectively. We say that S satisfies M_L^* [M_R^*] if and only if, for all $J \in S/\mathscr{J}$, the set of all $\mathscr{L}$-classes [$\mathscr{R}$-classes] of S contained in J contains a minimal member. The condition GB means that S is group-bound.

Let X and Y be semigroup conditions. We write $X \leqslant Y$ ('X implies Y') if and only if every semigroup that satisfies X also satisfies Y; furthermore, we write $X \Leftrightarrow Y$ ('X is equivalent to Y') if and only if $X \leqslant Y$ and $Y \leqslant X$. The relation $\leqslant$ is readily seen to be a partial ordering of any set of semigroup conditions. The conjunction of a finite family $(A_1, A_2, \ldots, A_n)$ of semigroup conditions will be denoted by $A_1 \wedge A_2 \wedge \ldots \wedge A_n$, and a semigroup S is said to satisfy this conjunction if S satisfies each of the conditions A_i ($1 \leqslant i \leqslant n$). Clearly, if A, B, and C are semigroup conditions such that $A \leqslant B$ then $A \wedge C \leqslant B \wedge C$.

For the remainder of this section we shall denote the family

$$(M_L, M_R, M_H, M_J, M_L^*, M_R^*, GB)$$

by Ω, and the set of all conjunctions of non-empty subfamilies of Ω by $\Lambda(\Omega)$. Clearly, $\Lambda(\Omega)$ is a finite lower semilattice with respect to $\leqslant$, the greatest lower bound of the pair (A, B) being their conjunction, $A \wedge B$. We shall establish that the Hasse diagram of the semilattice $\Lambda(\Omega)$ is as shown in Fig. 1.3. The verification of this, in which we shall follow Hall and Munn (1979), comprises two stages: first a theorem establishes all the implications that do hold along our seven conditions; and then a series of counter-examples shows that there are no implications other than those given by Fig. 1.3.

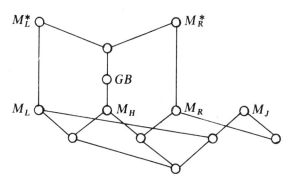

Fig. 1.3.

First we require an elementary property of the conditions M_L^* and M_R^*.

Lemma 1.2.16 *A semigroup S satisfies M_L^* if and only if each $\mathscr{L}$-class in each $\mathscr{J}$-class is minimal in the set of $\mathscr{L}$-classes of that $\mathscr{J}$-class.*

Proof Clearly, only the forward direction requires proof, so suppose that S satisfies M_L^* and that a and b are $\mathscr{J}$-related members of S such that $L_a \leqslant L_b$. By hypothesis there exists an $\mathscr{L}$-class, L_c say, minimal amongst the $\mathscr{L}$-classes in J_a. Now $L_a \leqslant L_b$ implies that there exists $x \in S^1$ such that $a = xb$. Since $J_b = J_c$, there exist $m, n \in S^1$ such that $b = mcn$; then $a = xb = xmcn$. Put $xmc = d$. If $L_d < L_c$, then since L_c is minimal in $J_a = J_c$, it follows that $J_d < J_c$, whence $J_a = J_{dn} \leqslant J_d < J_c$, a contradiction. Hence $L_d = L_c$. Thus there exist $l \in S^1$ such that $lxmc = c$. Then

$$b = mcn = m(lxmc)n = ml(xb) = (ml)a.$$

The equations $a = xb$ and $b = (ml)a$ together imply that $L_a = L_b$, as required. ∎

Theorem 1.2.17

(i) $M_L \leqslant M_L^*$; $M_R \leqslant M_R^*$;

(ii) $M_L \wedge M_R \leqslant M_J$;

(iii) $M_J \wedge M_L^* \leqslant M_L$; $M_J \wedge M_R^* \leqslant M_R$;
(iv) $M_L \wedge M_R^* \leqslant M_H$; $M_R \wedge M_L^* \leqslant M_H$;
(v) $M_H \leqslant GB$;
(vi) $GB \leqslant M_L^* \wedge M_R^*$.

Proof We note first that (i) follows from the definitions of the conditions involved. The statement (ii) is in earlier result of Green (1951) and is proved as follows.

Let $J_1 \geqslant J_2 \geqslant J_3 \geqslant \ldots \geqslant J_n \geqslant \ldots$ be a descending chain of $\mathcal{J}$-classes. Take $a_i \in J_i$ ($i = 1, 2, \ldots$). Then, since $J_i \geqslant J_{i+1}$, there exist x_i, y_i in S^1 such that $a_{i+1} = x_i a_i y_i$ ($i = 1, 2, \ldots$). Thus

$$a_{i+1} = x_i x_{i-1} \ldots x_1 a_1 y_1 y_2 \ldots y_i.$$

Now

$$R(a_1) \supseteq R(a_1 y_1) \supseteq \ldots \supseteq R(a_1 y_1 y_2 \ldots y_n) \supseteq \ldots,$$

and

$$L(x_1) \supseteq L(x_2 x_1) \supseteq \ldots \supseteq L(x_n x_{n-1} \ldots x_1) \supseteq \ldots.$$

Since $M_L \wedge M_R$ holds in S, and thus also in S^1, there exists a positive integer m such that

$$R(a_1 y_1 y_2 \ldots y_m) = R(a_1 y_1 y_2 \ldots y_m y_{m+1}) \quad \text{and}$$

$$L(x_m x_{m-1} \ldots x_1) = L(x_{m+1} x_m \ldots x_1).$$

Thus there exist $u, v \in S^1$ such that

$$a_1 y_1 y_2 \ldots y_{m+1} v = a_1 y_1 y_2 \ldots y_m \quad \text{and}$$

$$u x_{m+1} x_m \ldots x_1 = x_m x_{m-1} \ldots x_1.$$

Hence

$$u x_{m+1} \ldots x_1 a_1 y_1 \ldots y_{m+1} v = x_m \ldots x_1 a_1 y_1 \ldots y_m;$$

that is, $u a_{m+2} v = a_{m+1}$. This implies that $J_{m+1} \leqslant J_{m+2}$, whence these two $\mathcal{J}$-classes are equal, and therefore we conclude that S satisfies M_J.

(iii) Let S be a semigroup satisfying M_J and M_L^*. Consider a non-empty set $\mathcal{C}$ of $\mathcal{L}$-classes of S. Since S satisfies M_J there exists some a in S such that $L_a \in \mathcal{C}$ and, for all $x \in S$, if $L_x \in \mathcal{C}$ and $J_x \leqslant J_a$ then $J_x = J_a$. Suppose that $b \in S$ is such that $L_b \in \mathcal{C}$ and $L_b \leqslant L_a$. Then $J_b \leqslant J_a$ and so $J_a = J_b$. But since S satisfies M_L^* it follows from Lemma 1.2.16 that L_a is minimal in the set of all $\mathcal{L}$-classes contained in J_a. Hence $L_b = L_a$. Consequently, L_a is minimal in $\mathcal{C}$. This shows that S satisfies M_L: thus $M_J \wedge M_L^* \leqslant M_L$. A dual argument shows that $M_J \wedge M_R^* \leqslant M_R$.

(iv) Let S be a semigroup satisfying M_L and M_R^*. Consider a sequence $a_1, a_2, a_3, \ldots$ of elements of S such that $H_{a_1} \geqslant H_{a_2} \geqslant H_{a_3} \ldots$. We have $L_{a_1} \geqslant L_{a_2} \geqslant L_{a_3} \geqslant \ldots$ and so, since S satisfies M_L, there exists a positive integer k such that the elements $a_k, a_{k+1}, a_{k+2}, \ldots$ are $\mathcal{L}$-related. Thus

$a_k, a_{k+1}, a_{k+2}, \ldots$ are $\mathscr{J}$-related. But $R_{a_k} \geqslant R_{a_{k+1}} \geqslant \ldots$ and so, by the dual of Lemma 1.2.16, the elements $a_k, a_{k+1}, a_{k+2}, \ldots$ are $\mathscr{R}$-equivalent. It follows that $H_{a_k} = H_{a_{k+1}} = H_{a_{k+2}} = \ldots$. This proves that S satisfies M_H. Hence $M_L \wedge M_R^* \leqslant M_H$, and the other implication in the statement follows dually.

(v) Let S be a semigroup satisfying M_H. Consider any element $a \in S$. Since $H_a \geqslant H_{a^2} \geqslant H_{a^3} \geqslant \ldots$ there exists a positive integer n such that $H_{a^n} = H_{a^{n+1}} = H_{a^{n+2}} = \ldots$. But this means that $a^n \mathscr{H} a^{2n}$ and so, by Corollary 1.2.6, H_{a^n} is a group. Thus S is group-bound and therefore $M_H \leqslant GB$.

(vi) Let S be a group-bound semigroup. We shall show that S satisfies M_L^*. Let $a, b \in S$ be such that $a \mathscr{J} b$ and $L_b \leqslant L_a$. Then there exist elements $u, v, c \in S$ such that $a = ubv$ and $b = ca$. Thus $a = (uc)av$ and so $a = (uc)^n av^n$ for all positive integers n. Now S is group-bound, and so we can choose n such that $n > 1$ and $(uc)^n$ lies in a subgroup of S, with identity element e, say. Write $g = (uc)^n$ and let g^{-1} denote the inverse of g in the subgroup H_e of S. Then

$$ea = e(gav^n) = (eg).av^n = gav^n = a$$

and so

$$g^{-1}(uc)^{n-1}ub = g^{-1}(uc)^{n-1}uca = g^{-1}ga = ea = a.$$

Hence $L_a \leqslant L_b$, from which it follows that $L_a = L_b$. Consequently, S satisfies M_L^*. By duality, S also satisfies M_R^*. Thus $GB \leqslant M_L^* \wedge M_R^*$. ∎

Corollary 1.2.18 $M_L \wedge M_R = M_J \wedge M_H = M_J \wedge GB = M_J \wedge M_L^* \wedge M_R^*$.

Proof By (ii) $M_L \wedge M_R \leqslant M_J$ and, by (i) and (iv), $M_L \wedge M_R \leqslant M_L^* \wedge M_R \leqslant M_H$. Thus $M_L \wedge M_R \leqslant M_J \wedge M_H$. On the other hand, by (v), (vi), and (iii),

$$M_J \wedge M_H \leqslant M_J \wedge GB \leqslant M_J \wedge (M_L^* \wedge M_R^*)$$
$$= (M_J \wedge M_L^*) \wedge (M_J \wedge M_R^*) \leqslant M_L \wedge M_R.$$

These two statements together yield the result. ∎

It follows easily from Theorem 1.2.17 and Corollary 1.2.18 that $\Lambda(\Omega)$ consists of at most 13 members, and indeed it follows that if they are all distinct, the implications between them are all exhibited in Fig. 1.3. We proceed to show by means of examples that this is the case. We shall denote the dual of a semigroup S by S^*. A *0-direct union* of disjoint semigroups S_i^0 ($i \in I$) is the set-theoretic union of the S_i^0 with all zeros identified with a single zero symbol 0, where the product $\circ$ is defined by

$$a \circ b = \begin{cases} ab & \text{if } a, b \in S_i, \text{ for some } i \in I, \\ 0 & \text{otherwise.} \end{cases}$$

For any two semigroups S and T, let $S + T$ denote the 0-direct union of S and T. This idea is introduced at this point because of the following observation.

Lemma 1.2.19 *Let S and T be semigroups and let X be any member of Ω. Then $S + T$ satisfies X if and only if both S and T satisfy X.*

We now consider four semigroups S_i ($i = 1, 2, 3, 4$) defined as follows: S_1, S_2, and S_3 are, respectively, an infinite monogenic semigroup, an infinitely descending semilattice, and the Baer–Levi semigroup of Exercise 1.1.11, while S_4 is the semigroup with zero 0, non-zero elements the ordered pairs (i, j) of positive integers i, j such that $i < j$, and multiplication of non-zero elements according to the rule that

$$(i, j)(r, s) = \begin{cases} (i, s) & \text{if } j = r, \\ 0 & \text{otherwise.} \end{cases}$$

(Associativity of this product is easily seen or, alternatively, S_4 can be realized as the subsemigroup of the partial transformation semigroup on the positive integers consisting of all mappings $\{(i, j)\}$ with $i < j$.)

These semigroups, and four others derived from them, label the rows of Table 1.1, the columns of which are labelled by the members of Ω.

The entry in Table 1.1 corresponding to a semigroup S and a condition X is 1 or 0 according as S satisfies or fails to satisfy X. For S_4 see Exercise 1.2.18. By Lemma 1.2.19, the row corresponding to $S + T$ is the boolean product of the row of S with the row of T.

Table 1.1

	M_L	M_R	M_J	M_H	M_L^*	M_R^*	GB
S_1	0	0	0	0	1	1	0
S_2	0	0	0	0	1	1	1
S_3	0	1	1	0	0	1	0
S_4	1	0	0	1	1	1	1
$S_1 + S_3$	0	0	0	0	0	1	0
$S_3 + S_3^*$	0	0	1	0	0	0	0
$S_3 + S_4^*$	0	1	0	0	0	1	0
$S_4 + S_4^*$	0	0	0	1	1	1	1

The information contained in Table 1.1 is sufficient to verify that the 13 conditions indicated in Fig. 1.3 are distinct. For instance, S_3 shows that M_R^* does not imply M_L^*, whence M_R^* does not imply any of the seven conditions that lie below M_L^* in Fig. 1.3, and thus is distinct from each of them. Also, M_R^* differs from M_R, and from M_J by virtue of S_1, whence M_R^* differs from all the other conditions. By symmetry, the same can be said of M_L^*. We can then work down the diagram calling upon appropriate lines from the table as we go.

Note that there are semigroups that satisfy none of our seven conditions: for instance, $S_1 + (S_3 + S_3^*)$.

Theorem 1.2.20 *For any semigroup S satisfying $M_R^* \wedge M_L^*$, Green's relations $\mathcal{D}$ and $\mathcal{J}$ coincide. In particular, $\mathcal{D} = \mathcal{J}$ for any group-bound semigroup.*

Proof Suppose that $(a, b) \in \mathcal{J}$ in S. Then $b = xay$ for some $x, y \in S^1$. Now $L_{xa} \leq L_a$ and by Lemma 1.2.16 we infer that $L_{xa} = L_a$. From $R_b \leq R_{xa}$ and the dual of the same lemma, we infer that $R_b = R_{xa}$. Hence $a \mathcal{L} xa \mathcal{R} b$, whence $a \mathcal{D} b$. ∎

Remarks 1.2.21 Each of the conditions M_R^* and M_L^* is not sufficient to ensure that $\mathcal{D} = \mathcal{J}$: the counter-examples given by Hall and Munn (1979) are so-called Croisot–Teissier semigroups (see Clifford and Preston 1967; and for further reference consult Levi 1986).

Finally, suppose that S is regular. Then, for all $a, b \in S$, $L_a \geq L_b$ if and only if to each $e \in E(L_a)$ there is a corresponding $f \in E(L_b)$ such that $e \geq f$, by Lemma 1.2.10. It follows that on S each of the conditions M_L, M_R, and M_H is equivalent to the condition that every non-empty set of idempotents of S contains a minimal member with respect to the natural partial ordering.

We conclude this section by introducing the concepts of 0-simplicity and principal factors, ideas that are central to the Rees Theorem and applications which are the subject of the next section.

As already stated, a semigroup S is simple if S is its only ideal, or equivalently $\mathcal{J}$ is the universal relation of S. We say that S is *0-simple* if S has a zero, 0, $S^2 \neq \{0\}$, and the only ideals of S are $\{0\}$ and S (the condition that $S^2 \neq \{0\}$ serves only to exclude two-element null semigroups). These semigroups arise when the multiplicative structure of $\mathcal{J}$-classes within a semigroup is examined.

A semigroup S either has no minimal ideals or possesses a unique minimal ideal denoted by K, the *kernel of S* (as two minimal ideals M, N must contain the ideal MN, and thus each other). If I is an ideal of K, then $I \supseteq KIK \supseteq S^1 KIKS^1 \supseteq K \supseteq I$, where the third containment is so because $S^1 KIKS^1$ is an ideal of S. We have proved the following.

Proposition 1.2.22 *If a semigroup S has a kernel K, then K is a simple semigroup. In particular, all finite semigroups have a simple kernel.*

In case S has a zero then trivially $K = \{0\}$, and the question arises as to the existence of ideals intermediate between $\{0\}$ and S. The following fact is not quite as obvious as one might anticipate.

Lemma 1.2.23 *S is 0-simple if and only if $SaS = S$ for each $a \in S \setminus \{0\}$.*

Proof Suppose that S is 0-simple. Then S^2 is an ideal of S, $S^2 \neq \{0\}$ whence $S^2 = S = S^3$. For any $a \in S$, SaS is an ideal of S, which implies that $SaS = S$ or $SaS = \{0\}$. The subset $I = \{x \in S : SxS = \{0\}\}$ is an ideal of S containing 0, whence $I = \{0\}$, as otherwise $I = S$ and $S^3 = \{0\}$. Hence $SaS = S$ for all $a \in S\setminus\{0\}$. The converse is readily obtained. ∎

By a 0-*minimal ideal* M of S we mean $M \neq \{0\}$ and contains no ideals of S other than $\{0\}$ and itself. If $M^2 \neq \{0\}$ then $M^2 = M = M^3$. Take $a \in M\setminus\{0\}$. Then $S^1 a S^1$ is an ideal of S and is not $\{0\}$, whence $S^1 a S^1 = M$. Thus $MaM \subseteq S^1 a S^1 = M = M^3 = M(S^1 a S^1)M = (MS^1)a(S^1 M) \subseteq MaM$, and so $MaM = M$ which proves the following.

Proposition 1.2.24 *If M is a 0-minimal ideal of S then either $M^2 = \{0\}$ or M is a 0-simple semigroup.*

Corollary 1.2.25 *If I, J are ideals of S such that $I \subseteq J$ and there is no ideal of S lying strictly between I and J, then J/I is either 0-simple or null.*

Proof Apply Exercise 1.1.6(c) and Proposition 1.2.24 above. ∎

Let J be a $\mathcal{J}$-class of S. Let $I(a) = \{b \in J(a) : J_b < J_a\}$. Either $I(a)$ is empty, whereupon $J(a)$ is the kernel of S, or $I(a)$ is an ideal of S contained in $J(a)$. Moreover, suppose that B is an ideal of S such that $I(a) \subseteq B \subset J(a)$ and let $b \in B$. Then, clearly, $J_b < J_a$ and so $b \in I(a)$. Since b was arbitrary we infer that $B = I(a)$. This yields the following theorem.

Theorem 1.2.26 *If $J_a \in S/\mathcal{J}$ then either J_a is the kernel of S or the set $I = \{x \in S : J_x < J\}$ is an ideal of S (and hence of $J(a)$) and the factor $J(a)/I$ is either 0-simple or null.*

The semigroups K and $J(a)/I(a)$ are called the *principal factors* of S. A semigroup is called *semisimple* if none of its principal factors are null. A principal factor J/I can be thought of as the $\mathcal{J}$-class J together with 0 and for any a, b in J the product of a and b is ab if $ab \in J$ but is 0 otherwise.

Exercise 1.2

1. For $e, f \in E(S)$, $e \mathcal{R} f$ if and only if $ef = f$ and $fe = e$.
2. If L is a left ideal and R is a right ideal of S then $RL \subseteq R \cap L$, with equality if S is regular.
3. If S is a right cancellative semigroup without idempotents, then every $\mathcal{L}$-class of S is trivial.
4.* Show that Green's relation on $\mathcal{T}_X$ are as follows:

   (i) $\alpha \mathcal{L} \beta$ if and only if $\operatorname{ran} \alpha = \operatorname{ran} \beta$;
   (ii) $\alpha \mathcal{R} \beta$ if and only if $\ker \alpha = \ker \beta$;

(iii) $\alpha \mathcal{H} \beta$ if and only if $\operatorname{ran} \alpha = \operatorname{ran} \beta$ and $\ker \alpha = \ker \beta$;
(iv) $\alpha \mathcal{D} \beta$ if and only if $\operatorname{rank} \alpha = \operatorname{rank} \beta$;
(v) $\mathcal{D} = \mathcal{J}$.

5. Let Y be a subset of X and π a partition of X such that $|Y| = |X/\pi|$. Let H be the $\mathcal{H}$-class of $\mathcal{T}_X$ determined by (π, Y). Then H is a group if and only if Y is a transversal of π, in which case $H \simeq \mathcal{G}_Y$, the symmetric group on Y.

6.* (a) Describe Green's relations on $\mathcal{PT}_X$.
   (b) Describe Green's relations on $\mathcal{I}_X$.

7. Let $e \in E(S)$. H_e is a subgroup of $eSe = eS \cap Se$.

8. Let $S = \{0, e, f, a, b\}$ be the semigroup of matrices under multiplication

$$0 = \begin{bmatrix} 0 & 0 \\ 0 & 0 \end{bmatrix}, \quad e = \begin{bmatrix} 1 & 0 \\ 0 & 0 \end{bmatrix}, \quad f = \begin{bmatrix} 0 & 0 \\ 0 & 1 \end{bmatrix},$$

$$a = \begin{bmatrix} 0 & 1 \\ 0 & 0 \end{bmatrix}, \quad b = \begin{bmatrix} 0 & 0 \\ 1 & 0 \end{bmatrix}.$$

Let U be the subsemigroup $\{0, e, f\}$ of S. Show that U is $\mathcal{J}$-trivial, yet $e \mathcal{D}^S f$.

9 *(a) Reconcile our two definitions of simple: i.e. show that $SaS = S$ for all $a \in S$ if and only if S has a single $\mathcal{J}$-class.
  (b) The set product LR of an $\mathcal{L}$-class L and an $\mathcal{R}$-class R is contained within a single $\mathcal{D}$-class.

10.* The following are equivalent for a regular semigroup S:
   (i) S is an inverse semigroup.
   (ii) S contains no two-element right zero semigroup, nor its dual.
   (iii) There is a subsemilattice E of S such that for each a in S there exists some $a' \in V(a)$ such that $aa', a'a \in E$.
   (iv) There exists a subsemilattice E of S such that for each e in $E(S)$ there exist $e', e'' \in V(e)$ such that $ee', e''e \in E$.
   (v) There is a subsemilattice E of S that meets every $\mathcal{L}$-class and every $\mathcal{R}$-class of S.

Moreover, the condition in (iii) cannot be weakened to the one-sided condition that insists only that $aa' \in E$.

11. Let X be an infinite set and consider

$$S = \{\alpha \in \mathcal{I}_X : |\overline{\operatorname{dom} \alpha}| < |X| \text{ and } |\overline{\operatorname{ran} \alpha}| < |X|\},$$

where $\bar{A}$ denotes the complement of the set A in X.

(a) Show that S is an inverse submonoid of $\mathcal{I}_X$.
(b) Use Exercise 1.2.6(b) and the method of proof of Corollary 1.2.15 to prove that any inverse semigroup can be embedded in a bisimple inverse monoid.

12. The following is a simple semigroup which is not bisimple.
    (a) Show that if S is cancellative without identity there is no pair of elements e, a in S such that $ea = a$ or $ae = a$. Deduce that in S, Green's $\mathscr{D}$ is trivial.
    (b) Show that with respect to matrix multiplication
    $$S = \left\{ \begin{bmatrix} a & 0 \\ b & 1 \end{bmatrix}; a, b \text{ positive reals} \right\}$$
    is cancellative without identity.
    (c) Show that $\mathscr{J}$ is universal on S, and that S is not group-bound.
13.* The *bicyclic monoid* C is the monoid with presentation $\langle a, b | ab = 1 \rangle$.
    (a) Let $S = \langle \alpha, \beta \rangle$, where $\alpha, \beta \in \mathscr{T}_{\mathbb{N}^0}$ are the mappings defined by $n\alpha = n + 1$, $n\beta = \max \{n - 1, 0\}$ for all $n \in \mathbb{N}^0$. Show that $\alpha\beta = 1$, but that $\beta\alpha \neq 1$. Hence deduce that S is a morphic image of C.
    (b) Show that any member of S, and of C, can be uniquely represented in the form $b^m a^n$. Deduce that S is a faithful representaion of C.
    (c) $b^k a^l . b^m a^n = b^i a^j$, where $i = k + m - \min(l, m)$ and $j = l + n - \min(l, m)$.
    (d) $b^i a^j \in E(C)$ if and only if $i = j$.
    (e) C is bisimple and if we represent C as the array

    $$\begin{array}{cccc} 1 & a & a^2 & \ldots \\ b & ba & ba^2 & \ldots \\ b^2 & b^2 a & b^2 a^2 & \ldots \\ & \vdots & & \end{array}$$

    then the $\mathscr{R}$- [$\mathscr{L}$-] classes of C are the rows [columns] of the array; each $\mathscr{H}$ class is a singleton.
    (f) The semilattice of idempotents is an infinite descending chain.
    (g) C satisfies neither M_L^* nor M_R^*.
14. A *right unit* u of S is a member of S such that $uS = S$. Suppose that S is a monoid and denote the subsemigroups of right units and of left units by P and Q respectively, while $P \cap Q = U$ is called the *group of units* of S.
    (a) Show that S is bisimple if and only if $QP = S$.
    (b) Describe the group of units of the monoid eSe, where $e \in E(S)$.
15. Show that S is semisimple if and only if $I = I^2$ for each ideal I of S.
16. Show that the relation 'is an ideal of' is transitive for semisimple semigroups, but not necessarily otherwise.
17. For elements a, b of a semigroups S define the relation $\mathscr{L}^*$ by saying that $a \mathscr{L}^* b$ if, for all $x, y \in S^1$, the following condition is satisfied:
    $$ax = ay \text{ if and only if } bx = by.$$

Let $\mathscr{R}^*$ denote the dual relation, $\mathscr{H}^* = \mathscr{L}^* \cap \mathscr{R}^*$, and call S *abundant* if every $\mathscr{L}^*$- and $\mathscr{R}^*$-class of S contains at least one idempotent.

(a) By considering the extended regular representations of S, prove that $a \mathscr{L}^* b$ [$a \mathscr{R}^* b$] if and only if $a \mathscr{L} b$ [$a \mathscr{R} b$] in some containing semigroup T of S.

(b) Show that any $\mathscr{H}^*$-class of an abundant semigroup S that contains an idempotent is a cancellative monoid.

Remark The class of abundant semigroups includes in particular the class of cancellative monoids, and any subsemigroup of a regular semigroup S that contains $E(S)$ is abundant. For results on abundant semigroups consult Fountain (1982). The earliest reference to the above-mentioned *-relations is Lyapin (1974).

18. Consider the semigroup $S = S_4$ defined after Lemma 1.2.19.

    (a) Show that the left ideal, right ideal, and ideal generated by (i, j) are respectively given by

    $$L_{ij} = \{0\} \cup \{(r, j) : 1 \leq r \leq i\}; \qquad R_{ij} = \{0\} \cup \{(i, s) : s \geq j\},$$

    and $I_{ij} = \{0\} \cup \{(r, s) : 1 \leq r \leq i, s \geq j\}$, respectively.

    (b) Verify the entries for S_4 in the table in Fig. 1.3.

    (c) Show that $I_{ij}, I_{i,j+1}, I_{i,j+2}, \ldots,$ is an infinite strictly descending chain of left ideals of S. Thus the condition M_L is weaker than the minimal condition on the set of all left ideals of a semigroup.

19.* (Venkatesan 1976) A semigroup S is *right-unipotent* if every $\mathscr{R}$-class has a unique idempotent. Prove that every right-unipotent semigroup is orthodox.

20. (Higgins 1987) Say that S has the RT^* property if for every sequence $a_1, a_2, \ldots$ of members of S there exist some positive integers $m < n$ such that $a_1 a_2 \ldots a_m = a_1 a_2 \ldots a_n$. S has LT^* if, dually, there always exists $m < n$ such that $a_m a_{m-1} \ldots a_1 = a_n a_{n-1} \ldots a_1$. If S enjoys both properties we say that S has T^*.

    (a) Prove that S has RT^* if and only if S has M_R and each $\mathscr{R}$-class of S is finite.

    (b) Prove that S has T^* if and only if S has M_J and each $\mathscr{J}$-class of S is finite.

    (c) Prove that the class of RT^* semigroups is closed under the taking of subsemigroups, morphic images, and finite direct products.

    (d) Show that the RT^* semigroups are not closed under the taking of direct powers by considering $S \times S \times \ldots$, where S is a 0-direct union of infinitely many one-element groups (S is sometimes called a *Kronecker semilattice*).

(e) Prove that the class of RT^* semigroups is *locally finite*, meaning that for any S with RT^*, any finitely generated subsemigroup of S is finite. Hence deduce that a finitely generated semigroup S is finite if and only if it has RT^*.

Remark Another remarkable class of locally finite semigroups is the class of bands (Green and Rees 1952; see also Howie 1976). More generally, a lemma of I. Simon (1980) states that a finitely generated semigroup S is finite if and only if $S\backslash E(S)$ is finite. This result can be proved by an application of a celebrated combinatorial theorem of Ramsay (see Pin 1986).

1.3 Completely 0-simple and completely regular semigroups

Completely 0-simple semigroups

Let S be a semigroup with zero 0. We say that $e \in E(S)$ is *primitive* if e is *0-minimal* in the natural partial order on $E(S)$: that is, $f \leqslant e$ implies $f = 0$ or $f = e$ ($f \in E(S)$)). A semigroup S is *completely 0-simple* if it is 0-simple and contains a primitive idempotent. A semigroup is *completely simple* if it is simple with a primitive idempotent. The effect of adjoining a zero to a completely simple semigroup is to create a completely 0-simple semigroup without zero divisors, and so any proposition about completely 0-simple semigroups has a corollary for the completely simple case which will generally not be stated in the sequel.

A semigroup S is said to be a *union of groups* if every element of S lies in a subgroup of S. It is easy to show that unions of groups are precisely the completely regular semigroups introduced in Section 1.1. It follows from Corollary 1.2.6 that each $\mathscr{D}$-class D of S is a disjoint union of (isomorphic) groups, and so by Theorem 1.2.5, D is a regular subsemigroup of S, indeed a bisimple (and thus simple) subsemigroup of S by Corollary 1.2.14. By Theorem 1.2.5 one readily sees that all idempotents of D are primitive (in the semigroup D) and thus any $\mathscr{D}$-class of a union of groups is completely simple. On the other hand, recall that a completely regular semigroup S is one in which, for each a in S there exists some $a^* \in V(a)$ such that $aa^* = a^*a$. This condition implies that $a \mathscr{H} a^*$, whence by Theorem 1.2.8 and Corollary 1.2.6, it follows that H_a is a group, and so S is a union of groups. (For this reason completely regular elements are sometimes described as those possessing 'group inverses'.)

The ease with which these facts are revealed demonstrates the usefulness of Green's relations and motivates the study of completely [0-] simple semigroups, which occur as structural blocks throughout semigroup theory—the kernel of any finite semigroup and the principal factors of finite full transformation semigroups, to mention two examples. These semigroups can be described up to the structure of the group of their non-zero $\mathscr{D}$-class (for they do turn out to be 0-bisimple) by the Rees Theorem, which is why they form the basis of other structure theories of more complex semigroups (see, for example, Allen 1971; Birget 1988a).

Let S be a finite 0-simple semigroup. Clearly, $E(S)$ is non-empty and since $S^2 \neq \{0\}$ it follows that S does not consist entirely of nilpotents (elements which have zero as a power) as any such semigroup is $\mathscr{J}$-trivial. We conclude that S contains non-zero idempotents, and in particular must contain a 0-minimal idempotent. The Rees Theorem for finite semigroups dates back to Suschkewitsch, and a short proof of this result, which provides a model for the general theorem, is provided here.

Since S is group-bound, $\mathscr{D} = \mathscr{J}$. Furthermore, as the non-zero $\mathscr{D}$-class D of S contains idempotents, it is regular, whence we conclude that S is 0-bisimple and regular. We utilize the extended right regular representation of S in $\mathscr{T}_{S^1} = T$. Suppose that $a, b \in D$ and that $ab \neq 0$, whence $ab \in D$ and so in T, $|\operatorname{ran} ab| = |\operatorname{ran} a| = |\operatorname{ran} b|$ (see Exercise 1.2.4) as $a \mathscr{D}^T ab \mathscr{D}^T b$. Since S is finite, this is possible only if ran a is a transversal (cross-section) of ker b (otherwise $|\operatorname{ran} ab| < |\operatorname{ran} a|$). But in this case ran $ab = \operatorname{ran} b$ and ker $ab = \operatorname{ker} a$, that is $a \mathscr{R}^T ab \mathscr{L}^T b$, which implies $a \mathscr{R}^S ab \mathscr{L}^S b$ by Lemma 1.2.13, since S is regular. We infer that for $a, b \in D$, either $ab = 0$ or $ab \in R_a \cap L_b$, and by Theorem 1.2.5 the latter occurs if and only if $L_a \cap R_b$ is a group.

Index the rows and columns of D by the index sets I, Λ respectively, and without loss we assume that $(1,1) \in I \times \Lambda$ with H_{11} a group. For each $i \in I$ and $\lambda \in \Lambda$ choose $r_i \in H_{i1}$ and $q_\lambda \in H_{1\lambda}$. By Lemma 1.2.4 and Theorem 1.2.5 the map $\varphi_{i\lambda} : H_{11} \to H_{i\lambda}$, where $a\varphi_{i\lambda} = r_i a q_\lambda$ is a bijection which, by Theorem 1.2.7, is an isomorphism if $H_{i\lambda}$ is a group. It follows that $x \in H_{i\lambda}$ can be represented as a triple $(a; i, \lambda)$, where $a \in H_{11}$ is such that $x = r_i a q_\lambda$. Take $y \in H_{j\mu}$, say, with representation $(b; i, \mu)$,. The product

$$(a; i, \lambda)(b; j, \mu) = (c; i, \mu)$$

where it remains only to determine c (if the product is 0, which occurs if and only if $H_{j\lambda}$ is not a group, we take $c = 0$, and agree that $(0; i, \lambda) = 0$ for all i in I and λ in Λ). But $xy = r_i a q_\lambda r_j b q_\mu$ and if the product is not 0, $xy \in H_{i\mu}$, whence it is evident that $c = a q_\lambda r_j b$. However, if the product is 0 then $q_\lambda r_j = 0$ so that in any case $c = a q_\lambda r_j b$. We have shown that any finite 0-simple semigroup can be represented as a Rees matrix semigroup, a concept which we now define.

Definition 1.3.1 Let I, Λ be index sets and let G^0 be a group with adjoined zero (we call G a 'group with zero'). Let $P = (p_{\lambda i})$ be a $\Lambda \times I$ matrix with entries from G^0. The *Rees semigroup with sandwich matrix* $P = M^0[G; I, \Lambda; P]$ is the set

$$S = \{(a; i, \lambda) : a \in G^0, i \in I, \lambda \in \Lambda\} \cup \{0\},$$

with product
$$(a; i, \lambda)(b; j, \mu) = (ap_{\lambda j}b; i, \mu)$$
with the understanding that $(0; i, \lambda) = 0$ for all $i \in I$, $\lambda \in \Lambda$ and that any product involving 0 is itself 0.

A general Rees semigroup S (for one readily checks associativity of this product) need not be finite nor must it be 0-simple: for example, if P is the zero matrix then S is null. The above argument shows that our finite 0-simple semigroup is isomorphic to a Rees semigroup $M^0[H_{11}; I, \Lambda; P]$, where $p_{\lambda i} = q_\lambda r_i$. Since every row and every column of D contains at least one group $\mathscr{H}$-class it follows that P is *regular*, meaning that no row and no column of P consists entirely of zeros. It is easy to verify that a Rees semigroup is regular if and only if its matrix P is regular.

Theorem 1.3.2 (Rees–Suschkewitsch Theorem) *A semigroup S is completely 0-simple if and only if it is isomorphic to some regular Rees semigroup.*

The verification that a regular Rees semigroup is completely 0-simple is straightforward (Exercise 1.3.2), and we have proved the converse for the case where S is finite. The general case depends on the following lemma.

Lemma 1.3.3 *For any non-zero $\mathscr{R}$-class R of a completely 0-simple semigroup S the set $R \cup \{0\}$ is a 0-minimal right ideal.*

Proof It is sufficient to show that $R \cup \{0\}$ is a right ideal; the 0-minimality follows, for if R' is a non-zero right ideal such that $R' \subseteq R \cup \{0\}$ then taking $a \in R'$, $a \neq 0$, and $x \in R$, we have $x \mathscr{R} a$ and thus $x = au \in R'$, which shows that $R' = R \cup \{0\}$. First, we prove that if R denotes the $\mathscr{R}$-class of a primitive idempotent e of S, then $R \cup \{0\}$ is the right ideal eS. The inclusion $R \cup \{0\} \subseteq eS$ is clear. Let $b \in eS$, $b \neq 0$. Since S is 0-simple, $e = xby$ for some $x, y \in S$. The element $f = byexe$ is an idempotent. Furthermore, $ef = fe = f$. Since e is primitive, $f = 0$ or $f = e$. But $f = 0$ implies $xfby = xbyexeby = e = 0$—a contradiction. Therefore $f = e = byexe$. This shows that $e \in bS$, which together with $b \in eS$ yields $b \in R_e$. Hence $R \cup \{0\} = eS$. From $S = SeS = S(R \cup \{0\})$ we deduce that every $x \in S$ is in $c(R \cup \{0\})$ for some $c \in S$. We prove that for $x \neq 0$, $R_x \cup \{0\} = c(R \cup \{0\})$ by showing inclusions in both directions. Assume that $x = cr$ for some $r \in R \cup \{0\}$. If $y \in R_x \cup \{0\}$, then $y = xt$ for some $t \in S$; it follows that $y = crt$ and thus $y \in c(R \cup \{0\})$. If $y \in c(R \cup \{0\})$, then $y = cr'$ for some r' in $R \cup \{0\}$; in case $r' = 0$, y is 0 and thus $y \in R_x \cup \{0\}$; in case $r' \neq 0$, we have $r' \mathscr{R} r$, which implies that $cr' \mathscr{R} cr$ or $y \in R_x$. Consequently, $R_x \cup \{0\} = c(R \cup \{0\}) = ceS$. ∎

36 | Techniques of semigroup theory

A dual proof shows that for a non-zero $\mathscr{L}$-class L, $L \cup \{0\}$ is a 0-minimal left ideal of S.

Lemma 1.3.4 *A completely 0-simple semigroup S is regular and 0-bisimple with non-zero $\mathscr{D}$-class D. Moreover, for $a, b \in D$ either $a \mathscr{R} ab \mathscr{L} b$ or $ab = 0$.*

Proof Let $a, b \in S \setminus \{0\}$. Then $aSb \neq \{0\}$, for otherwise we would have $S = S^2 = SaSSbS = S(aSb)S = S\{0\}S = 0$. Take $c \in aSb \setminus \{0\}$. Since $c \in aS \cap Sb$ we have, by Lemma 1.3.3, $a \mathscr{R} c \mathscr{L} b$, whence $a \mathscr{D} b$ as required. The non-zero $\mathscr{D}$-class is regular since it contains an idempotent. Finally, if $ab \neq 0$, then $R_{ab} \leq R_a$, $L_{ab} \leq L_b$ and, again from Lemma 1.3.3, we infer that $R_{ab} = R_a$ and $L_{ab} = L_b$, as required. ∎

Given Lemmas 1.3.3 and 1.3.4, a Rees representation of an arbitrary completely 0-simple semigroup can be constructed just as in the proof of the finite case, thus completing the proof of the Rees Theorem.

Another characterization of completely 0-simple semigroups was given by Munn (see Clifford and Preston 1961).

Theorem 1.3.5 *A 0-simple semigroup S is completely 0-simple if and only if S is group-bound.*

Proof Let a be a member of a completely 0-simple semigroup S. Then if H_a is not itself a subgroup of S then $a^2 = 0$, and so S is group-bound.

Conversely, suppose that S is 0-simple and group-bound. Then, by Theorem 1.2.20, $\mathscr{D} = \mathscr{J}$ for S, so let D denote the non-zero $\mathscr{D}$-class of S. As before, D does not consist entirely of nilpotents (as otherwise $\mathscr{J}$ would be trivial, and S would be a two-element null semigroup) but since S is group-bound, D must contain a subgroup, and thus an idempotent, whereupon D, and therefore also S, is regular. By Theorem 1.2.17, S satisfies $M_L^* \wedge M_R^*$, which evidently implies that S satisfies $M_L \wedge M_R$. But, by Remark 1.2.21, since S is regular, it follows that $E(D)$ contains a minimal member; that is, D has a primitive idempotent, as required. ∎

As a corollary, all periodic 0-simple semigroups are completely 0-simple.

The proof of Theorem 1.3.5 in Clifford and Preston (1961) is based on the following fact, attributed there to Andersen.

Theorem 1.3.6 *If e is any non-zero idempotent of a 0-simple semigroup S which is not completely 0-simple then S contains a bicyclic submonoid (see Exercise 1.2.13) with identity element e.*

We do not repeat the proof of Theorem 1.3.6 here, but it relies on the curious fact that any semigroup generated by two generators p and q such that $pq = 1$ but $qp \neq 1$ is isomorphic to the bicyclic monoid $\langle a, b| \, ab = 1\rangle$: in effect one merely repeats Exercise 1.2.13, working throughout with $\langle p, q \rangle$, to verify this. For an account of analogues of Andersen's Theorem for 0-simple semigroups with no non-zero idempotent, see Joñes (1987). For some generalizations of the bicyclic monoid, see Nivat and Perrot (1970), and Cowan and Rankin (1987).

Another fundamental characterization of completely 0-simple semigroups is due to Clifford (1949) (see Clifford and Preston 1961).

Theorem 1.3.7 *A 0-simple semigroup is completely 0-simple if and only if it contains at least one 0-minimal left ideal and at least one 0-minimal right ideal.*

The Baer–Levi semigroup (Exercise 1.1.11) with adjoined zero is 0-simple with a 0-minimal right ideal, but is not completely 0-simple. However, any 0-simple semigroup with a 0-minimal right ideal and non-zero idempotent is completely 0-simple (Clifford and Preston 1961, Exercise 2.7.12).

The isomorphism problem for completely 0-simple semigroups was solved by Rees (1940) (see Clifford and Preston 1961; Howie 1976).

Theorem 1.3.8 *Two regular Rees semigroups $S = M^0[G; I, \Lambda; P]$ and $T = M^0[K; J, M; Q]$ are isomorphic if and only if there exist an isomorphism $\theta: G \to K$, bijections $\psi: I \to J$, $\chi: \Lambda \to M$ and elements u_i $(i \in I)$, $v_\lambda (\lambda \in \Lambda)$ in K such that $p_{\lambda i}\theta = v_\lambda q_{\lambda \chi, i\psi} u_i$ for all λ, i.*

Consult Howie (1976) for a description of all congruences on a completely 0-simple semigroup.

Completely regular semigroups

We now return to our investigation of completely regular semigroups. In the preamble to the Rees Theorem we found that S was completely regular if and only if every $\mathscr{D}$-class D of S was a completely simple subsemigroup of S. The Rees Theorem gives us a hold on the $\mathscr{D}$-classes of S but provides no insight into any relationship between the $\mathscr{D}$-classes. A fruitful way to decompose a semigroup S is sometimes provided by the congruence η, the *minimum semilattice congruence* on S, which always exists, and is the intersection of all congruences σ on S such that S/σ is a semilattice. Equivalently, S/η is the maximum semilattice homomorphic image of S. The importance of the maximum semilattice image S/η lies partly is the observation that each η-class of S is a subsemigroup (a feature that evidently characterizes those congruences σ on S such that S/σ is a band). If a semigroup S has a congruence σ such that S/σ is a band B and each σ-class is a semigroup of type T, say, we call S a 'band of type-T semigroups'. We may specialize this definition to special classes of bands, in particular to semilattices.

38 | Techniques of semigroup theory

It is always the case that $\mathcal{J} \subseteq \eta$: to see this suppose that $a \mathcal{J} b$ in some semigroup S so that there exist $x, y, u, v \in S^1$ such that $a = xby$ and $b = uav$. Then, since η is the congruence on S generated by

$$\eta_0 = \{(a, a^2), (ab, ba) : a, b \in S\},$$

we obtain

$$a = xby \,\eta\, xb^2y \,\eta\, xbyb = ab = auav \,\eta\, ua^2v \,\eta\, uav = b,$$

and so $\mathcal{J} \subseteq \eta$.

The reverse inclusion is generally false (for example, in $(\mathbb{N}, +)$ $\mathcal{J}$ is trivial while η is universal) but we can demonstrate equality for the case where S is completely regular by proving that $\mathcal{J}$ is a semilattice congruence on S. Let $a, b \in S$. Since H_a is a group we have $a^2 \,\mathcal{H}\, a$ and thus $a \mathcal{J} a^2$. Then

$$J_{ab} = J_{(ab)^2} = J_{a(ba)b} \leq J_{ba},$$

but this of course yields $J_{ab} = J_{ba}$, as we can equally infer that $J_{ba} \leq J_{ab}$. Thus we have $J_a = J_{a^2}$ and $J_{ab} = J_{ba}$. It follows that if $\mathcal{J}$ is a congruence then $S/\mathcal{J}$ is a semilattice. Now suppose that $a \mathcal{J} b$ and $c \in S$. Take $x, y, u, v \in S^1$ such that $b = xay$ and $a = ubv$. Then $J_{ca} = J_{cubv} \leq J_{cub} = J_{bcu} \leq J_{bc} = J_{cb}$, and similarly $J_{cb} \leq J_{ca}$, whence $J_{ca} = J_{cb}$. Dually, we can show that $\mathcal{J}$ is a right congruence, thus concluding that $\mathcal{J}$ is a congruence on any completely regular semigroup.

Bands are completely regular, so that the above argument shows that in their case $\mathcal{J} = \eta$. But bands are also periodic, and we know that $\mathcal{D} = \mathcal{J}$ in any such semigroup. It follows that any η-class of a band is a completely simple band, thus yielding the following structure theorem.

Theorem 1.3.9 *In any band B, $\mathcal{D} = \eta$ and B is a semilattice of rectangular bands.*

We can do better. The following celebrated result is due to Clifford (1941).

Theorem 1.3.10 *For any completely regular semigroup S, $\mathcal{D} = \eta$ and S is a semilattice of completely simple semigroups.*

Proof We know that in S each $\mathcal{D}$-class is completely simple and that $\mathcal{J} = \eta$. Thus it remains only to prove that $\mathcal{D} = \mathcal{J}$.

Each $J \in S/\mathcal{J}$ is a subsemigroup of S (since $\mathcal{J} = \eta$), so that no principal factor of S is null. Indeed, the kernel of S (if it exists) is simple, and all other principal factors are 0-simple with the zero an adjoined one. It follows that S is a semilattice, $S/\mathcal{J}$, of simple semigroups which are also unions of groups. Therefore Clifford's Theorem will follow once we show that any simple completely regular semigroup T is completely simple, and this is done by showing that every member of $E(T)$ is a primitive idempotent.

Suppose that $e, f \in E(T)$ with $f \leqslant e$. Take $z, t \in T$ such that $e = zft$. Put $x = ezf$ and $y = fte$ to obtain

$$xfy = ezf^3 te = e(zft)e = e^3 = e \quad \text{and} \quad ex = xf = x, \quad fy = ye = y.$$

Since S is completely regular, we have $x \in H_g$ for some $g \in E(T)$. Thus $gx = xg = x$ and there exists $x^* \in H_g$ such that $xx^* = x^*x = g$. From $xf = x$ it follows that $x^*xf = x^*x$; hence $gf = g$. We also have

$$gf = gef = gxfyf = xfyf = ef = f.$$

Hence $g = f$. Therefore $f = fe = ge = gxfy = xfy = e$, as required. ∎

Another consequence of Clifford's Theorem is a description of the intersection of two interesting classes of semigroup: completely regular semigroups and inverse semigroups. The $\mathscr{D}$-classes of such a semigroup are completely simple with commuting idempotents and thus it follows that S must be a semilattice of groups. Conversely, one observes that any semilattice of groups is a union of groups in which inverses are unique. The structure of semilattices of groups can be described in more detail than for an arbitrary completely regular semigroup. The multiplicative structure of a semilattice of groups is an instance of the following type of construction.

Let Y be a semilattice and let $\{S_\alpha : \alpha \in Y\}$ be a collection of disjoint semigroups of the same type T, indexed by Y. Suppose that for each $\alpha, \beta \in Y$ such that $\alpha \geqslant \beta$ there is a homomorphism $\varphi_{\alpha, \beta} : S_\alpha \to S_\beta$ such that:

(i) $\varphi_{\alpha, \alpha}$ is the identity map on S;
(ii) $\varphi_{\alpha, \beta} \varphi_{\beta, \gamma} = \varphi_{\alpha, \gamma}$ for every $\alpha, \beta, \gamma \in Y$ with $\alpha \geqslant \beta \geqslant \gamma$.

We can then define an associative multiplication $\circ$ on $S = \{S_\alpha : \alpha \in Y\}$ by

$$a_\alpha \circ b_\beta = (a_\alpha \varphi_{\alpha, \alpha\beta})(b_\beta \varphi_{\beta, \alpha\beta}); \quad a_\alpha \in S_\alpha, \quad b_\beta \in S_\beta.$$

Notice that the product on the right is taken in $S_{\alpha\beta}$. We call S a *strong semilattice of semigroups of type T*, and write

$$S = S(Y; \{S_\alpha; \alpha \in Y\}; \{\phi_{\alpha, \beta} : \alpha, \beta \in Y, \alpha \geqslant \beta\}).$$

Remarkably, every semilattice of groups is a strong semilattice of groups. Moreover, the idempotents of a semilattice of groups are central (an element a of S is *central* if $ab = ba$ for all b in S).

Theorem 1.3.11 *The following are equivalent:*

(i) *S is regular and every idempotent is central;*
(ii) *every $\mathscr{D}$-class of S has a unique idempotent;*
(iii) *S is a semilattice of groups;*
(iv) *S is a strong semilattice of groups.*

Proof (i) implies (ii). Let D be a $\mathscr{D}$-class of S. Since S is regular, D contains at least one idempotent. Suppose that $e \mathscr{D} f$ with e, f idempotents. Then there exists a in S such that $e \mathscr{L} a \mathscr{R} f$, and thus for some $x \in S^1$, $e = xa$ whence, since idempotents are central, $e = xa = xfa = fxa = fe = ef$; similarly, we can obtain $f = ef$, so that $e = f$.

(ii) implies (iii). The given condition implies that each $\mathscr{D}$-class is a group $\mathscr{H}$-class. Hence S is a semilattice of completely simple semigroups which themselves are $\mathscr{D}$-classes of S, and thus groups.

(iii) implies (iv). Let S be a semilattice Y of groups $G_\alpha (\alpha \in Y)$ and for each $\alpha \in Y$, let e_α be the identity of G_α. It is routine to verify that the mappings $\varphi_{\alpha,\beta} : G_\alpha \to G_\beta$ ($\alpha \geq \beta$) defined by $a_\alpha \varphi_{\alpha,\beta} = a_\alpha e_\beta$ ($a_\alpha \in G_\alpha$) are homomorphisms with the required properties (i) and (ii) above and that S is the strong semilattice of groups $S(Y; G_\alpha; \varphi_{\alpha,\beta})$.

(iv) implies (i). If S is a strong semilattice $S(Y; G_\alpha; \varphi_{\alpha,\beta})$ of groups, then it is certainly regular. The idempotents of S are the identity elements e_α, of the groups G_α. If $e_\alpha \in E(S)$ and $b_\beta \in G_\beta$ then, writing γ for $\alpha\beta$, we have

$$e_\alpha b_\beta = (e_\alpha \varphi_{\alpha,\gamma})(b_\beta \varphi_{\beta,\gamma}) = e_\gamma(b_\beta \varphi_{\beta,\gamma}) = b_\beta \varphi_{\beta,\gamma} = (b_\beta \varphi_{\beta,\gamma})e_\gamma$$
$$= (b_\beta \varphi_{\beta,\gamma})(e_\alpha \varphi_{\alpha,\gamma}) = b_\beta e_\alpha,$$

and so every idempotent of S is central. ∎

Since the idempotents of a commutative regular semigroup are certainly central, we obtain the next corollary.

Corollary 1.3.12 *A commutative semigroup S is regular if and only if it is a strong semilattice of abelian groups.*

Proof Only the reverse direction requires a line of proof. Suppose that S is a strong semilattice of abelian groups $S(Y; G_\alpha; \varphi_{\alpha,\beta})$. Take any a_α in G_α, b_β in G_β; then

$$a_\alpha b_\beta = (a_\alpha \varphi_{\alpha,\gamma})(b_\beta \varphi_{\beta,\gamma}) \quad \text{(where } \gamma = \alpha\beta\text{)}$$
$$= (b_\beta \varphi_{\alpha,\gamma})(a_\alpha \varphi_{\alpha,\gamma}) \quad \text{(as both are members of the abelian group } G_\gamma\text{)}$$
$$= b_\beta a_\alpha.$$
∎

An heroic attempt to describe the structure of arbitrary completely regular semigroups has been made by Petrich (1974), and this description has been used by, for example, Gerhard (1983). The corresponding construction for bands (Petrich 1971) can be found in Howie (1976). Free completely regular semigroups exist, and descriptions including solution to the corresponding word problem have been given by Gerhard (1983), Trotter (1984), and more recently by Polak (1986).

Fundamentals | 41

Recall that a semigroup is called semisimple if none of its principal factors are null. We shall call a semigroup *completely semisimple* if all of its principal factors are completely [0-] simple. This is a broad class of semigroup which includes the classes of completely regular and finite regular semigroups.

Theorem 1.3.13 *In a completely semisimple semigroup S, $\mathscr{D} = \mathscr{J}$ and every $\mathscr{R}$-class [$\mathscr{L}$-class] is minimal within all the $\mathscr{R}$-classes [$\mathscr{L}$-classes] within its $\mathscr{J}$-class. Conversely, any regular semigroup satisfying M_L^* or M_R^* is completely semisimple.*

Our proof goes by way of lemma.

Lemma 1.3.14 *Let $a \mathscr{J} b$ in a regular semigroup S and let P be the principal factor of S associated with a. Then $R_a^S \leqslant R_b^S$ if and only if $R_a^P \leqslant R_b^P$. Hence S satisfies M_R^* if and only if each principal factor of S does also.*

Proof If $R_a^P \leqslant R_b^P$ then there exists $x \in J$ such that $a = bx$, whence clearly $R_a^S \leqslant R_b^S$. Conversely, if $R_a^S \leqslant R_b^S$ then there exists $x \in S^1$ such that $bx = a$. Take $e \in E(L_b)$, whence $b = be$. Then

$$J_a = J_{bx} = J_{bex} \leqslant J_{ex} \leqslant J_e = J_b,$$

so that $ex \mathscr{J} a$ and $a = b(ex)$, whence $R_a^P \leqslant R_b^P$. □

Proof of Theorem 1.3.13 Let S be completely semisimple. Then for every principal factor P of S, P is completely 0-simple, whence it follows from Lemma 1.3.3 that every non-zero $\mathscr{R}$-class of P is 0-minimal and so P satisfies M_R^*, and dually M_L^* also. Hence, by Lemma 1.3.14, S satisfies $M_R^* \wedge M_L^*$, whence $\mathscr{D} = \mathscr{J}$ in S by Theorem 1.2.20. Furthermore, it follows from Lemma 1.2.16 and its dual that every $\mathscr{R}$-class and every $\mathscr{L}$-class of S is minimal within its $\mathscr{J}$-class.

Conversely, let S be a regular semigroup satisfying M_R^*. Then for every principal factor P of S, P is a regular 0-simple semigroup satisfying M_R^* by Lemma 1.3.14. Evidently, P then satisfies M_R, whence by Remark 1.2.21 it follows that the set of non-zero idempotents of P possesses a minimal member, and thus P is completely 0-simple, as required. ■

For further results relating semisimplicity to minimal conditions, see Clifford and Preston (1967), Chapter 6.6.

Exercises 1.3

1. Show that a Rees semigroup is regular if and only if its matrix P is regular.

2. Verify that any completely 0-simple semigroup S has the following properties:

   (i) every non-zero idempotent is primitive;
   (ii) $\mathcal{H}$ is a congruence on S;
   (iii) any non-trivial homomorphic image of S is completely 0-simple.
   (iv) the direct product of completely 0-simple semigroups is completely 0-simple.

3. It is always possible to represent a completely 0-simple semigroup $S = M^0[G; I, \Lambda; P]$ so that a given row λ and column i of P consists only of 0 and e, the identity of G.

4.* A completely 0-simple inverse semigroup is called a *Brandt semigroup*. A semigroup is Brandt if and only if S is isomorphic to a Rees semigroup of the form $M^0[G; I, I; \Delta]$, where Δ is an $I \times I$ identity matrix.

5. Show that a regular semigroup S has all non-zero idempotents primitive if and only if S is a 0-direct union of completely 0-simple semigroups. [Hint: suppose that $\{0\} \neq J_f \leq J_e$ $(e, f \in E)$. Then $f = xey$, say; put $g = eyfxe$ and show that $g = e$.]

6. The following are equivalent for a regular semigroup S:

   (i) S is completely simple;
   (ii) $aba = a$ implies $bab = b$ for all $a, b \in S$;
   (iii) S is *weakly cancellative*: i.e. $ax = bx$ and $ya = yb$ for some x, y in S implies $a = b$.

7. A completely 0-simple semigroup with trivial maximal subgroups is called a *0-rectangular band*. S is a 0-rectangular band if and only if S has a zero, and satisfies the two conditions:

   (i) $xyx = x$ or $xyx = 0$ for every $x, y \in S$;
   (ii) $xSy = \{0\}$ implies $x = 0$ or $y = 0$.

8. A completely simple semigroup is orthodox if and only if S is a *rectangular group*: i.e. a direct product of a group and a rectangular band.

9. Prove that the following are equivalent for a regular semigroup S:

   (i) S is orthodox;
   (ii) there exists a subband B of S such that for all $a \in S$ there exists $a' \in V(a)$ such that $aa', a'a \in B$;
   (iii) there exists a subband B of S such that for all $e \in E(S)$ there exists $e', e'' \in V(e)$ such that $ee', e''e \in B$;
   (iv) S has a subband B that meets every $\mathcal{L}$-class and every $\mathcal{R}$-class of S.

   Moreover, the condition of (ii) cannot be weakened to the one-sided condition that insists that only $aa' \in B$, even if S is completely simple. (Compare this exercise with Exercise 1.2.10.)

10. The following are equivalent for a semigroup S:

    (i) S is a rectangular band;
    (ii) S is *nowhere commutative*; i.e. $ab = ba$ implies $a = b$ $(a, b \in S)$;
    (iii) every pair of elements of S are mutually inverse;
    (iv) S is a member of the variety $[x = x^2, xyz = xz]$;
    (v) S is a member of the variety $[x = xyx]$.

11. Characterize those semigroups S which have the property that every non-empty subset of S is a subsemigroup.

12.* A regular semigroup S is called a *locally T-semigroup* if the local subsemigroups, eSe $(e \in E(S))$ are all semigroups of type T.

    (a) Show that every regular semigroup is a locally regular semigroup.
    (b) (Hall 1982) Show that for any regular semigroup S, $\mathrm{Reg}(eS)$ forms a subsemigroup of S $(e \in E(S))$.

13.* A semigroup is called *normal* or *medial* if $abcd = acbd$ for all $a, b, c, d \in S$. Show that the following are equivalent for a band B:

    (i) B is a strong semilattice of rectangular bands;
    (ii) B is normal;
    (iii) B satisfies the identity $xyzx = xzyx$;
    (iv) B satisfies the identity $xyxzx = xzxyx$;
    (v) B is a *local semilattice*: i.e. eBe is a semilattice for each $e \in B$;
    (vi) the natural partial order on B is compatible with multiplication, i.e. $e \leq f$ implies $ge \leq gf$ and $eg \leq fg$ for all $e, f, g \in B$.

14. The variety of *right normal bands* is defined as $[x = x^2, xyz = yxz]$. Show that a band is right normal if and only if it is a strong semilattice of right zero semigroups.

15. A completely regular semigroup is a band of groups if and only if $\mathcal{H}$ is a congruence.

16. If S is a commutative semigroup and $a, b \in S$, write $a|b$ (a divides b) if there exists x in S^1 such that $ax = b$.

    (a) η is the set of all pairs (a, b) in S such that a divides some power of b, and b divides some power of a.
    A commutative semigroup T is called *archimedean* if for all a, b in T, some power of a divides b and some power of b divides a.
    (b) Show that any commutative semigroup is a semilattice of archimedean components.
    (c) Prove that an archimedean semigroup has at most one idempotent.

17. Show that $\mathcal{T}_X$ with X infinite is an example of a semisimple semigroup that is not completely semisimple.

18. (Petrich 1987) Let X be a non-empty set and λ be an equivalence

relation on X all of whose classes are of the same cardinality and put

$$\mathcal{T}_\lambda(X) = \{\varphi \in \mathcal{T}_X : \text{ran } \varphi \text{ is a } \lambda\text{-class and for every } \lambda\text{-class } L, \varphi|L \text{ is a bijection of } L \text{ onto ran } \varphi\}.$$

(a) Show that $\mathcal{T}_\lambda(X)$ is a completely simple subsemigroup of $\mathcal{T}_X$. Let $\mathcal{T}_\lambda^*(X)$ be the dual of $\mathcal{T}_\lambda(X)$.

(b) For any completely simple semigroup S, the mapping $\tau : s \mapsto (\lambda_s, \rho_s)$ ($s \in S$), where $\lambda_s x = sx$ and $x\rho_s = xs$ for all $x \in S$, is a faithful representation $S \to \mathcal{T}_\lambda^*(S) \times \mathcal{T}_\lambda(S)$ and the latter is itself a completely simple semigroup.

(c) Construct a Rees semigroup representation of $\mathcal{T}_\lambda(X)$.

1.4 Properties of regularity

Partial orders

That various partial orders come to be associated with semigroups is expected in view of the fact that products tend to 'fall' into lower $\mathcal{J}$-classes, from which they cannot be retrieved by further multiplications. The partial orders introduced so far have been those defined on the Green's factor sets, $S/\mathcal{L}, S/\mathcal{J}$, etc., together with the natural partial order on $E(S)$. Mitsch (1986) introduced three partial orders on an arbitrary semigroup S, each of which reduce to the so-called natural partial order on a regular semigroup discovered independently by Nambooripad (1980) and Hartwig (1980). A partial order on a semigroup S is called *natural* if it is defined by means of the multiplication of S. The reason for the interest in such natural orders lies in the fact that such an ordering can provide additional information on a given semigroup since it reflects its multiplication in a particular way.

The search for natural partial orders on various classes of semigroup was inspired by the natural partial order for inverse semigroups which is defined by set containment of mappings. For this reason it merits special attention, although it can be regarded as a special case of the natural partial order on a regular semigroup that will be introduced later in this section.

The underlying idea is to formulate some natural relationship between mappings in the symmetric inverse semigroup in terms of equality of products in $\mathcal{I}_X$. This formulation can then be used as the definition of the relationship for abstract inverse semigroups, and provides a sound basis for a generalization to some wider class of semigroup.

We begin by considering the partial order on $\mathcal{I}_X$ defined by set inclusion: $\alpha \leqslant \beta$ ($\alpha, \beta \in \mathcal{I}_X$) if $\alpha \subseteq \beta$ when regarded as subsets of $X \times X$; that is, β is an extension of α or equivalently, $\alpha = \beta|\text{dom } \alpha$. It is easy to verify that $\alpha \leqslant \beta$ is equivalent to any of the following conditions: (i) $\alpha = \alpha\alpha^{-1}\beta$; (ii) $\alpha = \beta\alpha^{-1}\alpha$; (iii) $\alpha = \alpha\beta^{-1}\alpha$; (iv) $\alpha = \varepsilon\beta$ for some $\varepsilon \in E(\mathcal{I}_X)$; (v) $\alpha = \beta\varepsilon$ for some $\varepsilon \in E(\mathcal{I}_X)$.

Since any inverse semigroup S may be embedded in some symmetric inverse semigroup, it follows that any of these five conditions define a natural partial order, $\leqslant$, on S and, moreover, by referring to the set-theoretic formulation of $\leqslant$ in $\mathscr{I}_X$, it is clear that $\leqslant$ is compatible with multiplication ($a \leqslant b \Rightarrow ax \leqslant bx$, $xa \leqslant xb$ for all $x \in S$) and with inversion ($a \leqslant b \Rightarrow a^{-1} \leqslant b^{-1}$).

A second example of this approach is provided by the compatibility relation on $\mathscr{I}_X$. We call elements α and β of $\mathscr{I}_X$ *compatible* if $\alpha \cup \beta$ is a member of $\mathscr{I}_X$. Now α and β are compatible if and only if $\beta|\text{dom}\,\alpha = \alpha|\text{dom}\,\beta$ and $\alpha^{-1}|\text{ran}\,\beta = \beta^{-1}|\text{ran}\,\alpha$. These conditions are equivalent to the equality of the products $\alpha\alpha^{-1}\beta = \beta\beta^{-1}\alpha$ and $\alpha\beta^{-1}\beta = \beta\alpha^{-1}\alpha$ (in which case all four products equal $\alpha\alpha^{-1}\beta$). Thus we can introduce the compatibility relation between elements of an abstract inverse semigroup S: a and b are *compatible* if and only if $aa^{-1}b = bb^{-1}a$ and $ab^{-1}b = ba^{-1}a$.

Let α, β, and γ be elements of some inverse subsemigroup S of $\mathscr{I}_X$, such that $\alpha \subseteq \gamma$ and $\beta \subseteq \gamma$. Then $\alpha \cup \beta \subseteq \gamma$, so that α and β are compatible, and hence $\alpha \cap \beta \in S$. Thus we obtain an abstract property of any inverse semigroup S: if $a \leqslant c$ and $b \leqslant c$ for some $a, b, c \in S$, then a and b have a *glb* in S with respect to the natural partial order on S, which we shall denote for the moment by v. We have proved that if $(a, b) \in vv^{-1}$ (i.e. $a \leqslant c$ and $b \leqslant c$ for some c), then $d \leqslant a$ and $d \leqslant b$, for some d which we can take to be the *glb* of a and b in S. Thus $(a, b) \in v^{-1}v$ and it follows that $vv^{-1} \subseteq v^{-1}v$. Consequently, we see that the least equivalence relation ρ on S such that $v \subseteq \rho$ coincides with $v^{-1}v$. Since v is compatible with multiplication in S, ρ is a congruence. Now for any $e, f \in E(S)$, $ev^{-1}efvf$, so that $ev^{-1}vf$, whence we infer that $(e, f) \in \rho$ and so $\sigma \subseteq \rho$, where σ denotes the least group congruence on S. On the other hand, if $a\,v\,b$ then $b = ea$ for some $e \in E(S)$, so that $(a, b) \in \sigma$ and $v \subseteq \sigma$, whence we conclude that $\rho = \sigma$. It is now a simple matter to deduce the result of Munn (1961) that for elements a, b of an inverse semigroup S, $a\,\sigma\,b$ if and only if $ea = eb$ for some $e \in E(S)$ (see Exercise 1.4.8; for a description of the least group congruence on a regular semigroup, see Theorem 1.4.16).

Using the foregoing as motivation, we shall introduce partial orders on an arbitrary semigroup S, based on comparison of mappings in the extended regular representations of S.

Let $\varphi: S \to \mathscr{T}_{S^1}$ whereby $a\varphi = \rho_a$ (the right inner translation of S^1 by a) denote the extended right regular representation of S.

Definition 1.4.1 Say $a \leqslant_r b$ ($a, b \in S$) if $\ker\rho_b \subseteq \ker\rho_a$ and there exists a transversal (cross-section) T of the classes of $\ker\rho_a$ such that $\rho_a|T = \rho_b|T$. This relation is the *right partial order* on S.

The relation is clearly reflexive and antisymmetric. To see that $\leqslant_r$ is transitive suppose that $a \leqslant_r b \leqslant_r c$, with U and V the required transversals of the classes of $\ker\rho_a$ and $\ker\rho_b$ respectively. For any class K of $\ker\rho_a$ we can choose a required transversal member of K as follows: let $u = U \cap K$ and let K' be the class of $\ker \rho_b$ containing u; take $v = V \cap K'$, then $vc = vb = ub = va$, so v is as required. Note also that $a \leqslant_r b$ implies that $\operatorname{ran} \rho_a \subseteq \operatorname{ran} \rho_b$.

Definition 1.4.1* Say $a \leqslant_l b$ $(a, b \in S)$ if $\ker \lambda_b \subseteq \ker \lambda_a$ and there exists a transversal T of the classes of $\ker \lambda_a$ such that $\lambda_a|T = \lambda_b|T$. This relation is the *left partial order* on S.

Definition 1.4.2 The *natural partial order* on a semigroup S is the relation $\leqslant \; = \; \leqslant_r \cap \leqslant_l$.

These three partial orders all coincide with the natural partial order on idempotents (Exercise 1.4.1). The natural partial order $\leqslant$ reduces on regular semigroups to the ordering introduced by Nambooripad (1980).

On an arbitrary semigroup S define the following relation:

$$a \, v \, b \quad \text{if there exist } e, f \in E(S^1) \text{ such that } a = be = fb.$$

Then v is a reflexive relation on any semigroup, and if S is regular then v is the order of Nambooripad (1980) (although this formulation is due to P. R. Jones). The next result shows, in particular, that it is always the case that $v \subseteq \;\leqslant$.

Proposition 1.4.3 *Let $a, b \in S$. Then $a \leqslant b$ if and only if there exists $t, s \in S^1$ such that $tb = ta = a = as = bs$.*

Proof Suppose that $a \leqslant b$. Since $a \leqslant_r b$ there exists $t \in S^1$ such that $t = T \cap 1(\rho_a \circ \rho_a^{-1})$, where T is a required transversal of the classes of $\ker \rho_a$. Then $a = 1a = ta = tb$. The dual argument gives us an $s \in S^1$ such that $a = as = bs$.

Conversely, suppose that $xb = yb$ $(x, y \in S^1)$. Then $xa = xbs = ybs = ya$, which implies that $\ker \rho_b \subseteq \ker \rho_a$. Next let x be any member of S^1. Then $xa = (xt)a = (xt)b$ so that xt can act as the required transversal member of the class of $\ker \rho_a$ that contains x, thus proving that $a \leqslant_r b$. Combining this with the dual argument completes the proof. ∎

The preceding characterization of $\leqslant$ shows it to be a clear generalization of v. It also provides a form which can quickly yield properties of $\leqslant$: for instance, it is immediate that the natural partial order is preserved by homomorphisms. The conditions expressed in Proposition 1.4.3 can be taken as the definition of $\leqslant$, and indeed the natural partial order can be defined by the formally weaker conditions by which $a \leqslant b$ if there exist $t, s \in S^1$ such that $a = ta = tb = bs$, as these equations yield $as = tbs = ta = a$ also.

Theorem 1.4.4 *Let* $b \in \text{Reg}(S)$. *The following are equivalent for* $a \in S$:

(i) avb;   (ii) $a \leq b$;   (iii) $a \leq_r b$;   (iv) $a \leq_l b$;
(v) $a \in \text{Reg}(S)$ and $a = aa'b = ba'a$ for some $a' \in V(a)$;
(vi) $a \in \text{Reg}(S)$ and $a = aa'b \in bS$ for some $a' \in V(a)$;
(vii) $H_a \leq H_b$ and $a = ab'a$ for some [for all] $b' \in V(b)$;
(viii) $a \in bS$ and $a = eb$ for some $e \in E(R_a)$;
(ix) $a \in bS$ and $a = eb$ for some $e \in E(S)$.

Remark One can add the right–left dual conditions to (vi), (viii), and (ix) to the above list, and there are still a number of other similar equivalent formulations; see Exercise 1.4.1(b). Condition (v) of the theorem is interesting in that it shows that only regular elements lie below regular elements, and it is also true that only idempotents lie below idempotents (Exercise 1.4.1(a)).

Proof (i) $\Rightarrow$ (ii). This is immediate by Proposition 1.4.3.
(ii) $\Rightarrow$ (iii) and (iv) by definition. We show that (iii) $\Rightarrow$ (ii) and omit the dual proof that (iv) $\Rightarrow$ (ii). Suppose that $a \leq_r b$, and that $bx = by$ for some $x, y \in S^1$. Then, since $a \leq_r b$, there exists $t \in S^1$ such that $a = ta = tb$, whence $ax = tax = tbx = tby = ay$, and thus $\ker \lambda_b \subseteq \ker \lambda_a$. (Regularity of b has not been used to this stage.) To conclude that $a \leq_l b$ (and thus $a \leq b$) we must show that given $c \in S^1$ there exists d in S^1 such that $ac = ad = bd$. First note that $a \leq_r b \Rightarrow ab'a = a$ for all $b' \in V(b)$; to see this observe that $bb'a = a$ because $\ker \rho_b \subseteq \ker \rho_a$ whence, with t as before we obtain $ab'a = tab'a = tbb'a = ta = a$. Putting $d = b'ac$ we obtain

$$ad = ab'ac = ac \quad \text{and} \quad bd = bb'ac = ac,$$

as required.

(ii) $\Rightarrow$ (v). We saw above that $a \in \text{Reg}(S)$ so take $a' \in V(a)$. By Proposition 1.4.3 there exist $t, s \in S^1$ such that $tb = ta = a = as = bs$. Then $sa't \in V(a)$, which satisfies the requirements of (v).

(v) $\Rightarrow$ (i) is clear (thus establishing the equivalence of (i)–(v)), as is (v) $\Rightarrow$ (vi). To see that (vi) $\Rightarrow$ (i) suppose that $a = aa'b = bx$ ($x \in S^1$). Then $ax = aa'bx = aa'a = a$, whence $xa'a \in E(S)$ and $bxa'a = a$.

To deduce (vii), recall from above that $a \leq_r b \Rightarrow ab'a = a$ for each b' in $V(b)$, while clearly $a \leq_r b \Rightarrow H_a \leq H_b$. Conversely, suppose that $H_a \leq H_b$ and that $a = ab'a$ for some $b' \in V(b)$. Thus $a = xb = by$, say, and $b'ab' \in V(a)$. Then a is regular, a member of bS, and

$$a(b'ab')b = ab'b = xbb'b = xb = a,$$

which proves that (vii) implies (vi), thus establishing the equivalence of the first seven conditions.

Finally, note that the implications (v) $\Rightarrow$ (viii) $\Rightarrow$ (ix) are immediate, while given (ix) we certainly have $H_a \leq H_b$, and we may write $a = bx$ say ($x \in S$).

Then, for each $b' \in V(b)$ we obtain

$$ab'a = eb.b'.bx = e.bx = ea = a,$$

thus proving that (ix) $\Rightarrow$ (vii), and hence completing the proof. ∎

As mentioned in the introduction to this section, on an inverse semigroup, the order $\leq$ assumes a particularly simple form:

$$x \leq y \text{ if there exists } e \in E \text{ such that } x = ey.$$

From this formulation, taken together with its dual, it is immediate that $\leq$ is preserved under the taking of inverses (see Exercise 1.4.8). An analogue of this in our general setting is now easy to prove.

Proposition 1.4.5 *Let $a \leq b$ with $b \in \text{Reg}(S)$. Then for each $b' \in V(b)$ there exists $a' \in V(a)$ such that $a' \leq b'$.*

Proof From the proof of Theorem 1.4.4 we have $a = ab'a$ and so $b'ab' \in V(a)$. Take $a' = b'ab'$ whereupon, since $a'a = b'a$ and $aa' = ab'$, we obtain

$$a' = (a'a)a' = (b'a)b'ab' = (b'a)b' = b'(ab'),$$

from which it follows that $a' \leq b'$. ∎

Other basic properties of the natural partial order can be found in Exercise 1.4.1. It can also be shown that the three partial orders $\leq_r$, $\leq_l$, and $\leq$ are independent of the representation used to define them if S is regular. In general this is false, and the three partial orders can be distinct even for finite semigroups. One surprising result that emerges is that the relation v defines a partial order on any group-bound semigroup.

Corollary 1.4.6 *On any group-bound semigroup S, $\leq \; = v$.*

Proof Suppose that $a \leq b$. By Proposition 1.4.3 we take $t \in S^1$ such that $a = ta = tb$, whereupon $a = t^n a = t^n b$ for every positive integer n. Replacing t by a suitable power gives an element $t^n = u$ such that $a = ua = uu*ua = u*uua = u*ua = u*ub$ and $u*u \in E(S)$. Dually, we use $a \leq_l b$ to obtain f in $E(S)$ such that $a = bf$. Hence avb. The reverse implication is always true by Proposition 1.4.3. ∎

The natural partial order on a regular semigroup is not in general compatible with multiplication. We have already seen (Exercise 1.3.13) that the natural partial order is compatible with the multiplication on a band if and only if the band is normal. It follows that a necessary condition for compatibility on an orthodox semigroup S is that S be a *generalized inverse semigroup*, i.e. a regular semigroup in which $E(S)$ forms a normal band (or, equivalently, $E(S)$ is a local semilattice; see Exercises 1.3.12 and 1.3.13). This condition is also sufficient, and is indicative of the general situation where by a regular semigroup S has a compatible Nambooripad partial order if and only if eSe is inverse for all e in $E(S)$ (Nambooripad, 1980). Such semigroups are called *locally inverse* or *pseudo-inverse* semigroups, and include the classes of inverse semigroups, completely [0-] simple semigroups, and the generalized inverse semigroups which are exactly the orthodox locally inverse semigroups (see Exercise 1.4.2). Papers on locally inverse semigroups include those of Schein (1972), Pastijn (1982), Petrich (1982), and McAlister (1984); see also Hall (1986).

A structure theorem for generalized inverse semigroups is due to Yamada (1967); see also Madhaven (1978) and Gagnon (1981). The paper of Scheiblich (1982) is on generalized inverse semigroups with involution, while for amalgamation results on this topic consult Hall (1988) and Imaoka (1987).

The most general statement on compatibility is due to Blyth and Gomes (1983), which says that v is right compatible on a regular semigroup S if and only if eSe is an $\mathscr{L}$-unipotent semigroup (Exercise 1.2.19) for all e in $E(S)$ (see Theorem 1.5.10).

Hartwig (1980) defined a partial order on the regular elements of a semigroup, and in doing so independently discovered the natural partial order on a regular semigroup. Partial orders on transformation semigroups were studied by Kowol and Mitsch (1986). For other partial orders on certain classes of semigroup, see Koch (1984), Drazin (1986), Lawson (1987, 1989), and Stamenkovic and Protic (1987); see also Burgess and Raphael (1978), Hickey (1983), and Blyth and Hickey (1984). For fundamental order relations on inverse semigroups and many references, consult Goberstein (1980).

Congruences, idempotents and eventual regularity

There are substantial theories of regular semigroups and of finite semigroups. The latter is mainly due to the school of Rhodes, and is based on the use of wreath products to determine the structure of semigroups 'up to division' (homomorphic images of subsemigroups); as distinct from the approach in Clifford and Preston (1961, 1967) where the underlying goal is determination up to isomorphism (see, for example, Rhodes and Allen, 1976). There are certainly some aspects of the two theories which are similar. Our Corollary 1.4.6 reveals one feature that regular and finite semigroups have in common. One easily expressed property that has emerged in a number of papers (e.g. Birget 1988*b*; Protic 1987) was introduced in the Ph.D. thesis of P. Edwards as a generalization of both regularity and finiteness.

Definition 1.4.7 A member $a \in S$ is *eventually regular* if for some positive integer n (depending on a), a^n is regular. An *eventually regular semigroup* consists entirely of members with this property.

Clearly, all group-bound semigroups are eventually regular (see Edwards 1986). Eventual regularity is a useful context in which to describe idempotent-separating congruences. What follows is based on Edwards (1983).

We begin with a generalization of Lallement's Lemma, first proved for regular semigroups by Hall (1972).

Theorem 1.4.8 *Let $\varphi: S \to T$ be a surmorphism from an eventually regular semigroup S onto a semigroup T, and suppose that c and d are mutually inverse elements of T. Then there exist mutually inverse elements a and b of S such that $a\varphi = c$ and $b\varphi = d$.*

Proof As φ is onto T there exist $x, y \in S$ such that $x\varphi = c$ and $y\varphi = d$. As S is eventually regular there exists an integer $n > 1$ such that $((xy)^2)^n$ is regular. Let $z \in V((xy)^{2n})$ and put $a = (xy)^{2n-1}zxyx$ and $b = y(xy)^{2n-2}zxy$. It is easy to verify that $b \in V(a)$.

As $z \in V((xy)^{2n})$, $x\varphi = c$, $y\varphi = d$ and φ is a morphism it follows that $z\varphi \in V((cd)^{2n})$. As $d \in V(c)$, $(cd)^{2n} = cd$. Thus $z\varphi \in V(cd)$ and so will be denoted by $(cd)'$. As φ is a morphism, $a\varphi = (cd)^{2n-1}(cd)'cdc = cd(cd)'cdc = cdc = c$; and, $b\varphi = d(cd)^{2n-2}(cd)'cd = dcd = d$. Thus a and b meet the requirements of the theorem. ∎

A congruence on a semigroup is called *idempotent-consistent* (or in some places in the literature *idempotent-surjective*) if all of its idempotent congruence classes contain idempotents, and S itself is called *idempotent-consistent* if all its congruences enjoy this property. That regular semigroups are idempotent-consistent is the statement of Lallement's Lemma (Lemma 1.1.7).

Corollary 1.4.9 *An eventually regular semigroup S is idempotent-consistent. Indeed, if ρ is a congruence on S with $w\rho$ in $E(S/\rho)$ then there exists $e \in E(S) \cap w\rho$ such that $H_e \leqslant H_w$.*

Proof Since $w\rho \in V(w\rho)$ in S/ρ there exist mutually inverse a, b in S such that $a\rho = w\rho = b\rho$. Putting $e = ab$ yields $e = e^2$, $e\rho = (ab)\rho = a\rho b\rho = (w^2)\rho = w\rho$. Thus $e \in E \cap w\rho$. Finally, the proof of Theorem 1.4.8 shows that a and b can be taken to have the forms wrw and wsw respectively, whence $e = wrw^2sw$ and therefore $H_e \leqslant H_w$. ∎

An *idempotent-separating congruence* σ on a semigroup S is one in which no pair of distinct idempotents are contained within the same σ-class.

Lemma 1.4.10 *Suppose that $e\,\sigma\,b$, where e is an idempotent and σ is an idempotent-separating congruence on an eventually regular semigroup S. Then $H_e \leqslant H_b$.*

Fundamentals | 51

Proof Since $b\sigma = (b\sigma)^2$ it follows by Corollary 1.4.9 that there exist $f \in b\sigma \cap E$ such that $H_f \leqslant H_b$. Since σ is idempotent-separating, $f = e$. ∎

For an eventually regular semigroup S the maximum idempotent-separating congruence μ exists and can be introduced as the kernel of a certain representation of S.

Let S be any semigroup with Reg S non-empty. Let X and Y be the sets of regular $\mathscr{L}$- and regular $\mathscr{R}$-classes of S respectively together with a new symbol ∞. Define $\varphi: S \to \mathscr{T}_X \times \mathscr{T}_Y^*$ by $a\varphi = (\rho_a, \lambda_a)$, where

$$L_x \rho_a = \begin{cases} L_{xa} & \text{if } x \mathscr{R} xa, \\ \infty & \text{otherwise}, \\ \infty \mapsto \infty; \end{cases} \quad \lambda_a R_x = \begin{cases} R_{ax} & \text{if } x \mathscr{L} ax, \\ \infty & \text{otherwise}, \\ \infty \mapsto \infty. \end{cases}$$

Note that $\mathscr{T}_Y^*$ denotes the dual of the semigroup $\mathscr{T}_Y$: in $\mathscr{T}_Y^*$ mappings are written from left to right. That ρ_a, λ_a are well defined follows by Green's Lemma. Also, if $L_x \rho_a$ is L_{xa} then L_{xa} is regular as $x \mathscr{R} xa$, whence the common $\mathscr{D}$-class of x and xa is regular. To verify that φ is a representation we must check that $(\rho_{ab}, \lambda_{ab}) = (\rho_a, \lambda_a) \circ (\rho_b, \lambda_b)$. We check that $\lambda_{ab} = \lambda_a \lambda_b$, and omit the dual argument. Certainly, if either of λ_{ab} or $\lambda_a \lambda_b$ does not take the value ∞ on R_x the image is R_{abx}. For this to be so, we require that $x \mathscr{L} abx$ in the former case and that $x \mathscr{L} bx \mathscr{L} abx$ in the latter; these conditions are easily seen to be equivalent. Thus φ is a homomorphism, and $\ker \varphi$ can be formulated in terms of Green's relations.

Proposition 1.4.11 $\ker \varphi = \mu = \{(a, b) \in S \times S: \text{ if } x \in \text{Reg}(S) \text{ then each of } x \mathscr{R} xa, x \mathscr{R} xb \Rightarrow xa \mathscr{H} xb, \text{ and each of } x \mathscr{L} ax, x \mathscr{L} bx \Rightarrow ax \mathscr{H} bx\}.$

To see that μ is idempotent-separating take $(e, f) \in \mu$ with $e = e^2$ and $f = f^2$. Put $x = e$ in the given expression for μ to obtain $e \mathscr{R} e^2 \Rightarrow e \mathscr{H} ef$ and, similarly, $f \mathscr{L} f^2 \Rightarrow ef \mathscr{H} f$. Thus $e \mathscr{H} f$ and so $e = f$.

Theorem 1.4.12 *The following are equivalent for a congruence σ on an eventually regular semigroup S:*

(i) $\sigma \subseteq \mu$;
(ii) for all $e \in E(S)$ and for all $b \in S$, $(e, b) \in \sigma \Rightarrow H_e \leqslant H_b$;
(iii) for all $a \in \text{Reg}(S)$ and for all $b \in S$, $(a, b) \in \sigma \Rightarrow H_a \leqslant H_b$;
(iv) σ separates idempotents.

Proof (i) $\Rightarrow$ (ii) follows from Theorem 1.4.10.
(ii) $\Rightarrow$ (iii). Let $a \sigma b$ with $a \in \text{Reg}(S)$. Take $a' \in V(a)$. Since $aa' \sigma ba'$ and aa' is idempotent then $H_{aa'} \leqslant H_{ba'}$ by (ii). Therefore, $R_a = R_{aa'} \leqslant R_{ba'} \leqslant R_b$, and combining this with the dual argument yields $H_a \leqslant H_b$.

(iii) ⇒ (i). Take $a\,\sigma\,b$ and let $x\,\mathscr{R}\,xa$ with $x \in \text{Reg}(S)$. Then as $xa\,\sigma\,xb$ with xa regular we obtain $H_{xa} \leqslant H_{xb}$ by (iii). Hence $R_x = R_{xa} \leqslant R_{xb} \leqslant R_x$, whence $x\,\mathscr{R}\,xb$ and xb is regular. Again by (iii) $H_{xb} \leqslant H_{xa}$, and thus $H_{xa} = H_{xb}$. A dual argument shows that $x\,\mathscr{L}\,ax \Rightarrow ax\,\mathscr{H}\,bx$. These implications, together with symmetry, show that $a\,\mu\,b$. To finish the proof observe that (iv) ⇒ (ii) by Theorem 1.4.10, and (ii) ⇔ (i) ⇒ (iv), since μ separates idempotents. ∎

This theorem establishes μ as the maximum idempotent-separating congruence on an eventually regular semigroup S. Furthermore, $\mathscr{H}^b$, the maximum congruence contained in $\mathscr{H}$ (see Proposition 1.1.4) is contained in μ as $\mathscr{H}^b$ separates idempotents. If S is regular we have equality by Theorem 1.4.12(iii).

For any semigroup S, $a\,\mathscr{H}^b\,b$ if and only if $(xay, xby) \in \mathscr{H}$ for all x, y in S^1. Since $\mathscr{L}$ and $\mathscr{R}$ are right and left congruences respectively, this condition is equivalent to saying that $ay\,\mathscr{R}\,by$, $xa\,\mathscr{L}\,xb$ and $a\,\mathscr{H}\,b$ for all $x, y \in S^1$. If S is regular the condition that $a\,\mathscr{H}\,b$ is redundant as upon putting $y = a'a$ we obtain $a\,\mathscr{R}\,ba'a \Rightarrow R_a \leqslant R_b$; similarly, we obtain $R_b \leqslant R_a$, whence $R_a = R_b$, and combining with the dual argument yields $H_a = H_b$. The upshot of this is a similar representation for S to that given for eventually regular semigroups in Proposition 1.4.11.

Proposition 1.4.13 *Let S be a regular semigroup and define $\rho_a \in \mathscr{T}_{S/\mathscr{L}}$ by $L_x \rho_a = L_{xa}$ and $\lambda_a \in T^*_{S/\mathscr{R}}$ by $\lambda_a R_x = R_{ax}$. Then the map $\varphi: S \to \mathscr{T}_{S/\mathscr{L}} \times \mathscr{T}^*_{S/\mathscr{R}}$ whereby $a\varphi = (\rho_a, \lambda_a)$ is a representation of S with $\ker \varphi = \mathscr{H}^b = \mu$, the maximum idempotent-separating congruence on S.*

Meakin (1972b) and Hall (1973) showed that membership of μ is determined by equality of certain idempotents.

Theorem 1.4.14 *For a regular semigroup S, $\mu = \{(a, b) \in S \times S: \text{ for some } a' \in V(a), b' \in V(b), aa' = bb', a'a = b'b, \text{ and } a'ea = b'eb \text{ for each } e \in E(S) \text{ with } e \leqslant aa'\}$.*

Proof Take $(a, b) \in \mu = \mathscr{H}^b$ by above. By Theorem 1.2.8 we may take $a' \in V(a)$ and $b' \in V(a)$ such that $a'\,\mathscr{H}\,b'$ whereupon $aa' = bb'$ and $a'a = b'b$. Since $a'\mu$ is $\mathscr{H}$-related to $b'\mu$ in S/μ and both are inverses of $a\mu = b\mu$ in S/μ, we have, again by Theorem 1.2.8, $a'\mu = b'\mu$; i.e. $a'\mu b'$. Take any $e \in E(S)$ such that $e \leqslant aa'$. Since $\mu = H^b$ we obtain $ea\,\mathscr{H}\,eb$ and $a'e\,\mathscr{H}\,b'e$; but routinely one obtains $a'e \in V(ea)$, $b'e \in V(eb)$, whence $(a'e)(ea)$ and $(b'e)(eb)$ are $\mathscr{H}$-related idempotents, and so

$$a'ea = (a'e)(ea) = (b'e)(eb) = b'eb.$$

This proves that μ is contained in the given relation above.

Fundamentals | 53

Conversely, suppose that (a, b) is a member of the above relation. Then $a \mathcal{H} b$. Let $x \in S$. Since $L_{xaa'} \leqslant L_{aa'}$ by Lemma 1.2.10 there exists $e \in L_{xaa'}$ such that $e \leqslant aa'$. Then $a'e \in V(ea)$, $b'e \in V(eb)$ (easily checked) and so $L_x \rho_a = L_{xaa'a} = L_e \rho_a = L_{ea} = L_{a'ea} = L_{b'eb} = L_{x\rho_b}$, so that $\rho_a = \rho_b$. We can infer that $\lambda_a = \lambda_b$ by the dual argument provided that we show that $afa' = bfb'$ for each idempotent $f \leqslant a'a$. Since $e = afa'$ is idempotent and $e \leqslant aa'$ we obtain

$$afa' = aa'(afa')aa' = bb'(afa')bb' = b(b'eb)b' = b(a'ea)b'$$
$$= b(a'afa'a)b' = bb'bfb'bb' = bfb'.$$

Therefore $(a, b) \in \mu$, thus completing the proof. ∎

A semigroup S is called *fundamental* if $\mathcal{H}^\flat$ is the trivial congruence.

Theorem 1.4.15 *For any semigroup S the factor semigroup $S/\mathcal{H}^\flat$ is fundamental. In particular, if S is regular S/μ is fundamental, where μ is the maximum idempotent-separating congruence.*

Proof For convenience denote $S/\mathcal{H}^\flat$ by S' and denote the maximum congruence contained in $\mathcal{H}$ on S and on S' by μ and μ' respectively. Suppose $(a\mu, b\mu) \in \mu'$. Then, for all $x\mu, y\mu \in S'^1$, we have $((xay)\mu, (xby)\mu) \in \mu'$. Since $\mu' \subseteq R^{S'}$ there exist $r\mu, s\mu \in S'^1$ such that $(xayr)\mu = (xby)\mu$ and $(xbys)\mu = (xay)\mu$, whence

$$R_{xby} = R_{xayr} \leqslant R_{xay} = R_{xbys} \leqslant R_{xby} \Rightarrow xay \, \mathcal{R} \, xby.$$

Combine this reasoning with its dual to obtain

$$xay \, \mathcal{H} \, xby \quad \text{for all } x, y \in S^1 \Rightarrow a \, \mathcal{H}^\flat \, b,$$

that is $a\mu = b\mu$. ∎

In Edwards (1983, 1985a) the relation μ is defined on an arbitrary semigroup S as in Proposition 1.4.11, and this is shown to be an idempotent-separating congruence containing $\mathcal{H}^\flat$. Furthermore, $\mu(S/\mu)$ is always trivial and S/μ is fundamental. However, μ is not always even a maximal idempotent-separating congruence (Exercise 1.4.3).

If S is a generalized inverse semigroup S/μ can be embedded in the so-called symmetric generalized inverse semigroup of $\lambda - \mu$ relations on a set (Madhaven 1978).

Group congruences, which we now turn to, have the opposite nature to idempotent-separating congruences since they identify all idempotents. The least group congruence on a regular semigroup S has been characterized by Feigenbaum (1979).

54 | Techniques of semigroup theory

A subsemigroup T of a regular semigroup S is called *full* if $E(S) = E(T)$ and is *self-conjugate* if $xTx' \subseteq T$ for all $(x, x') \in V(S)$. By the standard intersection argument we see that the least full self-conjugate subsemigroup C of S exists.

Theorem 1.4.16 *The least group congruence σ on a regular semigroup S is the relation*

$$\rho = \{(a, b) \in S \times S : xa = by \text{ for some } x, y \in C\}.$$

Proof It is obvious that C consists of all members of S which can be formed from $E(S)$ by a finite number of operations, each of which is a product or a conjugation of a given $c \in C$ by x (by *conjugation* we mean formation of the product xcx' with $x' \in V(x)$). It follows that $c\sigma$ is the identity of the group S/σ for all $c \in C$, and thus $a \rho b \Rightarrow a \sigma b$. Conversely, if we prove that ρ is a congruence, it is immediate that ρ is the least group congruence as $E \times E \subseteq \rho$, whereupon we infer that $\sigma \subseteq \rho$.

Clearly, ρ is reflexive. Suppose that $xa = by$ for some $x, y \in C$. Then

$$a(a'xab'b) = (aa'byb')b$$

and since both bracketed products lie in C, we conclude that ρ is symmetric. Transitivity follows easily, for if $xa = by$, $zb = cw$ $(x, y, z, w \in C)$ then $(zx)a = zby = c(wy)$, whence $a \rho c$. To show that ρ is a right congruence suppose that $xa = by$ $(x, y \in C)$ and $c \in S$. Then $(bcc'b'x)ac = bc(c'b'byc)$, and both bracketed products belong to C. Invoking duality yields that ρ is a congruence, and thus we conclude that $\rho = \sigma$. ∎

An alternative characterization of group congruences on a regular semigroup has been provided by La Torre (1982); see also Masat (1973). Hanumantha Rao and Lakshmi (1988) have characterized group congruences on eventually regular semigroups.

The development of the theory of congruences on a regular semigroup has progressed largely along the lines of the successful theories for inverse semigroups. A congruence ρ on an inverse semigroup (in fact, on a regular semigroup) S is determined by the ρ-classes containing idempotents, and this approach leads to the characterization by the so-called kernel-normal systems of Preston (1954). As an alternative, Scheiblich (1974) developed the congruence-pair approach, which is based on the fact that ρ is determined by its *trace* ($\rho|E(S)$) and the union of all idempotent ρ-classes. A systematic account of the theory of inverse semigroup congruences is in Petrich (1984). Feigenbaum (1979) and Trotter (1978) studied the application of the congruence-pair approach to regular semigroups, a direction which has recently been explored further by Pastijn and Petrich (1986). A fundamental paper on the lattice of congruences on a regular semigroup is that of Reilly and Scheiblich (1967), which is also the subject of Hall (1969b) and Pastijn and Petrich (1988); see also Jones (1983). The lattice of congruences of an eventually regular semigroup is studied in Edwards (1985b). The survey on congruences of regular semigroups of Pastijn (1985) itself contains many references. Other relevant papers are Masat (1973, 1982), Nambooripad and Sitarman (1979), La Torre (1982, 1983), Trotter (1982), Jones (1984), Alimpic and Krgovic (1988), and Pastijn (1990).

A particularly crisp result concerning regular semigroups congruences is the following, due to Howie and Lallement (1966).

Theorem 1.4.17 *In a regular semigroup $\mathcal{D}^* = \mathcal{J}^* = \eta$.*

Proof Since $\mathcal{D}^* \subseteq \eta$ and η is generated by

$$\eta^0 = \{(a, a^2), (ab, ba): a, b \in S\}$$

it suffices to show that $\eta^0 \subseteq \mathcal{D}^*$. Let $a' \in V(a)$. Then $a \mathcal{D} aa' \Rightarrow a^2 \mathcal{D}^* aa'a = a$.

Next take $e, f \in E(S)$ and $y \in V(ef)$. Then $fye \in V(ef)$ and so $ef \mathcal{D} fye \Rightarrow fefe \mathcal{D}^* f^2 ye^2 = fye$. But then $fe \mathcal{D}^* (fe)^2 \mathcal{D}^* fye \mathcal{D}^* ef$. Finally, take $a, b \in S$. Then $ab \mathcal{D}^* aa'bb' \mathcal{D}^* bb'aa' \mathcal{D} ba$, as required. ∎

Further results concerning the congruences genefated by Green's relations can be found in the above paper, and this direction has been extended by Pastijn and Petrich (1987). The behaviour of Green's relations under inverse morphic images is dealt with in Hall (1972), and in Edwards (1987) for eventually regular semigroups.

The inverse fye used in Theorem 1.4.17 is an idempotent belonging to the 'sandwich set' $S(e, f)$, an idea introduced in Section 1.5 and developed in Chapter 3. Indeed, the choice of the 'special' inverse fye of ef has occurred twice previously in the text: in Theorem 1.1.5 [(iii) implies (i)], and in Theorem 1.1.9 [(iii) implies (i)].

What appears to be an innocuous problem in Clifford and Preston (1961) is an exercise, due to Miller and Clifford, which asks one to prove that an inverse of an idempotent e in a regular semigroup is always the product of two idempotents (take $y \in V(e)$; then $y = yey = (ye)(ey) \in E^2$). In generalizing this result, Fitzgerald (1972) discovered a construction fundamental to the theory of regular semigroups.

By $V(E^n)$ we mean the set $\{V(a): a \in E^n\}$. We shall follow Petrich in referring to the idempotent-generated subsemigroup $\langle E \rangle$ of a semigroup S as the *core* of S.

Theorem 1.4.18 *For a regular semigroup S, $V(E^n) = E^{n+1}$ for all $n = 1, 2, \ldots$. Thus the core of a regular semigroup is itself regular.*

Remark Another independent proof of this result is due to Eberhart *et al.* (1973); the case where $n = 2$ is in Howie and Lallement (1966).

Proof We have seen above that $V(E^n) \subseteq E^{n+1}$ if $n = 1$. Let $n > 1$ and take $x = e_1 e_2 \ldots e_n$ ($e_i \in E$) and $y \in V(x)$. Define $f_j = e_{j+1} \ldots e_n y e_1 \ldots e_j$, ($j = 1, \ldots, n-1$). Routinely we see that $f_j \in E$ and that $y = y x f_{n-1} \ldots f_1 x y$ is in E^{n+1}, as the product can be factored as

$$y(x.e_n y)(e_1 \ldots e_{n-1}.e_{n-1} e_n y)(e_1 \ldots e_{n-2}.e_{n-2} e_{n-1} e_n y) \ldots (xy)$$
$$= y(xy)^n = y.$$

Conversely, given $x = e_1 e_2 \ldots e_{n+1} \in E^{n+1}$, put $g_j = e_{j+1} \ldots e_{n+1} y e_1 \ldots e_j$, where $y \in V(x)$. Then $g_j \in E$ and $z = g_n \ldots g_1 \in E^n$ is inverse to x as $z = e_{n+1}(yx)^{n-1} y e_1 = e_{n+1} y e_1$, whence

$$zxz = e_{n+1} y e_1 x . e_{n+1} y e_1 = e_{n+1} y x y e_1 = z \quad \text{and}$$

$$xzx = x e_{n+1} y e_1 x = xyx = x. \qquad \blacksquare$$

Remark The above proof shows that $V(E^n) \subseteq E^{n+1}$ for an arbitrary semigroup.

The significant idea in the proof of this result is the introduction of the products f_j ($j = 1, \ldots, n-1$) which form the analogue of the single product, fye, introduced in the proof of Theorem 1.4.17. As we shall see in Chapter 3, the sequence $(f_1, \ldots, f_{n-1})$ is a typical member of the 'sandwich set' $S(e_1, \ldots, e_n)$.

Idempotents feature in virtually all our investigations involving regular semigroups, and are also important when studying specific semigroups such as $\mathcal{T}_X$, as they are easily recognized. A natural question is: To what extent do the idempotents of a regular semigroup S determine its nature? More precisely, what information can be elicited about S given its partial algebra $(E, \circ)$ of idempotents (by which we mean the set E together with the partial binary operation $\circ$, whereby $e \circ f = ef$ if ef is a member of E)? The answer is 'none' if S is a group, which suggests examination of the opposite end of the spectrum: idempotent-generated semigroups called *semibands* (Pastijn 1977). Let E be the set of idempotents of a semigroup S. From Theorem 1.4.18 it follows that there is a regular subsemigroup of S with E as its partial algebra of idempotents if and only if the core of S is such a semigroup. We shall show that a fundamental regular semiband is determined by its partial algebra of idempotents (Hall 1973). Our proof, which involves biordered sets, will be taken up in Chapter 3.

One of the main subjects of Chapter 2 is the study of so-called E-unitary inverse semigroups (see Exercise 1.4.18). In general, a subset A of a semigroup S is called *right unitary* if whenever $sa \in A$ then $s \in A$ ($a \in A$, $s \in S$), and A is *left unitary* if whenever $as \in A$ then $s \in A$. We call A *unitary* if it is both right and left unitary.

Fundamentals | 57

Theorem 1.4.19 (Howie and Lallement 1966) *Let E be the set of idempotents of a regular semigroup S. If E is a right unitary subset of S then E is a unitary subset of S.*

Proof Suppose that $es \in E$ with $e \in E$, $s \in S$. Then for each $s' \in V(s)$ $sess' \cdot sess' = s(es)^2 s' = sess'$; thus $se(ss') \in E$, and since $ss' \in E$ we can apply the right E-unitary property twice to obtain first, that $se \in E$, and then that $s \in E$, thus proving that E is also left unitary. ∎

Exercises 1.4

1. (a) On any semigroup S with set of idempotents E:

   (i) $\leqslant_r \cap (E \times E) = \{(e, f) \in E \times E : e = fe = ef\}$;

   (ii) if $a \leqslant_r e$ $(e \in E)$ then $a \in E$;

   (iii) $\leqslant_r \cap (\mathscr{R} \cup \mathscr{L}) = \iota$;

   (b) Prove that the following are equivalent to $a \leqslant b$ for any regular element b of a semigroup S:

   (i) $a = aa'b = ba''a$ for some $a', a'' \in V(a)$;

   (ii) $a'a = a'b$ and $aa' = ba'$ for some $a' \in V(a)$;

   (iii) $a = ab'b = bb'a$, $a = ab'a$ for some $b' \in V(b)$;

   (iv) $a = axb = bxa$, $a = axa$, $b = bxb$ for some $x \in S$;

   (v) for each $f \in E(R_b)$ there is some $e \in E(R_a)$ with $e \leqslant f$ and $a = eb$.

2. (Hickey 1983) Let $(S, .)$ be a semigroup and let $x \in S$. Define a binary operation $\circ$ on S by $a \circ b = axb$ $(a, b \in S)$. The resulting semigroup, denoted by (S, x) is called a *variant* of S. Now suppose that S is regular. Prove that $a \leqslant b$ if and only if there exists $x \in S$ such that $a, b \in E((S, x))$ and $a \leqslant b$ in the natural partial order on $E((S, x))$.

3. (a) (Yamada 1967) An orthodox semigroup is locally inverse if and only if it is a generalized inverse semigroup. [Hint: use Exercise 1.3.13.]

   (b) Show that the class of locally inverse semigroups is closed under the taking of regular subsemigroups, homomorphic images, and arbitrary direct products

4. Define μ on an arbitrary semigroup S as in Proposition 1.4.11 of the text. Show that μ is an idemopotent-separating congruence.

5. (Edwards 1983) Adjoin to a free semigroup F_X on X, an arbitrary generating set, two idempotents e and 1 such that $ew = we = 1w = w1 = w$ for all $w \in F_X$ and $e1 = 1e = e$. Show that μ as defined in Proposition 1.4.11 is strictly contained in the maximum idempotent-separating congruence on S.

58 | Techniques of semigroup theory

6.* (a) (Howie 1964) If S is an inverse semigroup with semilattice of idempotents E, then the relation

$$\mu = \{(a, b) \in S \times S : a^{-1}ea = b^{-1}eb \text{ for all } e \in E\}$$

is the maximum idempotent-separating congruence on S.

(b) (Meakin 1972a) Repeat (a) with S an orthodox semigroup. Then $(a, b) \in \mu$ if and only if

$$(\exists a' \in V(a))(\exists b' \in V(b))(\forall x \in E)[a'xa = b'xb \text{ and } axa' = bxb'].$$

7. (Tully 1961) Let S be any semigroup and for each element $a \in S$ define $\sigma_a \in T_{S^1/\mathscr{L}}$ by $L_x \sigma_a = L_{xa}$ for each $x \in S^1$. Then the mapping φ that maps each element $a \in S$ to σ_a is a homomorphism, and the congruence $\varphi \circ \varphi^{-1}$ is the maximum congruence contained in $\mathscr{L}$.

8.* Let σ denote the least group congruence on an inverse semigroup S.

(a) (Munn 1961) $a \sigma b$ if and only if there exists $e \in E$ such that $ea = eb$; also $a \sigma b$ if and only if $af = bf$ for some $f \in E$.

(b) Show that $a \leqslant b$ if and only if $a = eb$ for some $e \in E$, which is equivalent to the dual condition, $a = bf$, for some $f \in E$.

(c) Show that a and b are compatible elements of S if and only if $a^{-1}b$ and ab^{-1} are idempotents.

9. (Meakin 1972b) The least group congruence σ on an orthodox semigroup S is given by

$$\sigma = \{(a, b) \in S \times S : eae = ebe \text{ for some } e \in E(S)\}.$$

10. (a) (Howie and Lallement 1966) Let β denote the least band congruence on a regular semigroup S. Then $\mathscr{H}^* \subseteq \beta \subseteq \mathscr{R}^* \cap \mathscr{L}^*$.

(b) For the bicyclic monoid B (see Exercise 1.2.13), show that $\mathscr{H}^* = \iota$, while $\beta = \eta = \mathscr{R}^* = \mathscr{L}^* = \mathscr{D}^* = \mathscr{J}^* = B \times B$.

(c) Let F be the free band on three generators. Show that $\beta = 1_F \neq \mathscr{R}^* \cap \mathscr{L}^*$. [Hint: $(x_1 x_2 x_1 x_3 x_2 x_1, x_1 x_2 x_3 x_2 x_1) \in \mathscr{R}^* \cap \mathscr{L}^*$.]

11. (Hall 1973) Let e, f be $\mathscr{D}$-related idempotents of a semigroup S. Take any a in S and any $a' \in V(a)$ such that $aa' = e$ and $a'a = f$. Define $\theta_{a',a} : eSe \to fSf$ and $\theta_{a,a'} : fSf \to eSe$ by, for each x in eSe, $x\theta_{a',a} = a'xa$, and for each $y \in fSf$, $y\theta_{a,a'} = aya'$. Then $\theta_{a',a}$ is a $\mathscr{D}$-class preserving isomorphism from eSe onto fSf (meaning that $x\mathscr{D}^S x\theta_{a',a}$) and $\theta_{a',a}^{-1} = \theta_{a,a'}$.

Remark A generalized 'local' structure for arbitrary semigroups were provided by Hickey (1986).

12. $\text{Reg}(S)$ forms a subsemigroup of S if and only if $E(S)$ is non-empty and $ef \in \text{Reg}(S)$ for all $e, f \in E(S)$. [Hint: Exercise 1.2.9(b).]

13.* (Hall 1973) Let $A_1, \ldots, A_n$ be any elements of a regular semigroup S. Put $A_1 A_2 \ldots A_n = a$. Take $x \in V(a)$. Put $a_i = A_i A_{i+1} \ldots A_n x A_1 \ldots A_i$ for all $i = 1, 2, \ldots, n$. Show that:

(i) $a = a_1 a_2 \ldots a_n$ and $a_i \mathcal{D} a$ for all $i = 1, 2, \ldots, n$;
  [Hint: $a_i \mathcal{R} a_i A_{i+1} \ldots A_n = A_i \ldots A_n x \mathcal{L} a$.]
(ii) $a_i \leqslant A_i$, for all $i = 1, 2, \ldots, n$;
(iii) if $A_i \in E(S)$ then so is a_i;
(iv) S is orthodox if and only if each principal factor of S is orthodox;
(v) S is a semiband if and only if each principal factor of S is a semiband.

Remark Results along these lines are also in Nambooripad (1980).

14.* (Hall 1973; Fitzgerald unpublished) For a regular semigroup S, $\langle E \rangle$ is completely regular if and only if S is *E-solid*; i.e. for all $e, f, g \in E$ such that $e \mathcal{L} f \mathcal{R} g$ there exist $h \in E$ such that $e \mathcal{R} h \mathcal{L} g$. [Hint: use Theorem 1.4.18 and induction on n.]

15.* Let $\langle E \rangle$ be the core of an arbitrary semigroup S. Show that $V(x) \subseteq \langle E \rangle$ for any $x \in \langle E \rangle$ and thus prove that the core of any eventually regular [group-bound] semigroup is itself eventually regular [group-bound].

16. (Eberhart et al. 1973) An ideal of a regular semiband is itself a regular semiband.

17. (Johnston and Jones 1984) Let S be regular and let A be a full subset of S (meaning that $E(S) \subseteq A$) such that $V(a) \cap A$ is non-empty for each a in A. Prove that $\langle A \rangle$ is regular. [Hint: If $a = a_1 \ldots a_n$ then $a' = a' a a'_n e_n a'_{n-1} e_{n-1} \ldots a'_1 e_1$, where $e_i = a_i \ldots a_n a' a_1 \ldots a_{i-1}$.]

18.* Call a regular semigroup *E-unitary* if E is a unitary subsemigroup of S. Show that if S is E-unitary then $ab \in E \Rightarrow ba \in E$ for all $a, b \in S$. [Hint: show that $babb' \in E$.]

19. (Hall 1969a)
  (a) Let S be orthodox and denote the $\mathcal{D}$-class in E of $e \in E$ by D_e. If $a \in S$ and $a' \in V(a)$ then $V(a) = D_{a'a} a' D_{aa'}$.
  (b) Show that a regular semigroup is orthodox if and only if for all $a, b \in S$
  $$V(a) \cap V(b) \neq \emptyset \Rightarrow V(a) = V(b).$$
  [Hint: for converse let $x \in V(ef)$. Then $fxe, efxe \in E$ and $fxe \in V(fxe) \cap V(efxe)$. Then use $efe \in V(fxe)$ to verify orthodoxy.]
  (c) If S is orthodox then $\gamma = \{(a, b) \in S \times S : V(a) = V(b)\}$ is the least inverse congruence on S.

20. A regular semigroup S is completely semisimple if and only if no pair of distinct comparable elements are $\mathcal{D}$-related.

60 | Techniques of semigroup theory

1.5 Biordered sets

The previous section provided ample evidence to show that the partial algebra of idempotents, E, is central to the algebraic structure of a regular semigroup S, yet E is not generally a subsemigroup of S. In order to facilitate the study of such partial algebras, the notion of a biordered set was introduced by Nambooripad (1974, 1975) as a formal framework that captures the essential properties of E.

In this short introduction we follow Easdown (1986). The topic of biordered sets will be taken up again in depth in Chapter 3.

Let S be a semigroup and $E = E(S)$. As observed in Exercise 1.2.1,

$$e \mathscr{L} f \quad \text{if and only if } e = ef \text{ and } f = fe;$$

$$e \mathscr{R} f \quad \text{if and only if } e = fe \text{ and } f = ef.$$

To obtain the *right arrow* $\to$ and *left arrow* $\succ\!\!-$, collectively called the *biorder on E*, we split a $\mathscr{L}$ and $\mathscr{R}$ by defining

$$e \to f \quad \text{if and only if } fe = e \quad (e \text{ is a right zero for } f),$$

and

$$e \succ\!\!- f \quad \text{if and only if } ef = e \quad (e \text{ is a left zero for } f).$$

Both $\to$ and $\succ\!\!-$ are pre-orders (they are reflexive and transitive):

$$\leftrightarrow = \mathscr{R}|E \quad \text{and} \quad \succ\!\!-\!\!\prec = \mathscr{L}|E; \quad \text{and} \quad e \succ\!\!\to f \quad \text{if and only if } e \leqslant f.$$

We shall sometimes denote $E/\leftrightarrow [E/\succ\!\!-\!\!\prec]$ by $\mathscr{R}'$ [$\mathscr{L}'$] respectively and denote the $\mathscr{R}'$-[$\mathscr{L}'$-] class of e by $R'_e[L'_e]$ respectively. We have a natural partial order on $E/\mathscr{R}'$ where by $R'_e \leqslant R'_f$ if and only if $e \to f$ and dually for $E/\mathscr{L}'$.

Observe that if $e \to f$ then $efef = eef = ef$, and so $ef \in E$; the following relationships occur:

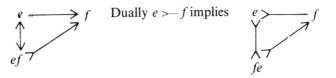

Dually $e \succ\!\!- f$ implies

Define the *biordered set of the semigroup S* to be the set $E = E(S)$ with a partial multiplication inherited from S with domain

$$D_E = \{(e, f) \in E \times E : e \to f \text{ or } e \succ\!\!- f \text{ or } e \leftarrow f \text{ or } e -\!\!\prec f\}.$$

Informally, we think of the biordered set as consisting of idempotents, arrows, and all products that arise from arrows. One may picture the biordered set as the 'skeleton' of the semigroup, the remnants when attention is paid only to idempotents and idempotent products arising from arrows.

A mapping $\alpha\colon E \to F$ is a *morphism* (called a *bimorphism* by Nambooripad) of the biordered set E into the biordered F if $(e,f) \in D_E$ implies that $(e\alpha, f\alpha) \in D_F$ and $(ef)\alpha = (e\alpha)(f\alpha)$. The biordered sets E and F are *isomorphic* (denoted $E \simeq F$) if there exists a bijection $\alpha\colon E \to F$ such that both α and α^{-1} are morphisms.

Theorem 1.5.1 *Let ρ be an idempotent-separating congruence on an eventually regular semigroup S. Then $E(S/\rho) \simeq E(S)$ and thus in particular $E(S/\mu) \simeq E(S)$.*

Proof Suppose that $A, B \in E(S/\rho)$ and $AB = B$. By Corollary 1.4.9, S is idempotent-consistent, so there exists $a \in A$, $b \in B$ with $a, b \in E(S)$. Now $(ab)\rho = b\rho$, and so $(ab, b) \in \rho \subseteq \mu \subseteq H^b$, from which it follows that $b = abx$ for some $x \in S^1$, whence $ab = a^2 bx = abx = b$. Dually, if $AB = A$, then for a and b as before, $ab = a$.

We now prove the statement of the theorem. Since S is idempotent-consistent, $\rho\,|\,E(S) = \alpha$, say, is a bijection of $E(S)$ upon $E(S/\rho)$. Clearly, α preserves left and right arrows, and the purpose of the preceding paragraph was to show that α^{-1} does likewise. Therefore $E(S) \simeq E(S/\rho)$. ∎

As mentioned previously, the nature of a regular semigroup depends to a large extent on its biordered set E, and thus a natural problem arises as to how to recognize whether or not a given biordered set could be that of some regular semigroup. The answer to this question involves the notion of a sandwich set of a pair of elements from a biordered set.

Definition 1.5.2 Let $e, f \in E(S)$. Define

$$M(e, f) = \{g \in E\colon ge = g = fg\}.$$

Define the *sandwich set of the ordered pair* (e, f), denoted by $S(e, f)$, as $\{g \in M(e, f)\colon eh \to eg$ and $hf \succ\!\!-\, gf$ for all $h \in M(e, f)\}$.

For a regular semigroup S, sandwich sets are always non-empty and offer two simple alternative definitions.

Theorem 1.5.3 *Let R be a regular semigroup. Then for any $e, f \in E = E(R)$, $S(e, f) = \{g \in E\colon ge = g = fg$ and $egf = ef\} = \{fye\colon y \in V(ef)\}$, and is non-empty.*

Proof We temporarily denote the three sets above by S, S', and S'' respectively. Clearly S'' is non-empty. For any $fye \in S''$ ($y \in V(ef)$) we see that $(fye)^2 = f(yefy)e = fye$, whence it is obvious that $fye \in M(e, f)$. Moreover, $e \cdot fye \cdot f = ef$, and so we conclude that $S'' \subseteq S'$. Conversely, take $g \in S'$. Then $g = fge$ and $g \in V(ef)$ (as is quickly checked), and thus $S' = S''$.

To see that $S'' \subseteq S$, take $fye \in S''$ as above and $h \in M(e,f)$. Then $efye.eh = efyefh = efh = eh$ and $hf.fyef = hefyef = hef = hf$, as required. Finally, let g denote any member of S; the proof is completed by verifying that $egf = ef$. Taking any $h \in S'$ and recalling that $eh \to eg$ and $hf \succ\!\!\!- gf$ we obtain

$$egf = eg^2 f = egefgf = egeh^2 fgf = (eg.eh)(hf.gf) = ehf = ef.\quad\blacksquare$$

Remark The proof shows that, in general, $S' = S'' \subseteq S(e,f)$ with equality throughout if all are non-empty; however, the inclusion can be strict: see Exercise 1.5.8.

In Chapter 3 we shall prove, conversely, that if every sandwich set of a biordered set E is non-empty, then E is the biordered set of some regular semigroup. The next definition anticipates this theorem.

Definition 1.5.4 A biordered set E is *regular* if $S(e,f)$ is non-empty for all $e, f \in E$. More generally, E is an *M-biordered set* if $M(e,f)$ is non-empty for all $e, f \in E$. A semigroup is an *M-semigroup* if $E(S)$ is an *M*-biordered set.

Suppose that S is eventually regular, so that $E(S) \neq \emptyset$, and take any $e, f \in E(S)$. There exists $n \geq 1$ such that $((fe)^2)^n \in \text{Reg}(S)$. Take x in $V((fe)^{2n})$ and put $k = (fe)^n x (fe)^n$. Then $k^2 = k$, $ke = k$, and $fk = k$, whence $k \in M(e,f)$. Thus we have proved the following.

Theorem 1.5.5 *Any eventually regular semigroup is an M-semigroup.*

In this case, however, the converse is false (see Exercise 1.6.4).

Before we give examples, we shall establish some conventions concerning biordered set diagrams. In a completed picture:

(1) different letters correspond to different objects;
(2) we omit the drawing of an arrow from an element to itself;
(3) $e \to f$ implies $e \leftrightarrow f$, and $e \succ\!\!\!- f$ implies $e \dashv\vdash f$;
(4) arrows not drawn may be deduced by *composition*, that is by transitivity.

For example,

and

both represent the same biordered set (which is in fact that of a three-element band in which $eg = f$). On the other hand, cannot be complete because the drawing implicitly implies that $e \not\to g$, which contradicts the existence of a right arrow from e to g, which results from composing right arrows from e to f to g. Hence the previous drawing necessarily becomes

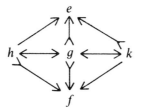

but even this cannot be complete because $g \mathrel{-\!\!\times\!} f$ (shorthand for $g \to f$ and $f \mathrel{\succ\!-} g$), which implies that $g = f$, and so the picture collapses to that of a two-element right zero band, $e \leftrightarrow f$, which is now complete.

Example 1.5.6

Since $h \to e$ and $k \to f$, both eh and kf are idempotents; so if this diagram is to represent the biordered set of some semigroup we must resolve the products eh and kf. Up to biordered set isomorphism there are three possibilities:

(i) E_1: $he = kf = g$.
(ii) E_2: $he = k$, $kf = h$.
(iii) E_3: $he = g$, $kf = h$.

The alternative $he = k$, $kf = g$ is isomorphic to E_3. All other candidates for the undetermined products introduce new arrows or collapse the biordered set: for example, if $he = f$ then $ehe = he = f$, so that $f \leqslant e$. This example will be thoroughly studied in Chapter 3. It turns out that E_1 is the biordered set of a six-element band, but by contrast E_1 comes from no eventually regular semigroup, while E_3 is the smallest biordered set to come from a finite, but not from a regular, semigroup. However, we can already see that E_2 and E_3 are not from regular semigroups: in both cases $M(e, f) = \{g, k\}$, whence $S(e, f)$ is empty.

The next result, the proof of which is set as an exercise, allows us to define the sandwich set of an arbitrary ordered pair from a regular semigroup.

Lemma 1.5.7 *Let S be a regular semigroup. Then:*

(i) $S(e,f)$ *is a rectangular band* $(e, f \in E(S))$;

(ii) $e \mathcal{L} f \Rightarrow S(e, g) = S(f, g)$, $e \mathcal{R} f \Rightarrow S(g, e) = S(g, f)$ *for all* $g \in E(S)$;

(iii) *if* $a' \in V(a)$, $b' \in V(b)$ *and* $g \in S(a'a, bb')$ *then* $b'ga' \in V(ab)$.

From part (ii) of this lemma we see that, for a regular semigroup S, the sandwich set, $S(e,f)$ of a pair of idempotents depends only on the $\mathcal{L}$-class of e and the $\mathcal{R}$-class of f. Thus for any members $x, y \in S$, we may define the sandwich set, $S(x, y)$ as $S(e,f)$, where $e \in E(L_x)$ and $f \in E(R_y)$. The following result of Nambooripad (1975) shows how sandwich sets can be used to 'locate' products in S.

Theorem 1.5.8 *If x, y are members of a regular semigroup S and $h \in S(x, y)$ then $h \in L_{xh} \cap R_{hy}$ and $(xh)(hy) = xy$.*

Remark The statement is illustrated by the egg-box diagram below:

	x					
						y

xh		(xh)(hy) = xy	
h		hy	

Proof For $h \in S(x, y)$ by Theorem 1.5.3 we obtain

$$(xh)(hy) = xhy = (xe)h(fy) = x(ehf)y = x(ef)y = (xe)(fy) = xy,$$

where $e \in E(L_x)$ and $f \in E(R_y)$. Taking $a \in S$ such that $ax = e$ we see that $xhaxh = xheh = xh^2 = xh$, whence it follows that $xh \in L_h$, and dually we can show that $hy \in R_h$. ∎

The reader should note that this result is effectively a reformulation of Exercise 1.4.13 for the case $n = 2$, expressed in the language of sandwich sets.

We close this section with an investigation of the compatibility with multiplication of the natural partial order on a regular semigroup. We begin with a simple lemma.

Lemma 1.5.9 *If $a \leqslant b$ and $b' \in V(b)$ then:*

(i) *there exists $a' \in V(a)$ such that $ab' = aa' \leqslant bb'$ and $a = aa'b$;*
(ii) *there exists $a'' \in V(a)$ such that $b'a = a''a \leqslant b'b$ and $a = ba''a$.*

Proof (i) Let $e, f \in E$ be such that $a = eb = bf$. Then

$$ab' \cdot ab' = ebb' \cdot bfb' = ebfb' = afb' = ab',$$

so $ab' \in E$. Since $ab'S = ebb'S = ebS = aS$, we infer that there exists $a' \in V(a)$ such that $ab' = aa'$. Since $ab' \cdot bb' = ab'$ and $bb'a \cdot b' = ab'$ we also have $ab' = aa' \leqslant bb'$. Finally, $aa'b = ab'b = a$.
(ii) This is proved similarly. ∎

If S is regular then so is the local submonoid eSe (Exercise 1.3.11). If each local submonoid of S is a semigroup of type T, say, we call S a *locally T-semigroup*. Recall from Exercise 1.2.19 the definition of an $\mathscr{L}$-unipotent semigroup as one in which each $\mathscr{L}$-class has a unique idempotent.

Theorem 1.5.10 (Blyth and Gomes 1983) *The following are equivalent for a regular semigroup S:*

(i) *S is locally $\mathscr{L}$-unipotent;*
(ii) *v is compatible on the right with multiplication;*
(iii) *$S(e, f)$ is a right zero semigroup for all $e, f \in E(S)$.*

Proof (i) $\Rightarrow$ (ii). Let $a \leqslant b$ and let $c \in S$. By Theorem 1.4.4(v) there exists $a' \in V(a)$ such that $a = aa'b$; choose $c' \in V(c)$ and $g \in S(a'a, cc')$. Then we have

$$ac \cdot c'ga' \cdot bc = aga'bc = aga'aa'bc = aga'ac = agc = ac$$

and so $ac = e \cdot bc$, where $e = ac \cdot c'ga' \in E$ since, by Lemma 1.5.7(iii), $c'ga' \in V(ac)$.

Now let $b' \in V(b)$ and choose $a'' \in V(a)$ such that $a''a \leqslant b'b$ and $a = ba''a$ by Lemma 1.5.9. Choose $c'' \in V(c)$ and $h \in S(a''a, cc'')$. Then we note first that $a''ah \in E$; and that $b'bh \cdot b'bh = b'bha''ab'bh = b'bha''ah = b'bh$, so that $b'bh \in E$. Now

$$a''ah \cdot b'bh = a''aha''ab'bh = a''aha''ah = a''ah, \qquad b'bh \cdot a''ah = b'bh,$$

and hence $a''ah \,\mathscr{L}\, b'bh$. Since $h = ha''a$ and $a''a \leqslant b'b$ we have that $a''ah$ and $b'bh$ are $\mathscr{L}$-related idempotents in the local submonoid $b'bSb'b$. It follows by (a) that $a''ah = b'bh$. Consequently,

$$bc \cdot c''ha'' \cdot ac = bhc = bb'bhc = ba''ahc = ahc = ac,$$

so that $ac = bc \cdot f$, where $f = c''ha'' \cdot ac \in E$ since $c''ha'' \in V(ac)$. Thus we see that $ac \leqslant bc$.

(ii) ⇒ (iii). If $g \in S(e,f)$ then we have $gf \in E$ with $gf \leqslant f$. If also $h \in S(e,f)$ then, by the given condition, we obtain $gh = gfh \leqslant fh = h$, whence $S(e,f)$ being a rectangular band yields $gh = hgh = h$. Thus $S(e,f)$ is a right zero semigroup.

(iii) ⇒ (i). If $e \in E$, $a \in eSe$, and $a' \in V(a) \cap eSe$ then clearly $a'a \in S(a'a, e)$. If also $a'' \in V(a) \cap eSe$ then, using Lemma 1.5.7(ii), we see that $a'a$ and $a''a$ belong to $S(a'a, e) = S(a''a, e)$ which, by (iii), is a right zero semigroup. Thus $a'a = a''a \cdot a'a = a''a$ and, consequently, the local submonoid eSe is $\mathscr{L}$-unipotent. ∎

Blyth and Gomes (1983) gives other characterizations of right compatibility, and constructs an example of a locally $\mathscr{L}$-unipotent semigroup which is not locally inverse.

Corollary 1.5.11 (Nambooripad 1980) *The following are equivalent for a regular semigroup S:*

(i) *S is locally inverse;*
(ii) *v is compatible with multiplication;*
(iii) *$S(e,f)$ is a singleton for all $e, f \in E(S)$.*

The idea of a regular biordered set can be thought of as a generalization of the semilattice of an inverse semigroup, with the sandwich set, $S(e, f)$, playing the role corresponding to the semilattice meet of a pair of idempotents e and f. The previous corollary shows that sandwich sets naturally lead to the consideration of locally inverse semigroups; indeed, one of Nambooripad's sources was the work of Schein (1972) on pseudosemilattices. Another important paper in this direction is that of McAlister (1984).

Exercises 1.5

1. (a) Prove the statements of Lemma 1.5.7.
   (b) Prove that if S is a regular semigroup then
   $$S(x, y) = \{yzx \text{ where } z \in V(xy)\}.$$

2. (a) Show that E, given by the picture

   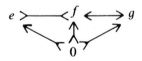

   is a regular biordered set, which is not that of any union of groups.
   (b) Find a completely 0-simple semigroup with E as its biordered set.

3. Show that the biordered set E_3 as given in Example 1.5.6 cannot be embedded in a band.
4. Let R be the free semigroup on idempotent generators e, f. Extend R to a semigroup P by adjoining idempotents g, h satisfying $gr = rg = g$, $hr = rh = h$ for all $r \in R$ and with $g \leftrightarrow h$. Let E be the biordered set of P.

   (a) Show that P is an M-semigroup.
   (b) E is not the biordered set of any eventually regular semigroup. [Hint: suppose such an eventually regular S exists. As in the proof of Theorem 1.5.5, construct $k \in M(e, f)$ and x. Deduce the contradiction $h = hg = g$.]

5. (Meakin 1980) Call a biordered set E *rectangular* if for all $e, f \in E$ there exists $g \in E$ such that $e \mathscr{R} g$ and $g \mathscr{L} f$. For an arbitrary biordered set E the following are equivalent:

   (i) E is a rectangular biordered set;
   (ii) E is an M-biordered set and $\leqslant \; = 1 | E$;
   (iii) E is the biordered set of some rectangular band.

6. (Edwards 1983) Let S be an eventually regular semigroup.

   (a) $E(S) \simeq E(T)$, for some finite semigroup T if and only if $E(S)$ is finite.
   (b) If $E(S)$ is finite, then for any congruence ρ on S, $E(S/\rho) \simeq E(S)$ if and only if $\rho \subseteq \mu$, the maximum idempotent-separating congruence on S.

7.* (Nambooripad 1979) Let E be a biordered set and $(e, f) \in D_E$. Then $ef \in S(f, e)$.

8. Let S be the free semigroup on two idempotent generators e and f. Then $E(S^0)$ is a three-element semilattice $\{e, f, 0\}$ and $S(e, f)$ is not empty, even though ef is not regular.

9. Consider Example 1.5.6 and show, as claimed, that there are, up to biordered set isomorphism, only three ways in which to resolve the products he and kf to yield a biordered set of some semigroup.

10. (Hall 1982b; see Exercise 1.3.12)

    (a) Show that the regular members of every right and every left principal ideal of a locally orthodox semigroup form an orthodox semigroup.
    (b) Show that the regular members of every right principal ideal of a locally $\mathscr{L}$-unipotent semigroup form an $\mathscr{L}$-unipotent semigroup.

1.6 Graph theory preliminaries

Graphical ideas will emerge in several places in our investigations into semigroups: in the diagram theory of the next section and of Chapter 5; in the

Munn birooted tree representation of free inverse semigroups in Chapter 2; and in the study of finite full transformation semigroups in Chapter 6. We take this opportunity to record some basic definitions and facts concerning graphs and digraphs (directed graphs). We shall not define all elementary concepts from graph theory (for a detailed account the reader can consult, for example, Harary 1969), but the use of some terms needs to be fixed for our own purposes as there are small variations of usage in the literature.

We shall understand a graph G to be finite but not necessarily simple; that is, loops and multiple edges will be permitted. The number of vertices of G will be denoted by p, and the number of edges by q. The respective sets of vertices and edges of G will be denoted by $V(G)$ and $E(G)$. A vertex may also be referred to as a point. A *walk* π in G is an alternating sequence of vertices and edges $v_0, e_1, v_1, \ldots, v_{n-1}, e_n, v_n$, beginning and ending with vertices, in which each edge is incident with the two vertices immediately preceding and following it, and the positive integer n is the *length* of π, denoted by $|\pi|$; it is convenient to admit the *null walk* of length zero at a vertex v. If a walk $\pi = \alpha\sigma\beta$ then σ is a *segment* of π, and σ is *proper* if not both α and β are null; π is *closed* if $v_0 = v_n$ and is *open* otherwise. We call π a *trail* if all the edges are distinct, and π is a *path* if all its vertices (and hence also its edges) are distinct. A closed trail is a *cycle* if its n vertices are distinct and it has at least one edge. A graph is *acyclic* if it has no cycles. A connected acyclic graph is a *tree*.

Theorem 1.6.1 *The following are equivalent for a graph G:*

(i) G is a tree;
(ii) every two points of G are joined by a unique path;
(iii) G is connected and $p = q + 1$;
(iv) G is acyclic and $p = q + 1$.

Proof (i) $\Rightarrow$ (ii). Let u and v be vertices. Suppose there are two distinct paths P and Q between vertices u and v. Let w be the first vertex we meet as we travel from u to v along P, the following edge of which is not an edge of Q, and let w' be the next vertex on P after w which is also on Q. Then the segments of P and Q running between w and w' would form a cycle, so this cannot occur. Since a path from u to v certainly exists, it must be unique.

(ii) $\Rightarrow$ (iii). Clearly, G is connected. We proceed by induction on p; the result being clear if $p = 1$ or 2. If G has p vertices, the removal of any edge of G disconnects G into two components. By the inductive hypothesis, each component has one more vertex than edge. Thus the total number of edges of G must be $p - 1$.

(iii) $\Rightarrow$ (iv). Assume that G has a cycle of length n. There are n vertices and n edges on the cycle, and for each of the $p - n$ vertices not on the cycle there is an edge incident on a shortest path to a vertex on the cycle. Each such edge is different, giving $q \geq p$, a contradiction.

(iv) ⇒ (i). Since G is acyclic it is a forest of a number of trees, k, say. Since each tree component has one more vertex than edge, we see that in G, $p = q + k$, whence $k = 1$. ∎

The *distance* between two-vertices u and v in a connected graph G, denoted by $d(u, v)$, is the length of a shortest path between them. The *eccentricity*, $e(v)$, of a vertex v in G is the maximum of $d(u, v)$ (u a vertex). The *radius* $r(G)$ is the minimum eccentricity of the vertices. A vertex v is *central* if $e(v) = r(G)$, and the *centre* of G is the set of all central vertices.

Theorem 1.6.2 *Every tree T has a centre consisting of one vertex or two adjacent vertices.*

Proof First note that T has a central endpoint if and only if T consists of one vertex or two adjacent vertices, and the theorem is true in these cases. Otherwise, consider the tree T' formed by deleting all endpoints from T. In passing from T to T' the eccentricity of all remaining vertices is reduced by one, and so T and T' share a common centre. The result now follows by induction on the number of vertices of T. ∎

Now let D be a digraph. If e is an edge of D with initial endpoint a and terminal endpoint b, we write e or e^+ to denote e considered with its given orientation in D, we refer to e^+ as the *positive edge* or *arc* associated with the edge e, and we write e^- to denote e considered with the reverse of the given orientation. We refer to e^- as the *negative edge* associated with e, and say that e^- has initial endpoint b and terminal endpoint a. An *oriented edge* is either a positive edge or a negative edge of D. A *walk* π in D of length n is a sequence of oriented edges $(f_1, f_2, \ldots, f_n)$ such that the terminal endpoint of f_i is the initial endpoint f_{i+1} for $1 \leq i < n$. A walk π is *directed* if each oriented edge of π is positive. The definitions of trail, path, and cycle are worded as for graphs; in particular, we shall refer to a directed path as a *dipath*. A walk in D is *two-sided* if it has the form $\alpha\beta^{-1}$, where α and β are positive walks and β^{-1} denotes the reverse of β.

A digraph D is *strong* if, given any two vertices i and j of D there exists a dipath from i to j. A digraph is *complete* if for all vertices $i, j \in V(D)$ either (i, j) or (j, i) is an arc of D. A complete digraph D is a *tournament* if whenever (i, j) is an arc of D then (j, i) is not. It is also usual to insist that tournaments do not have loops, in which case they do model the result of (say) a 'round-robin' tennis tournament by indicating that player i has defeated player j if and only if there is an arc from i to j in D.

A member α of T_n can be written as a sequence, in which $i\alpha$ is the ith member of the sequence. For example, $(2, 3, 1, 3, 4, 4, 8, 8)$ represents the mapping α in T_8 in which 1 is mapped to 2, 2 to 3, and so on. Members of PT_n can be written in a similar manner: for instance, the list $(-, 1, 2, -, -, 6, 6)$

would represent the member $\alpha \in PT_7$ with dom $\alpha = \{2, 3, 6, 7\}$, with images as shown.

It has often been observed (first in Suschkewitsch, 1928) that $\alpha \in T_n$ can be depicted as a digraph on n vertices, also called α, in which ij is an arc (positive edge) of α if $i\alpha = j$. For this reason, a digraph such as this one, in which every vertex has outdegree one, is called *functional* (which implies that there is at most one dipath between any two vertices).

The structure of functional digraphs can be formulated in several ways. First let us say that vertex v is *reachable* from a vertex u in a digraph D if there is a dipath from u to v. A *sink* in D is a vertex which can be reached from all others; a *source* is the dual concept. A *receiver* is a vertex of outdegree zero, while a *transmitter* is the dual concept. An *in-tree* is a digraph, the underlying graph of which is a tree, that possesses a (unique) sink, which is then necessarily the unique receiver. The dual concept is called an *out-tree*, and it possesses a (unique) source which is then the unique transmitter. A digraph D is called *weak* if it is *weakly connected*, meaning that its underlying graph is connected.

Proposition 1.6.3 *A weak digraph T is an in-tree if and only if exactly one vertex has outdegree zero and all others have outdegree one.*

Proof If T is an in-tree the sink must have outdegree zero in order that the graph of T be cycle-free. Suppose that u is a vertex with edges from u to v_1 and v_2. Let P and Q be dipaths from v_1 and v_2 to the sink. It follows that $v_1 = v_2$, as there is a unique dipath from u to the sink.

Conversely, T is connected, and since there is evidently a one-to-one correspondence between the edges of T and the vertices of T of outdegree one, it follows that $p = q + 1$ for T, whence the underlying graph of T is a tree by Theorem 1.6.1. Next take any vertex u of T, and consider the path in T formed by following successive out-edges beginning with u. This path must terminate when it reaches the vertex of outdegree zero, which is therefore the sink of T. ∎

Theorem 1.6.4 *The following are equivalent for a digraph D:*
  (i) *D is functional and weak;*
  (ii) *D has exactly one directed cycle, the removal of the edges of which results in a digraph in which each weak component is an in-tree with its sink in the cycle;*
  (iii) *D has exactly one directed cycle Z, the removal of any edge of which results in an in-tree.*

Proof (i) $\Rightarrow$ (ii). By finiteness of D, it is clear that D has at least one directed cycle Z. Suppose that Z' were another and let z, z' be vertices of Z and of Z'

respectively. Since D is weakly connected there is a path in the underlying graph of D, the vertices of which are $z, z_1, \ldots, z_n, z'$, say. Since each vertex has outdegree one, the edge zz_1 is negative, whence for the same reason, so are $z_1z_2, z_2z_3, \ldots, z_nz'$; but then z' has outdegree two, and so we conclude that the cycle Z of D is unique. Next suppose that the edges of Z are removed. The component of this new digraph containing z then has as vertices z together with all vertices not in Z from which z is reachable. The rest of (ii) now follows from Proposition 1.6.3.

(ii) $\Rightarrow$ (iii). Remove all edges from Z. By hypothesis the resulting digraph is a collection of $k = |Z|$ in-trees, each with its sink a vertex of Z. Observe that as each edge of Z is replaced, the resulting digraph retains the property that each weak component is an in-tree, but the number of components is reduced by one at each stage. Therefore, after any $k - 1$ of the edges of Z are replaced, the digraph we obtain is an in-tree, as required.

(iii) $\Rightarrow$ (i). Remove a directed edge uv from the directed cycle Z of D. Then the sink of the resulting graph must be u. It follows immediately that D is weakly connected, and that the outdegree of every vertex of D is one. ∎

Thus we see that the digraph of $\alpha \in T_n$ consists of a number of components, each with a unique cycle, together with a number of trees rooted around the vertices of the cycle. The direction of an arc within a tree is towards the sink of the associated in-tree (which is called the *root* of the in-tree), and we shall sometimes emphasize this by saying that the trees are *root-directed*. Note that, in general, a vertex in a cycle of α may be regarded as the root of a single tree, or of several trees; in other words, the root of an in-tree is not necessarily an endpoint, as shown in the example to follow. Every edge of an in-tree is a *bridge* of that component, meaning that its removal disconnects the component (as a graph).

We establish the convention that cycles are directed anticlockwise, and one-cycles (corresponding to fixed points) are shaded, so that the arrows may be deleted from the picture of α without ambiguity. A component is called *cyclic* if it is a cycle with more than one vertex. The *stable range* of α is the set of all points in some cycle of α (see Exercises 1.6.1, 1.6.2 and 1.6.4 below).

Example 1.6.5 Consider the digraph of a mapping $\alpha \in T_9$. Note that the cycle vertex 3 may be regarded as the root of one tree, or of a pair of trees:

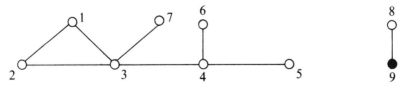

The enumeration of functional digraphs on n vertices has been carried out by Harary (1959).

We take this opportunity to record some basic observations concerning the digraph of a given map $\alpha \in T_n$.

For a given $\alpha \in T_n$ consider the relation ω on X_n whereby $x\omega y$ if and only if there exist $l, m \geq 0$ such that $x\alpha^l = y\alpha^m$. It is easily seen that ω is an equivalence relation on X_n and that the ω-classes $\Omega_1, \Omega_2, \ldots, \Omega_k$, say, called the *orbits* of α, correspond to the vertex sets of the components of the digraph of α. The labels of the cycle vertices of a component of Ω form a set sometimes called the *kernel* of Ω, denoted by $K(\Omega)$. The next lemma is a corollary of Theorem 1.6.4.

Lemma 1.6.6 *Let Ω be an orbit of α, with $|K(\Omega)| = r$. Then $y\alpha^r = y$ for all $y \in K(\Omega)$. Conversely, if $y \in \Omega$ is such that $y\alpha^k = y$ for some $k > 0$, then $y\alpha^r = y$ and $K(\Omega) = \{y, y\alpha, \ldots, y\alpha^{r-1}\}$.*

It will be useful to define, for each p in Ω, a subset $p\alpha^{-N}$ of Ω by

$$p\alpha^{-N} = \{x \in X_n : x\alpha^i = p \text{ for some } i > 0\}.$$

The preceding lemma then implies that

$$p \in K(\Omega) \quad \text{if and only if } p \in p\alpha^{-N}.$$

A further characterization of $K(\Omega)$, which is also a consequence of Theorem 1.6.4, is given by the following.

Lemma 1.6.7 *Let $p \in \Omega$. Then $p \in K(\Omega)$ if and only if $p\alpha^{-N} = \Omega$.*

Exercises 1.6

1. Identify the graphical counterparts of the following features of a given $\alpha \in T_n$:
   (a) the range of α, ran α;
   (b) the stable range of α, stran $\alpha = \bigcap_{k=0}^{\infty} \text{ran } \alpha^k$.

2. For a positive integer m and $\alpha \in T_n$, define the set of mth roots of α as $\alpha^{1/m} = \{\beta \in T_n : \beta^m = \alpha\}$. Show that stran α (see above) is invariant under the taking of mth powers and of mth roots.

3. (a) Prove that a subsemigroup S of T_n is completely regular if and only if ran α = stran α, for all $\alpha \in S$.
   (b) The condition ran α = stran α for all $\alpha \in S$ does not generally ensure that S is a union of groups if S is a subsemigroup of an infinite full transformation semigroup.

4. (a) For $\alpha \in T_n$ show that $\alpha|\text{stran }\alpha$ is a permutation (called the *main permutation* of α).

Fundamentals | 73

(b) Define the *quasi-inverse* of $\alpha \in T_n$ as α^m, where m is the least positive integer such that $\alpha^m |$ stran α is the inverse of the main permutation of α. Show that α is a completely regular element of T_n if and only if the quasi-inverse of α is an inverse of α.

(c) For a completely regular element α of T_n, describe the digraph of the quasi-inverse of α in terms of that of α.

5. Describe the index and period of $\alpha \in T_n$ in terms of features of the digraph of α.

1.7 Semigroup diagrams

In this section we introduce a geometric method, based on so-called semigroup diagrams, used as a tool in word arguments. For example, let S be the semigroup on four generators a, b, c, d with relations $a = cb, c = bd$, and $b = ac$. In S we have the sequence of equations:

$$a^2 d = acbd = ac^2 = bc.$$

To display this derivation pictorially, we can construct Fig. 1.4. Conversely, we could sequentially strip regions from the diagram to recover our sequence of equations. Equally, though, we can strip the regions from the diagram to obtain another derivation of $a^2 d = bc$,

$$a^2 d = acbd = b^2 d = bc,$$

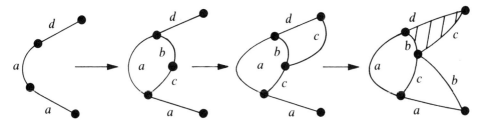

Fig. 1.4.

and our diagram may be constructed from this sequence of equations as well as from the first. Thus several different sequences of algebraic transitions may be subsumed by a single diagram.

Our diagram method, rudimentary as it is at this stage, clearly involves associating a labelled planar digraph with a transition sequence, the edges of which are directed upwards and labelled with words from the free semigroup on the given generators. If the reader constructs some diagrams himself he will soon notice that the diagrams have no interior transmitter or receiver, and that they have a unique transmitter and a unique receiver (which are the unique source and sink respectively) on the boundary of the diagram. Furthermore, each region is bounded by two 'sides' (a side being a directed path) and these sides are labelled by a pair of related words in our semigroup. Additionally, the boundary of the entire diagram also enjoys this 'two-sided' property. Our first goal is to prove that these features characterize this type of diagram.

Definition 1.7.1 A *semigroup diagram* (or more briefly, a diagram) is a planar connected digraph Γ embedded in $\mathbb{R}^2$, together with a labelling function which assigns to each oriented edge of Γ an element of F_X, the free semigroup on some set X. We also require the following:

(i) Each component of $S^2 - \Gamma = (\mathbb{R}^2 \cup \infty) - \Gamma$ is *two-sided*, meaning that it is bounded by a clockwise cycle of the form $\alpha\beta^{-1}$, where α and β are non-empty directed paths (that is, paths which traverse edges only in the direction of their orientation; β^{-1} denotes the reverse of β).
(ii) Γ has exactly one source, O, and one sink I, each of which lies on the unbounded component of $S^2 - \Gamma$.

If M is a diagram with underlying graph Γ, then the bounded components of $S^2 - \Gamma$ are the *regions* of M. The number of regions of M is denoted by $\|M\|$. If D is a region of M with clockwise boundary cycle $\alpha\beta^{-1}$ as in (i) above, then $\alpha[\beta]$ is called the *left* [*right*] *boundary* of D. The labelling function φ extends to a function of non-empty directed paths and takes values in the free semigroup F_X. The *left* and *right boundary labels* of α and β are then $r = \varphi(\alpha)$ and $s = \varphi(\beta)$ respectively, and we say that ∂D, the boundary of D, has label rs^{-1}.

A similar terminology applies to the whole diagram M. If D is the unbounded region of $S^2 - \Gamma$, and $\alpha\beta^{-1}$ is a clockwise boundary cycle of D as in (i) above, then $\beta\alpha^{-1}$ is a clockwise boundary cycle of $S^2 - \operatorname{int} D$. We call $\beta[\alpha]$ the *left* [*right*] *boundary* of M. Also, $\varphi(\beta)[\varphi(\alpha)]$ is the *left* [*right*] *boundary label* of M. In our previous diagram the shaded region has boundary labels bd and c respectively, while the boundary labels of the diagram are $a^2 d$ and bc, the source and sink being represented by the lowermost and uppermost vertices in the picture.

Note that since the source O of a semigroup diagram is unique, it must also be the unique transmitter of the diagram; dually, the unique sink is its unique receiver.

Let S be a semigroup with presentation $(X; R)$ where, without loss, we shall assume that R is symmetric. A *diagram* M *over* $(X; R)$ is one in which each interior region has left boundary label r and right boundary label s, where $\{r, s\} \in R$; we call M a (u, v)-*diagram* for the pair (u, v), where u and v are the left and right boundary labels of M respectively. We shall sometimes describe M as a *derivation diagram* over $(X; R)$ for the pair (u, v). We shall write $u \equiv v$ if u and v represent the same member of F_X and write $u = v$ if u and v are equal in S. A *subdiagram* M' of M is a diagram over $(X; R)$ the vertices, edges, and interior regions of which are themselves vertices, edges, and interior regions of M.

Having established these preliminaries, we may easily construct a (u, v)-diagram for any sequence of elementary R-transitions $u \equiv u_0 \to u_1 \to \ldots \to u_n \equiv v$ ($u, v \in F_X$). If $n = 0$, then $u \equiv v$ and M can be a simple directed open arc with label u. If $n > 0$ we assume inductively that N is a diagram for the pair (u, u_{n-1}) with $n - 1$ regions. Now $u_{n-1} = prq$ and $u_n = psq$ for some defining relation $\{r, s\}$. The right side of N can be written as $\beta\rho\gamma$, where the paths β, ρ, γ have labels p, r, and q respectively. Let σ be a simple arc with label s running from the initial to the terminal vertex of ρ in the exterior of N, chosen so that the cycle $\sigma\rho^{-1}$ is oriented anticlockwise. Then the diagram M obtained from N by adjoining σ as a labelled arc is as required.

Theorem 1.7.2 *Let $S = \langle X | R \rangle$ and $u, v \in F_X$. Then there is a sequence of elementary transitions $u \equiv u_0 \to \ldots \to u_n \equiv v$ if and only if there is a (u, v)-diagram over $(X; R)$ with exactly n regions.*

To establish the converse part of the theorem let M be a (u, v)-diagram over $(X; R)$, that is:

(i) each region of M is labelled in the clockwise direction by a word rs^{-1} for some $\{r, s\} \in R$;
(ii) ∂M, the boundary of M, carries a clockwise label uv^{-1} with boundary vertices O and I the unique source and sink respectively of M.

We preface the proof with a lemma.

Lemma 1.7.3 *Every vertex and edge of M lies on a dipath running from O to I.*

Proof Since we may assume that there are no isolated vertices, it is sufficient to prove the statement for a given edge $e = (a, b)$, and clearly only the case where e does not lie on ∂M requires proof. Suppose then that e is interior to ∂M (meaning that e does not meet ∂M, except perhaps at endpoints). Now if $b \neq I$ there is an edge $f = (b, c)$ in M because b is not a receiver. It we continue building a dipath from f in this way we shall eventually reach ∂M, or obtain a directed cycle lying in the interior of M. We show that the latter alternative is impossible.

Suppose that γ is a directed cycle strictly interior to ∂M so that, in particular, γ does not contain O or I. We show by induction on k, the number of edges enclosed by γ, that γ must contain a cyclic region—a contradiction. If $k = 0$ then γ is itself the boundary of a cyclic region. Suppose that $k > 0$: we assert that γ encloses a smaller directed cycle. The result then follows by induction. Take σ to be a maximal trail which is properly enclosed by γ (σ is enclosed by and is unequal to γ). Thus if no segment of σ is a directed cycle, the endpoints of σ divide γ into two dipaths, one of which, α say, runs from the terminal vertex to the initial vertex of σ. The directed cycle $\alpha\sigma$ is then properly enclosed by γ, as required. We thus infer that there is a dipath from b to some vertex on ∂M and, similarly, that there is a dipath from some vertex on ∂M to a. Bearing in mind that there are no directed cycles in M, one now readily obtains a required dipath from O to I that runs through e. ∎

Since a diagram M cannot contain a directed cycle, it follows that α and β, the left and right sides of ∂M respectively, are paths, and not merely trails. There are thus unique concatenations $\alpha = \alpha_1 \ldots \alpha_n$ and $\beta = \beta_1 \ldots \beta_n$ of non-trivial segments, where for each i either $\alpha_i = \beta_i$ or $\alpha_i \beta_i^{-1}$ is a clockwise-oriented cycle and for each $i \geq 2$ either $\alpha_i \beta_i$ or $\alpha_{i-1} \beta_{i-1}^{-1}$ is a cycle. The submap bounded by $\alpha_i \beta_i^{-1}$ is called the *ith simple component* of M, and is either a dipath or is bounded by a two-sided cycle as illustrated in Fig. 1.5 (where for convenience we have directed the graph from left to right instead of from bottom to top).

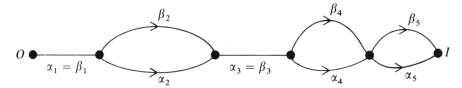

Fig. 1.5 The boundary of a diagram with five simple components.

Proof of Theorem 1.7.2 (converse direction) Assume that M is an n-region (u, v)-diagram with left side labelled $u = r_1 r_2 \ldots r_m$ and right side labelled $v = s_1 s_2, \ldots, s_m$, where $r_i s_i^{-1}$ is the label on the boundary of the ith simple component M_i of M, traversed clockwise starting at the initial vertex of M_i. If M has no interior edges, then for each i either $r_i \equiv s_i$ or M_i is a cycle and $\{r_i, s_i\}$ is a defining relation. Then v may be derived from u by a sequence of n non-overlapping elementary transitions of the form $pr_i q \to ps_i q$.

If M has an interior edge e, we extend e, by Lemma 1.7.3, to a dipath running from O to I. The submaps M' bounded by $\alpha \pi^{-1}$ and M'' bounded by $\pi \beta^{-1}$ are also semigroup diagrams, and each has fewer interior edges than does M. By induction we may assume that there are elementary transition sequences $u \to w$ and $w \to v$ ($w = \varphi(\pi)$) of lengths equal to the number of regions of M' and M'' respectively. By concatenating these sequences we obtain the desired sequence of elementary transitions from u to v of length n. ∎

The definition of semigroup diagram can be given in a formally weaker fashion.

Fundamentals | 77

Proposition 1.7.4 *In Definition 1.7.1, condition (ii) may be replaced by the following:*
(ii)* *the boundary of M is two-sided, and M has no interior transmitter or receiver.*

Proof Suppose that M satisfies the conditions of Definition 1.7.1. Then, as has been observed previously, O and I are the unique transmitter and receiver of M respectively, and so M has no interior transmitter or receiver; also, that the boundary of M is two-sided is equivalent to the assumption that the unbounded component of M is two-sided.

Conversely, suppose that M satisfies condition (i) of Definition 1.7.1 and condition (ii)*. Let O and I denote the initial and terminal vertices, respectively, of the two sides of M. Observe that the proof of Lemma 1.7.3 goes through under the weaker hypotheses that M satisfies conditions (i) and (ii)*. Hence O is a source in M and, further more, O must be the unique source in order to preclude the existence of directed cycles in M, which cannot occur, again by the proof of Lemma 1.7.3; similarly, I is the unique sink of M, and so condition (ii) of Definition 1.7.1 is also satisfied. ∎

We close this introductory section with one application: a proof that a semigroup $S = \langle X \mid R \rangle$ which has a 'cycle-free' presentation is cancellative (Remmers 1980). The argument involves the notion of angle in a semigroup diagram.

A two-sided cycle γ of a diagram M and a vertex P on γ determine an *interior angle* θ of M at P, the *sides* of which are the edges of γ incident with P; a region is *included in* θ if it is enclosed by γ and its boundary contains P. An interior angle is a *source-angle* [*sink-angle*] if its sides and included edges (edges interior to γ and incident with P) are all out-edges [in-edges] at P. A pair of out-edges [in-edges] are *coinitial* [*coterminal*] if they have the same initial [terminal] vertex.

Lemma 1.7.5 *Let M be a semigroup diagram. Every pair of distinct coinitial edges at P are the initial edges of a two-sided cycle γ of M. The angle at P determined by γ is a source-angle.*

Proof The first statement is a consequence of Lemma 1.7.3. Suppose that γ does not determine a source angle at P, but that $e = (Q, P)$ is an in-edge at P enclosed by γ. As in the proof of Theorem 1.7.2, we take a maximal trail, π enclosed in γ, which ends with e. This necessarily begins on the two-sided cycle γ, which yields a directed cycle enclosed by γ—a contradiction. ∎

The *left* [*right*] *graph* $LG(X; R)[RG(X; R)]$ associated with the presentation $(X; R)$ has vertex set X and edge set R. A defining relation $\{r, s\}$ joins the initial [terminal] letters of r and s. Loops and parallel edges may occur in either graph.

78 | Techniques of semigroup theory

Example 1.7.6 The left and right graphs associated with the presentation $S = \langle a, b, c, d; abd = dba, adb = cad, aba = ad^2, bab = c, cab = acb \rangle$ are, respectively:

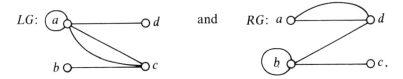

A walk of length $m \geq 0$ in an undirected graph is determined by a sequence of vertices $a_0, a_1, \ldots, a_m$ and edges $e_1, \ldots, e_m$ such that e_i runs between a_{i-1} and a_i ($i = 1, \ldots, m$). The walk is *closed* if $a_0 = a_m$ and is a *cycle* if it is a closed trail, $m \geq 1$; and the vertices $a_1, \ldots, a_m$ are all different. (In particular, any loop is a cycle.) Walks and cycles $LG(X, R)$ are called *left walks* and *left cycles* respectively for $(X; R)$. The pair $(X; R)$ is *cycle-free* if $(X; R)$ has no left and no right cycles.

Theorem 1.7.7 *If $(X; R)$ has no left [right] cycle and M is a (u, v)-diagram, then there exists a (u, v)-diagram M', having no more regions than does M, such that the labels on every pair of distinct coinitial [coterminal] edges of M' have distinct initial [terminal] letters. The modified diagram is then called source-reduced [sink-reduced]. If $(X; R)$ is cycle-free, then M' can be taken to satisfy both conditions simultaneously.*

Proof Suppose that $(X; R)$ has no left cycle and that there is at least one vertex P of M which possesses two out-edges bearing labels with the same initial letter. Since M has no directed cycles, we may choose P extremal, so that no other vertex on any dipath starting at P possesses two such out-edges. From Lemma 1.7.5 we infer that our two out-edges at P, e and f say, are the sides of a source angle θ at P. Let $D_1, D_2, \ldots, D_k$ be the regions included in θ, listed in clockwise cyclic order; that is, the order in which they are met when traversing a small circle centred as P in the clockwise direction. By choice of the ordering of the D_i, the right side of D_{i-1} and the left side of D_i share a common initial edge; let x_i be the initial letter of the label of this edge and define $x_0 \equiv x_k$ to be the common initial letter of the labels of the sides of θ. The region D_i is labelled by the relation $f_i \in R$ and the sequences $x_0, x_1, \ldots, x_k$ and $f_1, f_2, \ldots, f_k$ constitute a closed left walk of $(X; R)$, but $(X; R)$ has no left cycles. This can only occur if, for some i, f_i and f_{i+1} are the same relation, so that the right side β of D_i and the left side γ of D_{i+1} are identically labelled and are coinitial. Now β and γ must coincide for otherwise the extremal assumption on P would be violated. The deletion of β from M consolidates D_i and D_{i+1} into a single two-sided region E. We now continuously deform M to identify both the left and right sides of E, thus

Fundamentals | 79

reducing the number of regions of M by two. Iteration of this process leads to a diagram with the required properties. ∎

Corollary 1.7.8 *If $(X; R)$ has no left [right] cycles and there is a sequence of elementary transitions of length n from uv to uw [vu to vw] in $\langle X | R \rangle$, then there is a corresponding sequence from v to w in $\langle X | R \rangle$ of length no more than n. In particular, $\langle X | R \rangle$ is left [right] cancellative.*

Proof By Lemma 1.7.7, under the unbracketed assumptions, there exists a source-reduced diagram M for (uv, uw) with no more than n regions. The left and right sides of M have initial segments ρ and π each labelled u; since M is source-reduced, ρ and π coincide. Deletion of ρ from M yields the required diagram for (v, w). ∎

The theory of semigroup diagrams will be expanded in Chapter 5, where it will be applied to solve the word problem in certain so-called small cancellation semigroups and to show the group embeddability of semigroups with cycle-free presentations. The latter topic also serves to forge a link between our subject and that of the diagrams of combinatorial group theory.

Exercises 1.7

1. Derive a diagram which shows that $dbadcd = abdc^2a$ in the semigroup of Example 1.7.6:

   $S = \langle a, b, c, d; abd = dba, adb = cad, aba = ad^2, bab = c, cab = acb \rangle.$

2. A vertex v of a digraph is *hyperbolic* if there is an angle (e, f) at v such that v is the initial endpoint of e and f, but neither (e, f) nor (f, e) is a source angle. Find the minimal example of a hyperbolic vertex. Can a hyperbolic vertex occur in a semigroup diagram?

3. Show that a semigroup diagram has no interior separating vertex (a vertex the removal of which disconnects the underlying graph).

4. Let M be a semigroup diagram with at least one region, and let γ denote one of the sides of ∂M. Show that M has a region D, one of the sides of which is a segment of γ.

5.* For v, v' vertices of a semigroup diagram M, define $v \leq v'$ if there is a dipath in M from v to v'.

   (a) Show that this defines a partial order on the vertex set V.
   (b) Show that $(V \leq)$ is a complete lattice.
   (c) Use this idea to define a partial order on $V \cup E$, the set of vertices and edges of M.

6. Let M be a semigroup diagram. The *reflection dual*, M^r, is the map obtained from M by reflecting M in any straight line in the plane not meeting M; while a distinct dual, the *directional dual*, M^d, is obtained by reversing the direction of each arrow.

   (a) Show that the two operations generate an abelian group of order four under composition.
   (b) Let $S = \langle X | R \rangle$ and let $\langle X | R^* \rangle$ be the corresponding presentation of the dual semigroup S^*. What is the relationship between a derivation diagram over $(X; R)$ and the corresponding diagram over $(X; R^*)$?

2 Free inverse semigroups and the theorems of McAlister

The existence of the free inverse semigroup FI_X was established in Chapter 1 (Theorem 1.1.10). A canonical form for the elements of FI_X is afforded by their representation as the birooted word trees of Munn (1974). This in turn allows us to construct an alternative representation of the members of FI_X due to Scheiblich (1972, 1973). Consideration of the semilattice of FI_X leads to a further description of FI_X as a so-called P-semigroup. This motivates the study of arbitrary P-semigroups which are then shown to comprise a class of E-unitary inverse semigroups. The famous theorem of McAlister (1974b) is that the converse is true: any E-unitary inverse semigroup is isomorphic to a P-semigroup. In proving this theorem we follow Munn (1976). Finally, we prove that any inverse semigroup is the image under an idempotent-separating homomorphism of an E-unitary semigroup. Again we do not record the original proof of McAlister (1974a), but establish the result by using a semidirect product construction of Wilkinson (1983), although the result can also be quickly obtained from the knowledge gained concerning FI_X (Exercise 2.2.3).

2.1 Free inverse semigroups

The existence of free inverse semigroups has long been known, and an explicit representation of the free monogenic inverse semigroup dates back to Gluskin (1957) (see Exercise 2.1.5(a); for a recent account of monogenic inverse semigroups, see Preston 1986a). Little further progress was made until the early 1970s. The main obstacle to development was the lack of a canonical form for the elements of FI_X. Simple-minded approaches, such as representing elements of FI_X by words of minimum length, were not fruitful. The difficulty was overcome by Scheiblich (1972, 1973) and independently by Munn (1974), ushering in a period of rapid progress, of which the McAlister 'P-theorem' is the centrepiece.

Recall from Section 1.1 that FI_X can be realized as $F/\sim$, where F is the free semigroup on $Y = X \cup X'$ and ρ is the least inverse congruence on F. To provide a convenient method for handling the $\sim$-classes of F we employ some graph-theoretic concepts. We shall make use of the properties of trees recorded in Section 1.6.

Definition 2.1.1 A *word tree* T on a non-empty set X is a finite tree, with at least one edge, satisfying the following two conditions:

(WT1) Each edge is oriented and labelled by an element of X.

(WT2) T has no subgraph of the form ○$\xrightarrow{x}$○$\xleftarrow{x}$○ or

○$\xleftarrow{x}$○$\xrightarrow{x}$○ $(x \in X)$.

The WT2 property is the key feature of our trees, and will be used repeatedly in the sequel to establish certain uniqueness properties.

We extend the set of labels from X to Y by making the convention that:

○$\xrightarrow{x'}$○ means the same as ○$\xleftarrow{x}$○ $(x \in X)$.

The set of vertices of a word tree will be denoted by $V(T)$, and the number of vertices of T will be denoted by $|T|$. Note that $|T| > 1$. If α, β are adjacent vertices of T, then $\alpha\beta$ will denote the edge connecting α and β, with orientation $\alpha \to \beta$.

Definition 2.1.2 Let T be a word tree on X and let $(\alpha, \beta) \in V(T) \times V(T)$.

(i) An (α, β)-walk on T is a sequence $(\alpha = \gamma_0, \gamma_1, \ldots, \gamma_n = \beta)$ of vertices of T such that γ_{i-1}, γ_i are adjacent for $1 \leq i \leq n$.
(ii) An (α, β)-walk $\Gamma = (\alpha = \gamma_0, \gamma_1, \ldots, \gamma_n = \beta)$ on T is said to *span* T, or to be a *spanning (α, β)-walk on T*, if and only if each vertex of T occurs at least once amongst the γ_i.
(iii) An (α, β)-*path* on T is the unique (α, β)-walk $(\alpha = \gamma_0, \gamma_1, \ldots, \gamma_n = \beta)$ on T such that no vertex of T occurs more than once amongst the γ_i. We denote it by $\Pi(\alpha, \beta)$. The integer n is called the *length* of $\Pi(\alpha, \beta)$.

It is convenient also to introduce the notion of the *null walk* at α. This is the walk consisting of the single vertex α; it will be denoted by $\Pi(\alpha, \alpha)$ and it has length 0.

Example 2.1.3 The word tree T:

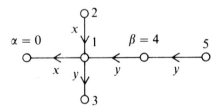

has spanning paths $\Gamma = (\alpha = 0, 1, 2, 1, 3, 1, 4, 5, 4 = \beta)$ and $\Delta = (\alpha = 0, 1, 4, 5, 4, 1, 3, 1, 2, 1, 4 = \beta)$. The (α, β)-path on T is of length 2 and is given by $\Pi(\alpha, \beta) = (\alpha = 0, 1, 4 = \beta)$.

Let T, T^* be word trees on X. An *isomorphism* $\theta: T \to T^*$ is a bijection from $V(T)$ to $V(T^*)$ which preserves adjacency, orientation of edges, and labelling of edges. Thus if α, β are adjacent vertices of T such that $\alpha\beta$ has label $y \in Y$ then $\alpha\theta$, $\beta\theta$ are adjacent vertices of T^* and $(\alpha\theta)(\beta\theta)$ has label y. An injective mapping from $V(T)$ to $V(T^*)$ that preserves adjacency, edge orientation, and labelling is called a *monomorphism* from T to T^*, and an isomorphism from T to itself is an *automorphism* of T.

Lemma 2.1.4 *Let T, T^* be word trees on X and let $\theta: T \to T^*$ and $\varphi: T \to T^*$ be monomorphisms such that $\alpha\theta = \alpha\varphi$ for some $\alpha \in V(T)$. Then $\theta = \varphi$.*

Proof Choose a spanning (α, β)-walk $\Gamma = (\alpha = \gamma_0, \gamma_1, \ldots, \gamma_n = \beta)$ on T for any vertex β of T. By hypothesis, $\gamma_0 \theta = \gamma_0 \varphi$. Suppose that $\gamma_{i-1}\theta = \gamma_{i-1}\varphi$. Then if x is the label on $\gamma_{i-1}\gamma_i$ it is also the label on both $(\gamma_{i-1}\theta)(\gamma_i\theta)$ and $(\gamma_{i-1}\varphi)(\gamma_i\varphi)$. Hence, by (WT2), $\gamma_i\theta = \gamma_i\varphi$. Thus by induction of i, $\gamma_i\theta = \gamma_i\varphi$ for $i = 0, 1, \ldots, n$. Every vertex of T occurs amongst the γ_i and so $\theta = \varphi$. ∎

Corollary 2.1.5 *The only automorphism of a word tree T on X is the identity automorphism.*

Proof Let θ be an automorphism of T. By Lemma 2.1.4, the result will be established if we can find a vertex α of T such that $\alpha\theta = \alpha$. Clearly, θ must induce a permutation on the centre C of T which consists of a single vertex or two adjacent vertices (Theorem 1.6.2). In the latter case (and obviously in the former) C is fixed pointwise as θ preserves the orientation of the edge joining the two central vertices, whereupon the result follows. ∎

This corollary shows that the vertices of a given word tree are 'distinguishable' in relation to the tree as a whole.

Let $\Gamma = (\alpha = \gamma_0, \gamma_1, \ldots, \gamma_m = \beta)$ and $\Delta = (\beta = \delta_0, \delta_1, \ldots, \delta_n = \gamma)$ be, respectively, an (α, β)-walk and a (β, γ)-walk and on a word tree T on X. Then we define an (α, γ)-walk $\Gamma\Delta$ on T by

$$\Gamma\Delta = (\alpha = \gamma_0, \ldots, \gamma_{m-1}, \beta, \delta_1, \ldots, \delta_n = \gamma).$$

Clearly, we may form products of more than two walks, and the associative law holds for such products. If Γ is an (α, α)-walk, then Γ^r has the obvious meaning for a positive integer r, and we take $\Gamma^0 = \Pi(\alpha, \alpha)$. We also define the inverse Γ^{-1} of an (α, β)-walk $\Gamma = (\alpha = \gamma_0, \gamma_1, \ldots, \gamma_n = \beta)$ to be the (β, α)-walk that results from reversing Γ. For a non-null (α, β)-walk $\Gamma = (\alpha = \gamma_0, \gamma_1, \ldots, \gamma_n = \beta)$ on T, we define an element $w(\Gamma)$ of F by the rule that $w(\Gamma) = y_1 y_2 \ldots y_n$, where $y_i \in Y$ is the label on the edge, $\gamma_{i-1}\gamma_i (i = 1, \ldots, n)$. We call $w(\Gamma)$ the *label* of Γ. We also make the convention that $w(\Pi(\alpha, \alpha)) = 1$, the empty word.

84 | Techniques of semigroup theory

We now prove a series of five lemmas which together establish the correspondence between the members of the free inverse semigroup and so-called birooted word trees. First we show that the labels of any two (α, β)-walks are in the same $\sim$-class (Lemma 2.1.7), then in Lemma 2.1.9 we show that every $u \in F$ is the label of some (α, β)-walk of some word tree T, and finally that the labels of (α, β)-walks of distinct birooted word trees are in distinct $\sim$-classes (Lemma 2.1.12).

Lemma 2.1.6 *Let Γ be an (α, β)-walk and Δ a (β, α)-walk on a word tree T on X. Then $w(\Gamma \Delta) = w(\Gamma) w(\Delta)$. Also $\Delta = \Gamma^{-1}$ if and only if $w(\Delta) = (w(\Gamma))'$.*

Proof The only part which is not quite immediate is the proof that if $w(\Delta) = (w(\Gamma))'$ then $\Delta = \Gamma^{-1}$. Accordingly, let $w(\Delta) = (w(\Gamma))' = y_1 y_2 \ldots y_n$ ($y_i \in Y$), and $\Gamma = (\alpha = \gamma_0, \gamma_1, \ldots, \gamma_n = \beta)$, $\Delta = (\beta = \delta_0, \delta_1, \ldots, \delta_n = \gamma)$. We have $\delta_0 = \gamma_n$. Suppose that we have shown that $\delta_{i-1} = \gamma_{n-i+1}$. Then y_i is the label of both $\delta_{i-1} \delta_i$ and $\gamma_{n-i+1} \gamma_{n-i}$. Hence, by (WT2), $\delta_i = \gamma_{n-i}$. It follows by induction on i that $\delta_i = \gamma_{n-i}$ for $0 \leqslant i \leqslant n$; that is, $\Delta = \Gamma^{-1}$. ∎

Lemma 2.1.7 *Let Γ and Δ be spanning (α, β)-walks on a word tree T on X. Then $w(\Gamma) \sim w(\Delta)$.*

Proof First consider the case in which $|T| = 2$. Let γ denote the vertex of T other than α, let $\Pi = \Pi(\alpha, \gamma)$, and let $y = w(\Pi)$ ($y \in Y$). If $\beta = \alpha$ then $\Gamma = (\Pi \Pi^{-1})^r$, $\Delta = (\Pi \Pi^{-1})^s$ for some positive integers r, s and so, by Lemma 2.1.6,

$$w(\Gamma) = (yy')^r \sim yy' \sim (yy')^s = w(\Delta).$$

On the other hand, if $\beta = \gamma$ then $\Gamma = (\Pi \Pi^{-1})^r \Pi$, $\Delta = (\Pi \Pi^{-1})^s \Pi$ for some non-negative integers r and s, whence

$$w(\Gamma) = (yy')^r y \sim y \sim (yy')^s y = w(\Delta).$$

Hence the result holds for $|T| = 2$.

Let n be a positive integer greater than 2. We make the inductive hypothesis that if Γ^*, Δ^* are spanning (α, β)-walks on any word tree T^* on X such that $|T^*| < n$, then $w(\Gamma^*) \sim w(\Delta^*)$.

Consider a word tree T on X such that $|T| = n$. The inductive hypothesis has the following consequence.

(A) If Γ_0 is a (γ, γ)-walk on a subtree T^* of T such that $|T^*| < n$, then $(w(\Gamma_0))^2 \sim w(\Gamma_0)$.

To see this, let T_0^* denote the subtree of T^* spanned by Γ_0. Then $|T^*| < n$, and Γ_0, Γ_0^2 are spanning (γ, γ)-walks on T^*. Hence $w(\Gamma_0) \sim w(\Gamma_0^2) = (w(\Gamma_0))^2$.

Now let Γ, Δ be spanning (α, β)-walks on T. We distinguish two cases.

(i) α is an endpoint of T. Let γ denote the unique vertex of T adjacent to α and let T^* denote the subtree of T obtained by deleting α and the edge $\alpha\gamma$ from T. Suppose first that $\beta = \alpha$. Then, for some (γ, γ)-walks $\Gamma_1, \Gamma_2, \ldots, \Gamma_k$ on T^* and some non-negative integers r_i ($i = 0, 1, \ldots, k$),

$$\Gamma = \Pi(\Pi^{-1}\Pi)^{r_0}\Gamma_1(\Pi^{-1}\Pi)^{r_1}\Gamma_2 \ldots \Gamma_k(\Pi^{-1}\Pi)^{r_k}\Pi^{-1},$$

where $\Pi = \Pi(\alpha, \gamma)$. By Lemma 2.1.6,

$$w(\Gamma) = y(y'y)^{r_0}u_1(y'y)^{r_1}u_2 \ldots u_k(y'y)^{r_k}y',$$

where $y = w(\Pi)$ and $u_i = w(\Gamma_i)$ for $i = 1, \ldots, k$. But by (A) above we have $u_i^2 \sim u_i$, whence, since idempotents commute with one another,

$$w(\Gamma) = y(y'y)^{r_0 + r_1 + \cdots + r_k}u_1u_2 \ldots u_k y' \sim yu^*y',$$

where $u^* = u_1 u_2 \ldots u_k$. Now $u^* = w(\Gamma^*)$, where $\Gamma^* = \Gamma_1\Gamma_2 \ldots \Gamma_k$. Moreover, since Γ spans T, it follows that Γ^* is a spanning (γ, γ)-walk on T^*. A similar argument shows that $w(\Delta) \sim yv^*y'$, where $v^* = w(\Delta^*)$ for some spanning (γ, γ)-walk Δ^* on T^*. But $|T^*| = n - 1$ and so, by the inductive hypothesis, $u^* \sim v^*$. Hence

$$w(\Gamma) \sim yu^*y' \sim yv^*y' \sim w(\Delta).$$

Next suppose that $\beta \neq \alpha$. Then $\beta \in T^*$, and a parallel argument to that above shows that

$$w(\Gamma) \sim yu^*, \qquad w(\Delta) \sim yv^*,$$

where y is as before and $u^* = w(\Gamma^*)$, $v^* = w(\Delta^*)$ for some spanning (α, β)-walks Γ^*, Δ^* on T^*. By the inductive hypothesis, $u^* \sim v^*$ and so $w(\Gamma) \sim w(\Delta)$.

(ii) α is not an endpoint of T. In this case we split T into two subtrees, T_1^*, T_2^*, such that $|T_1^*| < n$, $|T_2^*| < n$ and $V(T_1^*) \cap V(T_2^*) = \{\alpha\}$. Suppose first that $\beta = \alpha$. Then we can write $\Gamma = \Gamma_1\Gamma_2 \ldots \Gamma_r$, where $\Gamma_1, \Gamma_3, \Gamma_5 \ldots$ are (α, α)-walks on one of the subtrees T_k^* ($k = 1, 2$) and $\Gamma_2, \Gamma_4, \Gamma_6 \ldots$ are (α, α)-walks on the other subtree. Let $u_i = w(\Gamma_i)$ for $i = 1, \ldots, r$. Then $u_i^2 \sim u_i$, by condition (A), and so $w(\Gamma) \sim u_1^* u_2^*$, where $u_k^* = w(\Gamma_k^*)$ and Γ_k^* is the product of all the (α, α)-walks Γ_i on T_k^* ($k = 1, 2$). Moreover, since Γ spans T, it follows that Γ_k^* spans T_k^* ($k = 1, 2$). Similarly, we can show that $w(\Delta) \sim v_1^* v_2^*$, where $v_k^* = w(\Delta_k^*)$ for some spanning (α, α)-walk Δ_k^* on T_k^* ($k = 1, 2$). By the inductive hypothesis, $w(\Gamma_k^*) \sim w(\Delta_k^*)$ ($k = 1, 2$). Thus

$$w(\Gamma) \sim u_1^* u_2^* \sim v_1^* v_2^* \sim w(\Delta).$$

Now suppose that $\beta \neq \alpha$. Without loss of generality, we can assume that $\beta \in V(T_2)$. By an argument similar to that above we can show that $w(\Gamma) \sim u_1^* u_2^*$, $w(\Delta) \sim v_1^* v_2^*$, where $u_1^* = w(\Gamma_1^*)$, $v_1^* = w(\Delta_1^*)$ for some spanning (α, α)-walks Γ_1^*, Δ_1^* on T_1^* and $u_2^* = w(\Gamma_2^*)$, $v_2^* = w(\Delta_2^*)$ for some spanning (α, β)-walks Γ_2^*, Δ_2^* on T_2^*. By the inductive hypothesis, $u_1^* \sim v_1^*$ and $u_2^* \sim v_2^*$. Hence $w(\Gamma) \sim w(\Delta)$.

In each case we have shown that $w(\Gamma) \sim w(\Delta)$. The result now follows by induction on n. ∎

Example 2.1.8 Consider again the word tree T of Example 2.1.3 and the spanning (α, β)-walks Γ and Δ:

$\Gamma = (0, 1, 2, 1, 3, 1, 4, 5, 4)$

$w(\Gamma) = x'x'x\,y\,y\,y'y'y'y$

$\Delta = (0, 1, 4, 5, 4, 1, 3, 1, 2, 1, 4)$

$w(\Delta) = x'y'y'y\,y\,y\,y'x'x\,y'$

To show that $w(\Gamma) \sim w(\Delta)$ we use the fact that any subword of $w(\Gamma)$ corresponding to a closed subpath of Γ represents an idempotent. The commuting of idempotents then corresponds to exchanging the order of a pair of successive closed subpaths; the square of an idempotent in $w(\Gamma)$ corresponds to successive repetition of closed subpaths. The calculation to verify that $w(\Gamma) \sim w(\Delta)$ then is:

$$w(\Delta) = x'\underline{y'y'yyyy'}\,x'xy' = x'x'\underline{xy'y'yyyy'}\,y' = x'x'xyy'y'\,\underline{y'yyy'}$$
$$= x'x'xyy'y'yy'y'y = x'x'xyy'y'y'y = w(\Gamma),$$

where the underlined terms represent idempotents to be commuted.

Lemma 2.1.9 *Let $u \in F$. Then there exists a word tree T on X and a spanning (α, β)-walk Γ on T such that $u = w(\Gamma)$.*

Proof Let $u = y_1 y_2 \ldots y_n$, where $y_i \in Y$. We construct a sequence of word trees $T_1 \subseteq T_2 \subseteq T_3 \subseteq \ldots \subseteq T_n$ on X and a sequence of vertices $\gamma_0, \gamma_1, \ldots, \gamma_n$ of T_n such that T_i is spanned by $\Gamma_i = (\gamma_0, \gamma_1, \ldots, \gamma_i)$ and $y_1 y_2 \ldots y_i = w(\Gamma_i)$ ($i = 1, \ldots, n$).

First let T_1 denote the word tree with two vertices γ_0, γ_1 in which $\gamma_0 \gamma_1$ has label y_1. Now suppose that we have constructed the sequences as far as T_i and γ_i. Consider the $(i + 1)$th step. There are two possibilities:

(i) There exists a vertex δ in T_i, adjacent to γ_i and such that $\gamma_i \delta$ has label y_{i+1}. Then we take $T_{i+1} = T_i$ and $\gamma_{i+1} = \delta$.

(ii) There exists no vertex δ in T_i with the property stated in (i). In this case we adjoin a new vertex γ_{i+1} to T_i and a new edge $\gamma_i \gamma_{i+1}$ which we label y_{i+1}. Let T_{i+1} denote the word tree so formed.

In either case T_{i+1} is spanned by $\Gamma_{i+1} = (\gamma_0, \gamma_1, \ldots, \gamma_{i+1})$ and $y_1 y_2 \ldots y_{i+1} = w(\Gamma_{i+1})$. By induction on i, the sequences can be constructed as far as T_n and γ_n. The result follows by taking $T = T_n$, $\Gamma = (\gamma_0, \gamma_1, \ldots, \gamma_n)$, and $\alpha = \gamma_0$, $\beta = \gamma_n$. ∎

Example 2.1.10 Let $u = xy'yx'yy'xyy' \in F$. The sequence of nine word trees is given below, together with the endpoints α and β_i of the T_i:

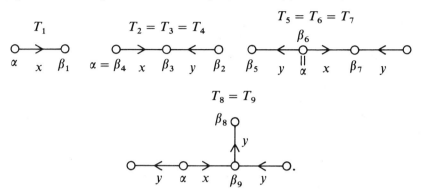

Lemma 2.1.11 *Let T, T^* be word trees on X. Let Γ be a spanning (α, β)-walk on T and Γ^* an (α^*, β^*)-walk on T^* such that $w(\Gamma) = w(\Gamma^*)$. Then there exists a monomorphism $\theta: T \to T^*$ such that $\alpha\theta = \alpha^*$, $\beta\theta = \beta^*$. If, in addition, Γ^* spans T^* then θ is an isomorphism.*

Proof Let $w(\Gamma) = w(\Gamma^*) = y_1 y_2 \ldots y_n$ ($y_i \in Y$) and let

$$\Gamma = (\alpha = \gamma_0, \gamma_1, \ldots, \gamma_n = \beta), \quad \Gamma^* = (\alpha^* = \gamma_0^*, \gamma_1^*, \ldots, \gamma_n^* = \beta^*),$$

where y_i is the label on both $\gamma_{i-1}\gamma_i$ and $\gamma_{i-1}^*\gamma_i^*$ ($1 \leq i \leq n$). Let T_i and T_i^* denote the subtrees of T and T^* spanned by the walks $(\gamma_0, \gamma_1, \ldots, \gamma_i)$ and $(\gamma_0^*, \gamma_1^*, \ldots, \gamma_i^*)$ respectively. Note that $T_n = T$ since Γ spans T.

Clearly, there exists an isomorphism $\theta_1: T_1 \to T_1^*$ such that $\gamma_j\theta_i = \gamma_j^*$ ($j = 0, 1$). Suppose that there exists an isomorphism $\theta_i: T_i \to T_i^*$ such that $\gamma_j\theta_i = \gamma_j^*$ ($j = 0, 1, \ldots, i$). We show that this can be extended to an isomorphism $\theta_{i+1}: T_{i+1} \to T_{i+1}^*$ such that $\gamma_j\theta_{i+1} = \gamma_j^*$ ($j = 0, 1, \ldots, i+1$).

Case (i): $\gamma_{i+1} \in V(T_i)$. In this case, $T_{i+1} = T_i$. Since γ_{i+1} is adjacent to γ_i and $\gamma_i\gamma_{i+1}$ has label y_{i+1}, there exists $\delta^* \in V(T_i^*)$ such that $\delta^* = \gamma_{i+1}\theta_i$, δ^* is adjacent to γ_i^*, and $\gamma_i^*\delta^*$ has label y_{i+1}. But, in T^*, γ_{i+1}^* is adjacent to γ_i^* and $\gamma_i^*\gamma_{i+1}^*$ has label y_{i+1}. Hence, by (WT2), $\delta^* = \gamma_{i+1}^*$. Thus $T_{i+1}^* = T_i^*$. We therefore define $\theta_{i+1}: T_{i+1} \to T_{i+1}^*$ by $\theta_{i+1} = \theta_i$. Then $\gamma_j\theta_{i+1} = \gamma_j^*$ ($j = 0, 1, \ldots, i+1$).

Case (ii): $\gamma_{i+1} \notin V(T_i)$. An argument similar to that of case (i), but with T and T^* interchanged, shows that $\gamma_{i+1}^* \notin V(T_i^*)$. We can therefore extend the bijection $\theta_i: V(T_i) \to V(T_i^*)$ to a bijection $\theta_{i+1}: V(T_{i+1}) \to V(T_{i+1}^*)$ by taking $\gamma_j\theta_{i+1} = \gamma_j^*$ ($j = 0, 1, \ldots, i+1$). This is evidently an isomorphism from T_{i+1} to T_{i+1}^*.

By induction on i, it follows that there exists an isomorphism $\theta_n: T \to T_n^*$ such that $\alpha\theta_n = \alpha^*$, $\beta\theta_n = \beta^*$ and we obtain the required monomorphism $\theta: T \to T^*$ by putting $\gamma_j\theta = \gamma_j\theta_n$ ($j = 0, 1, \ldots, n$).

88 | Techniques of semigroup theory

Finally, we note that if Γ^* spans T^* then $T_n^* = T^*$, and so θ is an isomorphism. ∎

Lemma 2.1.12 *Let T, T^* be word trees on X. Let Γ be a spanning (α, β)-walk on T and Γ^* an (α^*, β^*)-walk on T^* such that $w(\Gamma) \sim w(\Gamma^*)$. Then there exists an isomorphism $\theta: T \to T^*$ such that $\alpha\theta = \alpha^*$, $\beta\theta = \beta^*$.*

Proof By Lemma 2.1.11 it is enough to consider the case in which $w(\Gamma^*)$ is obtained from $w(\Gamma)$ by a single elementary $\sim$-transition, where $\sim$ is the least inverse congruence on F, as introduced in Section 1.6. Thus we have the following cases:

(i) $w(\Gamma) = pyp$, $\quad w(\Gamma^*) = pyy'yq \quad (y \in F; p, q \in F^1)$;

(ii) $w(\Gamma) = pyy'zz'q$, $\quad w(\Gamma^*) = pzz'yy'q \quad (y, z \in F; p, q \in F^1)$.

Case (i): for some $\xi, \eta \in V(T)$ there exist (α, ξ)-, (ξ, η)-, (η, β)-walks $\Gamma_1, \Gamma_2, \Gamma_3$ on T such that $\Gamma = \Gamma_1 \Gamma_2 \Gamma_3$, where $w(\Gamma_1) = p$, $w(\Gamma_2) = y$, and $w(\Gamma_3) = q$. (Note that one or both of Γ_1, Γ_3 may be null.) Let Δ denote the (α, β)-walk $\Gamma_1 \Gamma_2 \Gamma_2^{-1} \Gamma_2 \Gamma_3$ on T. Then Δ spans T since Γ spans T; also, by Lemma 2.1.11, there is an isomorphism $\theta: T \to T^*$ with $\alpha\theta = \alpha^*$, $\beta\theta = \beta^*$, since $w(\Delta) = w(\Gamma^*)$.

Case (ii): by Lemma 2.1.6, for some $\xi, \eta, \zeta \in V(T)$ there exist (α, ξ)-, (ξ, η)-, (η, ζ)-, (ζ, β)-walks $\Gamma_1, \Gamma_2, \Gamma_3, \Gamma_4$ on T such that

$$\Gamma = \Gamma_1 \Gamma_2 \Gamma_2^{-1} \Gamma_3 \Gamma_3^{-1} \Gamma_4,$$

where $w(\Gamma_1) = p$, $w(\Gamma_2) = y$, $w(\Gamma_3) = z$, and $w(\Gamma_4) = q$. Let Δ denote the (α, β)-walk $\Gamma_1 \Gamma_3 \Gamma_3^{-1} \Gamma_2 \Gamma_2^{-1} \Gamma_4$ on T. Then Δ spans T since Γ spans T; also $w(\Delta) = w(\Gamma^*) = pzz'yy'q$. Hence, by Lemma 2.1.11, there exists an isomorphism $\theta: T \to T^*$, such that $\alpha\theta = \alpha^*$, $\beta\theta = \beta^*$. This completes the proof. ∎

Now let $\mathcal{S}_X$ denote a transversal of the isomorphism classes of word trees on X, and let $\mathcal{BS}_X$ denote the set of all ordered triples (T, α, β) where $T \in \mathcal{S}_X$ and $(\alpha, \beta) \in V(T) \times V(T)$. We refer to such a triple as a *birooted word tree* on X.

Theorem 2.1.13 *If Γ, Δ are spanning (α, β)-walks on a word tree T on X then $[w(\Gamma)] = [w(\Delta)]$ and the mapping $\varphi: \mathcal{BS}_X \to FI_X$ defined by*

$$(T, \alpha, \beta)\varphi = [w(\Gamma)],$$

where Γ is any spanning (α, β)-walk on T, is a bijection. Furthermore, $\{(T, \alpha, \beta)\varphi\}^{-1} = (T, \beta, \alpha)$ and (T, α, β) is an idempotent if and only if $\alpha = \beta$.

Proof By Lemma 2.1.7, if Γ, Δ are spanning (α, β)-walks on a word tree T on X then $[w(\Gamma)] = [w(\Delta)]$; hence we may define a mapping φ as stated.

By Lemma 2.1.9, φ is surjective. To show that it is injective, suppose that $(T, \alpha, \beta)\varphi = (T^*, \alpha^*, \beta^*)\varphi$, where $(T, \alpha, \beta), (T^*, \alpha^*, \beta^*) \in \mathcal{BS}_X$. Then, by

Lemma 2.1.12, there exists an isomorphism $\theta: T \to T^*$ such that $\alpha\theta = \alpha^*$, $\beta\theta = \beta^*$. Thus $T = T^*$, by definition of $\mathscr{S}_X$. But then θ is an automorphism of T and so, by Corollary 2.1.5, $\alpha = \alpha^*$ and $\beta = \beta^*$, as required. Thus φ is a bijection.

Let $(T, \alpha, \beta) \in \mathscr{BS}_X$ and let Γ be a spanning (α, β)-walk on T. Then Γ^{-1} is a spanning (β, α)-walk on T and hence, by Lemma 2.1.6,

$$(T, \beta, \alpha)\varphi = [w(\Gamma^{-1})] = [(w(\Gamma))'] = [w(\Gamma)]^{-1} = \{(T, \alpha, \beta)\varphi\}^{-1}.$$

Suppose that (T, α, β) is an idempotent. Then $\{(T, \alpha, \beta)\varphi\}^{-1} = (T, \alpha, \beta)\varphi$ and so, by the previous result, $(T, \beta, \alpha)\varphi = (T, \alpha, \beta,)\varphi$. Thus $\alpha = \beta$. Conversely, if Δ is a spanning (α, α)-walk on T then so also is $\Delta\Delta^{-1}$ and therefore

$$(T, \alpha, \alpha)\varphi = [w(\Delta\Delta^{-1})] = [w(\Delta)(w(\Delta))'] = [w(\Delta)][w(\Delta)]^{-1},$$

which is an idempotent. This completes the proof. ∎

Theorem 2.1.13 justifies the use of the following notation. Let $u \in F$, write $[u]$ for the $\sim$-class of u, and let $[u]\varphi^{-1} = (T, \alpha, \beta) \in \mathscr{BS}_X$. Then we write

$$T(u) = T, \qquad \alpha(u) = \alpha, \qquad \beta(u) = \beta.$$

We have seen in Lemma 2.1.9 that there exists a spanning $(\alpha(u), \beta(u))$-walk Γ on $T(u)$ such that $u = w(\Gamma)$; moreover, by virtue of (WT2), Γ is uniquely determined by u. We denote it by $\Gamma(u)$. The mapping $x \mapsto \Gamma(x)$ is a bijection from $[u]$ to the set of all spanning $(\alpha(u), \beta(u))$-walks on $T(u)$ with inverse $\Gamma \to w(\Gamma)$.

With the aid of Lemma 2.1.9 we can easily construct the birooted word tree $(T(u), \alpha(u), \beta(u))$ for any given $u \in F$. Indeed, the product of birooted word trees (identifying them with the corresponding members of FI_X) can be effectively constructed. For example, the birooted word trees corresponding to $u = xx'yy'yyy'zz'$ and to $v = y'xx'yyy'zzz'$ are, respectively,

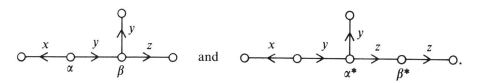

and

The birooted tree of uv is obtained by taking the word uv and following the algorithm in the proof of Lemma 2.1.9:

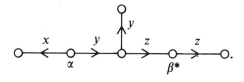

The product $(T, \alpha, \beta)(T^*, \alpha^*, \beta^*)$ can, of course, be formulated without reference to labels of specific spanning (α, β)-walks: one 'pastes' the two trees together, identifying β with α^* and then further identifying any isomorphic paths from the common vertex $\beta = \alpha^*$; again, (WT2) ensures that this can be done unambiguously.

Many properties of FI_X are now easily revealed.

Lemma 2.1.14 *Let $u, v \in F$. Then:*

(i) *$R_{[v]} \geqslant R_{[u]}$ if and only if there exists a monomorphism $\theta: T(u) \to T(v)$ such that $\alpha(u)\theta = \alpha(v)$;*

(ii) *$L_{[u]} \geqslant L_{[v]}$ if and only if there exists a monomorphism $\theta: T(u) \to T(v)$ such that $\beta(u)\theta = \beta(v)$;*

(iii) *$J_{[u]} \geqslant J_{[v]}$ if and only if there exists a monomorphism $\theta: T(u) \to T(v)$.*

Proof (i) Let $R_{[u]} \geqslant R_{[v]}$ Then there exists $y \in F^1$ such that $v \sim uy$. Hence, by Theorem 2.1.13, $T(v) = T(uy)$, $\alpha(v) = \alpha(uy)$, and $\beta(v) = \beta(uy)$. Let $\Gamma = \Gamma(u)$ and $\Delta = \Gamma(uy)$. Then there exists a vertex ξ of $T(v)$ such that $\Delta = \Gamma^*\Delta^*$, where Γ^* is an $(\alpha(v), \xi)$-walk on $T(v)$ such that $u = w(\Gamma^*)$, and Δ^* is a $(\xi, \beta(v))$-walk on $T(v)$ such that $y = w(\Delta^*)$. Thus $w(\Gamma) = w(\Gamma^*) = u$ and so, by Lemma 2.1.11, there exists a monomorphism $\theta: T(u) \to T(v)$ such that $\alpha(u)\theta = \alpha(v)$.

Conversely, suppose that there exists a monomorphism $\theta: T(u) \to T(v)$ such that $\alpha(u)\theta = \alpha(v)$. Let $\Gamma = \Gamma(u) = (\alpha(u) = \gamma_0, \gamma_1, \ldots, \gamma_n = \beta(u)))$. Define an $(\alpha(v), \gamma_n\theta)$-walk Γ^* on $T(v)$ by $\Gamma^* = (\alpha(v) = \gamma_0\theta, \gamma_1\theta, \ldots, \gamma_n\theta)$. Then, since θ is a monomorphism, we have $w(\Gamma^*) = w(\Gamma) = u$. Now let Δ^* be any $(\gamma_n\theta, \beta(v))$-walk on $T(v)$ such that $\Gamma^*\Delta^*$ is a spanning $(\alpha(v), \beta(v))$-walk on $T(v)$, and let $y = w(\Delta^*)$. Then $w(\Gamma^*\Delta^*) = uy$ and so, by Thoerem 2.1.13, $uy \sim v$. Thus $R_{[u]} \geqslant R_{[v]}$.

Proofs of (ii) and (iii) are provided by similar arguments. ∎

Green's relations correspond to natural features of the birooted word trees.

Theorem 2.1.15 *Let $u, v \in F$. Then:*

(i) *$[u] \mathscr{R} [v]$ if and only if $T(u) = T(v)$ and $\alpha(u) = \alpha(v)$;*

(ii) *$[u] \mathscr{L} [v]$ if and only if $T(u) = T(v)$ and $\beta(u) = \beta(v)$;*

(iii) *$[u] \mathscr{H} [v]$ if and only if $[u] = [v]$;*

(iv) *$[u] \mathscr{D} [v]$ if and only if $T(u) = T(v)$;*

(v) *$\mathscr{J} = \mathscr{D}$.*

Proof (i) Let $[u] \mathscr{R} [v]$. Then, by Lemma 2.1.14(i), there exists a monomorphism $\theta: T(u) \to T(v)$ such that $\alpha(u)\theta = \alpha(v)$ and a monomorphism $\varphi: T(v) \to T(u)$ such that $\alpha(v)\varphi = \alpha(u)$. Since $T(u)$ and $T(v)$ are finite trees, it follows that θ is an isomorphism. Thus $T(u) = T(v)$. Then, by Lemma 2.1.4, θ is the identity automorphism, and so $\alpha(u) = \alpha(v)$.

Conversely, if $T(u) = T(v)$ and $\alpha(u) = \alpha(v)$ then, by Lemma 2.1.14(i), $R_{[u]} \geqslant R_{[v]}$ and $R_{[v]} \geqslant R_{[u]}$, from which we obtain $[u] \mathscr{R} [v]$.

(ii) The proof is similar to (i).

(iii) By (i) and (ii), $[u] \mathscr{H} [v]$ if and only if $(T(u), \alpha(u), \beta(u)) = (T(v), \alpha(v), \beta(v))$, and the result follows.

(iv) If $[u] \mathscr{D} [v]$ it follows at once from parts (i) and (ii) that $T(u) = T(v)$. Conversely, let $T(u) = T(v) = T$, say, let Γ be a spanning $(\alpha(u), \beta(v))$-walk on T, and let $z = w(\Gamma)$. Then, by Theorem 2.1.13,

$$(T(z), \alpha(z), \beta(z)) = (T, \alpha(u), \beta(v)).$$

Thus $T(u) = T(z)$, $\alpha(u) = \alpha(z)$, and so, by (i), $[u] \mathscr{R} [z]$ and, similarly, we show by using (ii) that $[z] \mathscr{L} [v]$, whereupon the result follows.

(v) If $[u] \mathscr{J} [v]$ it follows from Lemma 2.1.14(iii) that there are monomorphisms from $T(u)$ and $T(v)$ into one another, and again, since we are dealing with finite trees, it follows that $T(u)$ and $T(v)$ are isomorphic, and thus equal. This completes the proof. ∎

Other features of FI_X are outlined in the exercises that follow.

Exercises 2.1

We use, without further comment, the notation of Section 2.1.

1. For all $u \in F$ show that:
   (i) $|R_{[u]}| = |L_{[u]}| = |T(u)|$;
   (ii) $|D_{[u]}| = |J_{[u]}| = |T(u)|^2$.

2. FI_X is $\mathscr{H}$-trivial and completely semisimple. In fact, if P is a principal factor of FI_X associated with the word tree T, with $|T| = n$, then P is isomorphic to the semigroup M_n, where $M_n = \{(i,j) : 1 \leqslant i, j \leqslant n\} \cup \{0\}$, with multiplication defined by

$$(i,j)(k,l) = \begin{cases} (i,l) & \text{if } j = k, \\ 0 & \text{otherwise.} \end{cases}$$

3. Prove that the word problem for FI_X is solvable; that is, show that there exists an algorithm for deciding whether or not $u \sim v$, where $u, v \in F$.

4. A semigroup S is called *residually finite* if and only if for any two distinct elements u, v of S there exists a congruence ρ on S such that $u\rho \neq v\rho$ and $|S/\rho| < \infty$. Prove that FI_X is residually finite as follows:
   (i) prove that there are only finitely many J-classes above $J_{[uv]}$, for any $u, v \in F$;

92 | Techniques of semigroup theory

    (ii) prove that the set $B = \{[y] : y \in F \text{ and } J_{[y]} \geqslant J_{[uv]}\}$ is finite;
    (iii) show that FI_X/C is finite, where C is the ideal $FI_X \backslash B$, and from this deduce that FI_X has the required property.

5. Show that the free monogenic inverse semigroup is isomorphic to each of the following:

    (i) $\{(p, q, r) \in \mathbb{Z}^3 : p \geqslant 0, p + q \geqslant 0, q + r \geqslant 0, r \geqslant 0, p + q + r > 0\}$ with multiplication
$$(p, q, r)(p', q', r') = (\max\{p, p' - q\}, q + q', \max\{r', r - q'\}).$$
    (ii) (Scheiblich 1971) The inverse subsemigroup of $C \times C$ generated by (b, a), where C denotes the bicyclic monoid (Exercise 1.2.13).
    (iii) The inverse subsemigroup $\langle \alpha \rangle$ of $\mathscr{I}(\mathbb{Z})$, where α is defined by
$$\alpha = \begin{pmatrix} \ldots & -3 & -2 & -1 & 0 & 1 & 2 & 3 & \ldots \\ \ldots & -2 & -1 & 0 & & 2 & 3 & 4 & \ldots \end{pmatrix}.$$

Remark More generally, FI_X is the inverse subsemigroup of $\mathscr{I}(G_X)$ (the symmetric inverse subsemigroup on the free group on X) generated by Xf, where $f: X \to \mathscr{I}_{G(X)}$ is defined by

$$w(xf) = \begin{cases} w \cdot x & \text{if } w \neq 1 \text{ and } w \neq x^{-1}; \\ \text{undefined} & \text{otherwise,} \end{cases}$$

where $w \in G$, and the product $w \cdot x$ is in G_X (Reynolds 1984). For a survey on the general topic of monogenic inverse semigroups, see Preston (1986a).

2.2 The Scheiblich representation and P-semigroups

We now use the birooted tree description of the free inverse semigroup to derive the representation of Scheiblich which furnishes a description of FI_X in terms of the free group G_X on X. Both descriptions are useful: that of Scheiblich makes the nature of the semilattice of FI_X particularly transparent.

Other representations of the free inverse semigroup can be found in the papers of Reynolds (1984) and Margolis and Pin (1987b). Inverse subsemigroups of a free inverse semigroup are not necessarily free, but Reilly (1972) determined when a subset of an arbitrary inverse semigroup generates a free inverse subsemigroup. For an embedding theorem for free inverse semigroups, see Munn (1981).

An element $u \in F^1$ is said to be *reduced* if it contains no subword of the form xx' or $x'x$ ($x \in X$). In particular, the empty word 1 is reduced. Given any word $u \in F$, we can obtain a unique reduced word $\hat{u}$ by successively deleting from u all subwords of the form xx' or $x'x$ ($x \in X$) until no such subword remains. Let G denote the set of all reduced words in F^1, and define a multiplication

(.) in G by the rule that $u \cdot v = \widehat{uv}$, the reduced word obtained from the product uv in F^1 ($u, v \in G$). Then $(G, .)$ is the free group on X.

Recall that $\Pi(\alpha, \beta)$ represents the unique path between vertices α and β.

Lemma 2.2.1 *Let T be a word tree on X.*

(i) *Let Γ be any (α, β)-walk on T. Then $w(\Gamma) \in G$ if and only if $\Gamma = \Pi(\alpha, \beta)$.*
(ii) *Let $\alpha, \beta, \gamma \in V(T)$. Then $w(\Pi(\alpha, \gamma)) = w(\Pi(\alpha, \beta)) \cdot w(\Pi(\beta, \gamma))$.*

Proof (i) This is an immediate consequence of (WT2).

(ii) Let δ be the vertex on $\Pi(\alpha, \gamma)$ that is also on $\Pi(\alpha, \beta)$, but no subsequent vertex on $\Pi(\alpha, \gamma)$ has this property. Then the right-hand side of the equation in (ii) can be written

$$w(\Pi(\alpha, \delta))w(\Pi(\delta, \beta)) \cdot w((\Pi(\delta, \beta))^{-1})w(\Pi(\delta, \gamma)) = w(\Pi(\alpha, \delta))w(\Pi(\delta, \gamma))$$
$$= w(\Pi(\alpha, \gamma)). \qquad \blacksquare$$

We next define a partial ordering $\leq$ on G by the rule that $u \leq v$ if and only if u is an initial segment of v ($u, v \in G$). By convention we take $1 \leq v$ for all $v \in G$. For any subset P of G, we write

$$\max P = \{a \in P : a \text{ is maximal in } P \text{ under } \leq \}.$$

Now let M denote the set of all non-empty finite subsets P of $G \setminus \{1\}$ such that $P = \max P$. Define S_X by

$$S_X = \{(P, g) \in M \times G : g \leq u \text{ for some } u \in P\}$$

with multiplication $(\circ)$ on S defined by the rule that

$$(P, g) \circ (Q, h) = (\max(P \cup g \cdot Q), g \cdot h),$$

where

$$g \cdot Q = \{g \cdot q : q \in Q\}.$$

Theorem 2.2.2 $(\mathscr{S}_X, \circ) \simeq FI_X$.

Remark This version of FI_X is a modification due to Munn (1974) of Scheiblich's original construction and is based on a result of Schein (1975b).

Proof First we verify that S_X is closed under $\circ$. Let $(P, g), (Q, h)$ be members of S_X. Clearly, $\max(P \cup g \cdot Q) \in M$. Now consider $g \cdot h$; if $g = g_0 h'$ for some $g_0 \in F^1$ then $g \cdot h = g_0 \leq g \leq p$ for some $p \in P$. On the other hand, if g does not have this form then $g \cdot h \leq g \cdot q$, where $q \in Q$ is such that $h \leq q$. Thus, in either case, $g \cdot h \leq u$ for some u in $\max(P \cup g \cdot Q)$. Hence $(P, g) \circ (Q, h) \in S_X$.

94 | Techniques of semigroup theory

We prove the theorem by showing that the mapping $\theta: (S_X, \circ) \to FI_X$ defined by

$$(\{u_1, \ldots, u_r\}, g)\theta = [u_1 u'_1 u_2 u'_2 \ldots u_r u'_r g]$$

is an isomorphism.

(i) θ is surjective. Let $u \in F$. Let $\varepsilon_1, \ldots, \varepsilon_r$ denote the endpoints of $T(u)$, omitting $\alpha(u)$ if this happens to be an endpoint. Write $\alpha = \alpha(u)$, $\beta = \beta(u)$, and $\Pi_i = \Pi(\alpha, \varepsilon_i)$ $(i = 1, \ldots, r)$. Then

$$\Gamma = \Pi_1 \Pi_1^{-1} \Pi_2 \Pi_1^{-1} \ldots \Pi_r \Pi_r^{-1} \Pi(\alpha, \beta)$$

is a spanning (α, β)-walk on $T(u)$ and so, by Theorem 2.1.13,

$$[u] = [w(\Gamma)] = [u_1 u'_1 u_2 u'_2 \ldots u_r u'_r g],$$

where $u_i = w(\Pi_i)$ $(i = 1, \ldots, r)$ and $g = w(\Pi(\alpha, \beta))$. Since the ε_i are endpoints, u_i cannot be an initial segment of u_j if $i \neq j$; also $g \leqslant u_k$ for some k. (Note that $g = 1$ if and only if $\alpha = \beta$.) Thus $(\{u_1, \ldots, u_r\}, g) \in S_X$ and $[u] = (\{u_1, \ldots, u_r\}, g)\theta$.

(ii) θ is injective. Let $(\{u_1, \ldots, u_r\}, g)$, $(\{v_1, \ldots, v_s\}, h) \in S_X$, where the u_i are distinct and the v_j are distinct. Let

$$u = u_1 u'_1 u_2 u'_2 \ldots u_r u'_r g, \quad v = v_1 v'_1 v_2 v'_2 \ldots v_s v'_s h,$$

and let $[u] = [v]$. We have to show that $g = h$, $r = s$, and, with a suitable ordering of the v_i, $u_i = v_i$ $(i = 1, \ldots, r)$. By Theorem 2.1.13,

$$(T(u), \alpha(u), \beta(u)) = (T(v), \alpha(v), \beta(v)) = (T, \alpha, \beta), \quad \text{say}.$$

Consider the spanning (α, β)-walk $\Gamma(u)$ on T. By Lemmas 2.2.1(i) and 2.1.6, there exist vertices $\varepsilon_1, \ldots, \varepsilon_r$ of T such that

$$\Gamma(u) = \Pi_1 \Pi_1^{-1} \Pi_2 \Pi_2^{-1} \ldots \Pi_r \Pi_r^{-1} \Pi(\alpha, \beta),$$

where $\Pi_i = \Pi(\alpha, \varepsilon_i)$, $u_i = w(\Pi_i)$, and $g = w(\Pi(\alpha, \beta))$. Since u_i is not an initial segment of u_j when $i \neq j$ and since $g \leqslant u_k$ for some k, it follows that $\varepsilon_1, \varepsilon_2, \ldots, \varepsilon_r$ are the endpoints of T, excluding α if α is an endpoint. Now apply a similar argument to v. It follows that $r = s$, $g = w(\Pi(\alpha, \beta)) = h$ and, with a suitable ordering of the v_i, $u_i = w(\Pi_i) = v_i$ $(i = 1, \ldots, r)$, as required.

(iii) θ is a homomorphism. Again let $(\{u_1, \ldots, u_r\}, g)$, $(v_1, \ldots, v_s\}, h)$ be members of S_X and let $u = u_1 u'_1 u_2 u'_2 \ldots u_r u'_r g$, $v = v_1 v'_1 v_2 v'_2 \ldots v_s v'_s h$. Write

$$(T, \alpha, \beta) = (T(uv), \alpha(uv), \beta(uv)).$$

By considering the spanning (α, β)-walk $\Gamma(uv)$ on T we see, as in (ii), that there exist vertices $\eta_1, \ldots, \eta_r, \gamma, \zeta_1, \ldots, \zeta_s$ of T such that

$$u_i = w(\Pi(\alpha, \eta_i)) \quad (i = 1, \ldots, r), \quad g = w(\Pi(\alpha, \gamma));$$
$$v_j = w(\Pi(\gamma, \zeta_j)) \quad (j = 1, \ldots, s), \quad h = w(\Pi(\gamma, \beta)).$$

Moreover, the set of endpoints of T (excluding α should it happen to be an endpoint) is a subset of $\{\eta_1, \ldots, \eta_r, \zeta_1, \ldots, \zeta_s\}$. Hence Γ, defined by

$$\Gamma = Y_1 Y_1^{-1} Y_2 Y_2^{-1} \ldots Y_r Y_r^{-1} Z_1 Z_1^{-1} Z_2 Z_2^{-1} \ldots Z_s Z_s^{-1} \Pi(\alpha, \beta),$$

where $Y_i = \Pi(\alpha, \eta_i)$ and $Z_j = \Pi(\alpha, \zeta_j)$, is a spanning (α, β)-walk on T. It follows from Theorem 2.1.13 that

$$[u][v] = [w(\Gamma)].$$

But by Theorem 2.2.1(ii),

$$w(Z_j) = w(\Pi(\alpha, \gamma)) \cdot w(\Pi(\gamma, \zeta_j)) = g \cdot v_j \qquad (j = 1, \ldots, s)$$

and

$$w(\Pi(\alpha, \beta)) = w(\Pi(\alpha, \gamma)) \cdot w(\Pi(\gamma, \beta)) = g \cdot h.$$

Hence

$$w(\Gamma) = (u_1 u_1') \ldots (u_r u_r')(g \cdot v_1)(g \cdot v_1)' \ldots (g \cdot v_s)(g \cdot v_s)' g \cdot h$$

$$\sim (w_1 w_1') \ldots (w_t w_t') g \cdot h,$$

where $\{w_1, \ldots, w_t\} = \max\{u_1, \ldots, u_r, g \cdot v_1, \ldots, g \cdot v_s\}$. Thus

$$[u][v] = [(w_1 w_1') \ldots (w_t w_t') g \cdot h] = ((\{u_1, \ldots, u_r\}, g) \circ (\{v_1, \ldots, v_s\}, h))\theta.$$

This completes the proof. ∎

Remark The proof that θ is a bijection entails showing that, up to the order of the factors $u_i u_i'$, there is a unique word in each $\sim$-class of F of the form

$$(u_1 u_1')(u_2 u_2') \ldots (u_r u_r') g,$$

where $(\{u_1, \ldots, u_r\}, g) \in S_X$. Any such word is said to be in *Schein left canonical form*. Clearly, a dual argument proves the existence of a unique *Schein right canonical form*

$$h(v_1 v_1')(v_2 v_2') \ldots (v_s v_s'),$$

and Theorem 2.2.2 can be dualized in the obvious way.

From this description of FI_X we may recover the original formulation of Scheiblich

Once again let $G = FG_X$ denote the free group on X. For $a \in G$ we shall write $\bar{a} = \{x \in G: x \leqslant a\}$ and for a subset A of G define

$$\bar{A} = \bigcup_{a \in A} \bar{a}.$$

A subset A of G is called *convex* if whenever $a, b \in A$ with $a \leqslant b$ and $a \leqslant c \leqslant b$ ($c \in G$) then $c \in A$. It is readily seen that ($\bar{}$) is a closure operator on the power set of G ($A \subseteq \bar{A}$, $\bar{\bar{A}} = \bar{A}$, and $A \subseteq B$ implies $\bar{A} \subseteq \bar{B}$ for all $A, B \subseteq G$) and that a subset A containing 1 is convex if and only if A is *closed*; that is, $A = \bar{A}$. Note also that for any two subsets A, B of G have $\overline{A \cup B} = \bar{A} \cup \bar{B}$.

96 | Techniques of semigroup theory

Let Y denote the set of all finite convex subsets of G containing 1 and at least one other element. The mapping θ which sends $(A, g) \in S_X$ to $(\bar{A}, g)$ is a bijection of S_X upon

$$F = \{(A, g) \in Y \times G : g \in A\},$$

for θ is injective because distinct maximal sets have distinct closures, and θ is surjective since, for a typical member (A, g) of F, we see that $(A, g) = (\overline{\max A}, g)$. The set F can then be regarded as FI_X if we define the multiplication on F in such a way so as to make θ an isomorphism; that is, by choosing the multiplication on F that makes the following diagram commute:

$$((\max A, g), (\max B, h)) \stackrel{\circ}{\to} (\max(\max A \cup g . \max B), g . h)$$
$$\theta \downarrow \qquad\qquad\qquad \downarrow \theta$$
$$((A, g), (B, h)) \to (\overline{\max(\max A \cup g . \max B)}, g . h).$$

The first entry in our product in F can be simplified. First consider $g . b$, where $b \in B$. If g has the form $g_0 b'$ then $g . b = g_0 \in A$. Otherwise, g can be factored as $g = g_0 b'_0$, where $b = b_0 b_1$ and $g . b = g_0 b_1$ is reduced, in which case any initial segment of $g . b$ is either an initial segment of g (and is thus in A) or has the form $g_0 b_2 \leqslant g_0 b_1 = g . b$, whence

$$g_0 b_2 = g . b_0 b_2 \in g . B \cap \overline{g . \max B}.$$

We conclude that any member of $\overline{g . B}$ is either in A or is in $g . B \cap \overline{g . \max B}$. From this we see that

$$\overline{A \cup g . B} = \overline{\bar{A} \cup g . B}.$$

Also, for any set A we have $\bar{A} = \overline{\max A}$. Thus we obtain:

$$\overline{\max(\max A \cup g . \max B)} = \overline{\max A \cup g . \max B} \subseteq \overline{A \cup g . B}$$
$$= \overline{A \cup g . B} \subseteq \overline{\bar{A} \cup g . B}$$
$$\subseteq \overline{\max A \cup A \cup g . \max B} = \overline{\max A}$$
$$\cup \overline{g . \max B} = \overline{\max A \cup g . \max B},$$

giving equality throughout. This yields the next result.

Theorem 2.2.3 (Scheiblich 1973) *The set*

$$F = \{(A, g) \in Y \times G : g \in A\},$$

with multiplication

$$(A, g)(B, h) = (A \cup g . B, g . h),$$

is isomorphic to FI_X. *The idempotents of F are evidently the elements* $(A, 1)$ *$(A \in Y)$ and the product of two typical idempotents* $(A, 1)(B, 1) = (A \cup B, 1)$.

Corollary 2.2.4 *The semilattice of FI_X is isomorphic to the set Y of all finite closed subsets of G containing at least two elements under the operation of set union.*

The Scheiblich representation of FI_X involves a certain type of multiplication on a particular subset of the cartesian product of a semilattice and a group. As we shall show, abstraction of the ideas encountered here will be rewarded with a structure theorem for E-unitary inverse semigroups. First we shall re-examine F in detail.

The group entries in a given pair of members of F are simply multiplied by components, but the formation of the first entry in the product is more subtle. The group and semilattice interact in that the first group entry g acts on the left of the second semilattice entry B in such a way that the 'join' $A \cup g.B$ is in Y. The complication to note is that the set $g.B$ may not itself be in Y, as $g.B$ is not necessarily closed. Thus it will become desirable to introduce a third object into our abstract formulation apart from a group G and a semilattice Y; namely, a partially ordered set X corresponding in F to the product $FG_X . Y$ of the group and semilattice. Following the state of affairs in F, we shall require that Y is a subsemilattice of X, and that the meet in X of a member of Y with a member of X exists and is in Y (the semilattice Y of F is more naturally regarded as an upper semilattice as its operation is one of set union; however, when working with abstract semilattices it is the usual practice to consider them to be lower semilattices and denote their operation as a meet). Finally, the underlying set of F is not all of $Y \times G$, but consists of those pairs (A, g) with $g \in A$. Since we intend our abstract construction modelled on F to involve an arbitrary group, semilattice pair, the condition that $g \in A$ ($g \in G$, $A \in Y$) is unsuitable. An appropriate formulation is at hand, however, based on the observation that, in F:

2.2.5 $g \in A \Leftrightarrow g^{-1}.A \in Y$.

Clearly, $g^{-1}.A \in Y \Rightarrow 1 \in g^{-1}.A \Leftrightarrow g \in A$. Conversely, if $g \in A$, note that all initial segments of g^{-1} lie in $g^{-1}.A$ because A is closed. Secondly, take any $a \in A$ and factor a as $g_0 a_0$, where $g = g_0 g_1$ and $g^{-1}.a = g_1^{-1} a_0$. That any initial segment of $g_1^{-1} a_0$ is in $g^{-1}.A$ follows from the facts that it is either an initial segment of g_1^{-1}, and thus of g^{-1}, or has the form $g_1^{-1} a_1$, where $a_0 = a_1 a_2$, say; now $g_0 a_1 \leqslant g_0 a_0 = a \in A$, and so $g_0 a_1 \in A$, whence $g_1^{-1} a_1 = g^{-1}.(g_0 a_1) \in g^{-1}.A$.

The foregoing serves to motivate the notion of a *P*-semigroup.

Definition 2.2.6 By a *McAlister triple* we shall mean one of the form $(G, \mathscr{X}, \mathscr{Y})$ consisting of a group G, a partially ordered set $(\mathscr{X}, \leqslant)$ and a nonempty subset $\mathscr{Y}$ and $\mathscr{X}$ satisfying these conditions:

(P1) $\mathscr{Y}$ is a lower semilattice under $\leq$ and an *order ideal* of X, meaning that if $A \leq B$ with $A \in \mathscr{X}$ and $B \in \mathscr{Y}$, then $A \in \mathscr{Y}$, and if $A, B \in \mathscr{Y}$ then $A \wedge B$ exists and is in $\mathscr{Y}$ (where $A \wedge B$ denotes the glb of A and B in $\mathscr{X}$, if it exists).

(P2) G acts on X, on the left, by order automorphisms; meaning that there is a homomorphism $\theta: G \to \operatorname{Aut} \mathscr{X}$, the group of automorphisms of $(\mathscr{X} \leq)$, such that for all $g, h \in G, (hg)\theta$ is the composition of $g\theta$ followed by $h\theta$; consequently if we write gA for $A(g\theta)(A \in \mathscr{Y}, g \in G)$ then $(hg)A = h(gA)$.

(P3) $G\mathscr{Y} = \mathscr{X}$.

(P4) For all $g \in G$, there exists $A \in \mathscr{Y}$, such that $gA \in \mathscr{Y}$.

The first part of (P1) states that if A, B are any two elements of $\mathscr{Y}$, then they have a glb C in $\mathscr{Y}$ under $\leq$; hence, by the second part of (P1), $A \wedge B$ exists and equals C. Also note, by (P2), if $A, B \in \mathscr{X}$ and $A \wedge B$ exists, then, for all $g \in G, gA \wedge gB$ exists and equals $g(A \wedge B)$.

Recall the definition of a right unitary subset of a semigroup. An inverse semigroup S is called *E-unitary* if $E(S)$ is right unitary (and thus also left unitary by Theorem 1.4.19).

Theorem 2.2.7 *The set $S = \{(A, g) \in \mathscr{Y} \times G: g^{-1}A \in \mathscr{Y}\}$, with multiplication given by*

$$(A, g)(B, h) = (A \wedge gB, gh) \tag{1}$$

is an E-unitary inverse semigroup. Moreover, if E is the semilattice of S and σ its least group congruence, then $E \simeq \mathscr{Y}$ and $S/\sigma \simeq G$. We denote S by $P(G, \mathscr{X}, \mathscr{Y})$ and call S a P-semigroup.

Proof First we verify that the operation on S is well-defined. Since $g \in A, B \in \mathscr{Y}, g^{-1}A \wedge B \in \mathscr{Y}$, which yields $A \wedge gB \in \mathscr{Y}$ upon pre-multiplying by g. Next, $(gh)^{-1}(A \wedge gB) = h^{-1}g^{-1}(A \wedge gB) = (h^{-1}g^{-1})A \wedge h^{-1}B \in \mathscr{Y}$ as $h^{-1}B \in \mathscr{Y}$ and $\mathscr{Y}$ is an order ideal of $\mathscr{X}$. Thus (1) defines a binary operation on S, the associativity of which is checked without difficulty.

If $(A, g) = (A, g)^2$ we obtain $g = g^2$ so that $g = 1$, the identity of G. Since the automorphism 1θ is necessarily the identity map on $\mathscr{X}$, we do indeed have $(A, 1)^2 = (A \wedge 1A, 1) = (A, 1)$, so that $E(S) = \{(A, 1): A \in \mathscr{Y}\}$; moreover, $E \simeq \mathscr{Y}$ under the mapping whereby $(A, 1) \mapsto A$ because $(A, 1)(B, 1) = (A \wedge B, 1)$ $(A, B \in \mathscr{Y})$. By direct calculation one verifies that any $(A, g) \in S$ has an inverse in $(g^{-1}A, g^{-1})$, so that S is regular, and since idempotents commute with one another, S is an inverse semigroup. Furthermore, since the product of an idempotent with a non-idempotent is evidently also not an idempotent, it follows that S is E-unitary. Finally, suppose that $(A, g)\sigma(B, h)$

so that, for some idempotent $(C,1)$,

$$(C, 1)(A, g) = (C, 1)(B, h) \qquad (2)$$
$$\Leftrightarrow (C \wedge A, g) = (C \wedge B, h) \Rightarrow g = h.$$

Conversely, if $g = h$ then (2) is satisfied if we put $C = A \wedge B$; thus $(A, g)\sigma(B, h)$ if and only if $g = h$. In addition, for each $g \in G$ there exists, by (P4), an $A \in \mathcal{Y}$ such that $g^{-1}A \in \mathcal{Y}$, whereupon $(A, g) \in S$; it follows easily that the mapping whereby $((A, g))\sigma \mapsto g$ is a well-defined isomorphism of S/σ onto G. ∎

It is clear from our preamble that FI_X corresponds to the triple $(FG_X, G_X\mathcal{Y}, \mathcal{Y})$ where Y is the set of all finite closed subsets of FG_X containing more that one element, and G acts on the left of $\mathcal{Y}$ by set products; in fact it only remains to check (P4) in order to verify that we have a P-semigroup. To see this, observe that for a given $g \in G\setminus\{1\}$ we have $\overline{g^{-1}} \in Y$ and $g \cdot (\overline{g^{-1}}) = \bar{g} \in Y$.

In summary, the class of P-semigroups is a class of E-unitary inverse semigroups containing our prototype, the free inverse semigroup on X. However, their significance lies in two theorems of McAlister: every E-unitary inverse semigroup is isomorphic to a P-semigroup, and every inverse semigroup S has an E-unitary 'cover' P, meaning that S is a morphic image of P by an idempotent-separating homomorphism, and so S and P essentially share the same semilattice of idempotents.

Before proving these results we shall record some basic facts about E-unitary inverse semigroups. The following characterization of an E-unitary inverse semigroup, essentially due to Saito (1965), shows that it can be 'coordinatized' by the pair $(S/\sigma, E)$, where σ denotes the least group congruence on S.

Theorem 2.2.8 *The following are equivalent for an inverse semigroup S:*

(i) S is E-unitary;
(ii) if $e \in E$ and $a \in S$ are such that $ae \in E$ then $a \in E$;
(iii) E is a σ-class of S;
(iv) if $e \in E$ and $e \leqslant a$ then $a \in E$;
(v) $\sigma \cap \mathcal{R} = 1$;
(vi) the map $\varphi: S \to S/\sigma \times E$ defined by $a\varphi = (a\sigma, aa^{-1})$ is injective.

Proof The equivalence of (i) and (ii) was proved in Theorem 1.4.19. The equivalence of the first four listed conditions is now an immediate consequence of the standard characterizations of σ and $\leqslant$ on an inverse semigroup (see Exercise 1.4.8). Also (v) and (vi) are obviously equivalent. We complete the proof by showing that conditions (iii) and (v) are equivalent.

(iii) ⇒ (v). Suppose that $(a,b) \in \sigma \cap \mathscr{R}$, so that $aa^{-1} = bb^{-1}$. Then $b^{-1}a\sigma b^{-1}b \Rightarrow b^{-1}a \in E$ by (iii). Also, $a\mathscr{R}b \Rightarrow b^{-1}a\mathscr{R}b^{-1}b$, whence $b^{-1}a = b^{-1}b$. Thus $a = aa^{-1}a = bb^{-1}a = bb^{-1}b = b$.

(v) ⇒ (iii). Suppose that $a\sigma e$ $(e \in E)$. Then $a^{-1}\sigma e$, whence $a\sigma aa^{-1}$ and $a\mathscr{R}aa^{-1}$, and so by (v) $a = aa^{-1} \in E$. ∎

We now embark on the proof of the theorem that any E-unitary inverse semigroup is isomorphic to a P-semigroup. To this end let S forthwith denote an E-unitary inverse semigroup with least group congruence σ, denoting the group S/σ by G, and semilattice of idempotents by E. We follow the proof of Munn (1976). The candidates for the group and semilattice of our McAlister triple are obviously G and E, so the first difficulty is identifying a suitable choice for our partially ordered set $\mathscr{X}$. In our proof $\mathscr{X}$ appears as the quotient of $E \times G$ by a certain equivalence.

We proceed by means of nine short lemmas. The product of elements g, h of G will be denoted by $g * h$, while g^{-1} will denote the inverse of g. Since S is E-unitary, it follows from Theorem 2.2.8(iii) that E is the identity of G.

First, we define a relation ρ on $E \times G$ by the rule:

2.2.9 $((a,g), (b,h)) \in \rho \Leftrightarrow a = xx^{-1}, b = x^{-1}x$ *for some* $x \in g^{-1} * h$.

Lemma 2.2.10 *The relation ρ is an equivalence on $E \times G$.*

Proof Reflexivity is seen by taking $x = a$ in 2.2.9. To see that ρ is symmetric, suppose that we are given x satisfying 2.2.9. Then

$$b = x^{-1}x, \quad a = (x^{-1})^{-1}x, \quad \text{where } x^{-1} \in h^{-1} * g,$$

giving $(b,h)\rho(a,g)$. Finally, for transitivity, suppose we have x as above and that for some $(c,k) \in E \times G$ we have

$$b = yy^{-1}, \quad c = y^{-1}y, \quad \text{for some } y \in h^{-1} * k.$$

Then $(xy)(xy)^{-1} = xyy^{-1}x^{-1} = xbx^{-1} = xx^{-1}xx = xx^{-1} = a$; while, similarly, $(xy)^{-1}(xy) = c$. Since $xy \in g^{-1} * k$, it follows that $(a,g)\rho(c,k)$, as required. ∎

We shall write $\mathscr{X}$ for $(E \times G)/\rho$, and denote the ρ-class of $E \times G$ that contains (a,g) by $(a,g)\rho$. The following facts are immediate.

Lemma 2.2.11 *For all $x \in S$,*

$$(xx^{-1}, E)\rho = (x^{-1}x, x\sigma)\rho, \quad (x^{-1}x, E)\rho = (xx^{-1}, (x\sigma)^{-1}\rho.$$

Lemma 2.2.12 *Let $a, b \in E$ and $g \in G$ be such that $(a,g)\rho = (b,g)\rho$. Then $a = b$.*

Free inverse semigroups and the theorems of McAlister | 101

Proof There exists $x \in g^{-1} * g$ such that $a = xx^{-1}$, $b = x^{-1}x$. But $g^{-1} * g = E$, and so $x^{-1} = x$. Hence $a = x = b$. ∎

To enable us to define a partial ordering of $\mathscr{X}$ in terms of the natural partial ordering of E we prove the following.

Lemma 2.2.13 Let $a, b, d \in E$ and $g, h \in G$ be such that
$$a \leqslant b, \qquad (b, g)\rho = (d, h)\rho.$$
Then there exists $c \in E$ such that
$$c \leqslant d, \qquad (a, g)\rho = (c, h)\rho.$$

Proof There exists $x \in g^{-1} * h$ such that $b = xx^{-1}$, $d = x^{-1}x$. Write $c = (ax)^{-1}ax$. Then $c \in E$ and $c \leqslant d$. Also, since $a \in E$, $ax \in g^{-1} * h$. Furthermore, $ax(ax)^{-1} = axx^{-1}a^{-1} = aba = a$, since $a \leqslant b$. Hence $(a, g)\rho = (c, h)\rho$. ∎

We now define a relation $\leqslant$ on $\mathscr{X}$ by the rule that, for $A, B \in \mathscr{X}$, $A \leqslant B$ if there exist $a, b \in E$ and $g \in G$ such that
$$a \leqslant b, \qquad (a, g) \in A, \qquad (b, g) \in B.$$
It follows from Lemma 2.2.13 that if $A, B \in \mathscr{X}$ are such that $A \leqslant B$ then, for all $(d, h) \in B$, there exists $c \in E$ such that $c \leqslant d$ and $(c, h) \in A$.

Lemma 2.2.14 The relation $\leqslant$ is a partial order on $\mathscr{X}$.

Proof Obviously $\leqslant$ is reflexive. Let $A, B \in \mathscr{X}$ be such that $A \leqslant B$ and $B \leqslant A$. Since $A \leqslant B$ there exist $a, b \in E$ and $g \in G$ such that $a \leqslant b$, $(a, g) \in A$ and $(b, g) \in B$. Hence, by the comment following Lemma 2.2.13, since $B \leqslant A$ and $(a, g) \in A$, there exists $c \in E$ such that $c \leqslant a$ and (c, g) is in B. Thus $(b, g)\rho = (c, g)\rho$. Hence, by Lemma 2.2.12, $b = c$ and so $b \leqslant a$. It follows that $a = b$, and thus $A = B$.

Next, let $A, B, C \in \mathscr{X}$ be such that $A \leqslant B$ and $B \leqslant C$. Since $B \leqslant C$ there exist $b, c \in E$ and $g \in G$ such that $b \leqslant c$, $(b, g) \in B$ and $(c, g) \in C$. Hence, again as a consequence of Lemma 2.2.13, since $A \leqslant B$ and $(b, g) \in B$ there exists $a \in E$ such that $a \leqslant b$ and $(a, g) \in A$. Thus $a \leqslant c$ and so $A \leqslant C$. ∎

Let us now write $\mathscr{Y} = \{(a, E)\rho : a \in E\}$. We shall show that, with an appropriate action of G on $\mathscr{X}$, $(G, \mathscr{X}, \mathscr{Y})$ is a McAlister triple.

Lemma 2.2.15 The mapping $\varphi : E \to \mathscr{Y}$ defined by $a\varphi = (a, E)\rho$ is an order isomorphism.

Proof By Lemma 2.2.12, φ is injective and is therefore a bijection. Furthermore, it is order-preserving, by the definition of the partial ordering of $\mathscr{X}$. It remains to show that φ^{-1} is order-preserving. Suppose that $(a, E)\rho \leqslant (b, E)\rho$. By Lemma 2.2.13, there exists $c \in E$ such that $c \leqslant b$ and $(a, E)\rho = (c, E)\rho$. Hence, by Lemma 2.2.12, $a = c$ and so $a \leqslant b$, as required. ∎

If $(a, g)\rho$ and $(b, h)\rho$ have a greatest lower bound in $\mathscr{X}$, this will be denoted by $(a, g)\rho \wedge (b, h)\rho$.

Lemma 2.2.16 $\mathscr{Y}$ *forms an order ideal of* $\mathscr{X}$ *and a lower semilattice under* $\leqslant$. *Moreover, for all* $a, b \in E$,

$$(a, E)\rho \wedge (b, E)\rho = (ab, E)\rho.$$

Proof Let $A \in \mathscr{X}$ and $B \in \mathscr{Y}$ be such that $A \leqslant B$. Since $B \in \mathscr{Y}$ there exists $b \in E$ such that $B = (b, E)$. Hence, by Lemma 2.2.13, there exists a in E such that $a \leqslant b$ and $(a, E)\rho = A$. Thus $A \in \mathscr{Y}$. This shows that Y is an order ideal of $\mathscr{X}$. Furthermore, by Lemma 2.2.15, $\mathscr{Y}$ is a lower semilattice under $\leqslant$. But, since $\mathscr{Y}$ is an order ideal of $\mathscr{X}$, the greatest lower bound in $\mathscr{Y}$ of two elements of $\mathscr{Y}$ must also be their glb in $\mathscr{X}$. The final statement follows from Lemma 2.2.15. ∎

Next, we define an action of G on $\mathscr{X}$ by order automorphisms. Suppose first that $(a, g)\rho = (b, h)\rho$. This means that there exists $x \in g^{-1} * h$ such that $a = xx^{-1}, b = x^{-1}x$. Let $k \in G$. Then $x \in (k*g)^{-1} * (k*h)$ and so $(a, k*g)\rho = (b, k*h)\rho$. We can therefore define $\circ : G \times \mathscr{X} \to \mathscr{X}$ by the rule:

$$k \circ (a, g)\rho = (a, k*g)\rho.$$

Lemma 2.2.17 *The mapping* $\circ$ *is an action of* G *on* $\mathscr{X}$ *on the left by order automorphisms.*

Proof It suffices to verify that, for all $a, b \in E$ and all $g, h \in G$, the following conditions hold, the first two of which are immediate:

(i) $E \circ (a, g)\rho = (a, g)\rho$;
(ii) $(h*k) \circ (a, g)\rho = h \circ (k \circ (a, g)\rho)$;
(iii) if $(a, g)\rho \leqslant (b, h)\rho$ then $k \circ (a, g)\rho \leqslant k \circ (b, h)\rho$.

To verify the third, suppose that $(a, g)\rho \leqslant (b, h)\rho$. Then, by Lemma 2.2.13, there exists $c \in E$ such that $c \leqslant b$ and $(a, g)\rho = (c, h)\rho$. Thus

$$k \circ (a, g)\rho = k \circ (c, h)\rho = (c, k*h)\rho \leqslant (b, k*h)\rho = k \circ (b, h)\rho. \qquad \blacksquare$$

So far we have proved that the triple $(G, \mathscr{X}, \mathscr{Y})$ satisfies conditions (P1) and (P2). The final lemma shows that it also satisfies (P3) and (P4).

Lemma 2.2.18

(i) $G \circ \mathcal{Y} = \mathcal{X}$;
(ii) for all $g \in G$ there exists $A \in \mathcal{Y}$ such that $g \circ A \in \mathcal{Y}$.

Proof (i) Let $(a, g)\rho \in \mathcal{X}$. Then $(a, g)\rho = g \circ (a, E)\rho \in G \circ \mathcal{Y}$. Thus $\mathcal{X} \subseteq G \circ \mathcal{Y}$.
(ii) Let $g \in G$. Choose $x \in g$ and let $A = (x^{-1}x, E)\rho$. Then $A \in \mathcal{Y}$ and $g \circ A = (x^{-1}x, x\sigma) = (xx^{-1}, E)\rho \in \mathcal{Y}$, by Lemma 2.2.11. ∎

Thus $(G, \mathcal{X}, \mathcal{Y})$ is a McAlister triple and so we can form the P-semigroup $T = P(G, \mathcal{X}, \mathcal{Y})$. We now prove that S is isomorphic to T.

First, we note that, for all $x \in S$,

$$(x\sigma)^{-1} \circ (xx^{-1}, E)\rho = (xx^{-1}, (x\sigma)^{-1})\rho = (x^{-1}x, E)\rho \in \mathcal{Y},$$

by Lemma 2.2.11.

Hence we can define a mapping $\theta: S \to T$ by the rule that

$$x\theta = ((xx^{-1}, E)\rho, x\sigma).$$

We shall show that θ is an isomorphism.

Let $x, y \in S$ be such that $x\theta = y\theta$. Then $(xx^{-1}, E)\rho = (yy^{-1}, E)\rho$ and so, by Lemma 2.2.12, $xx^{-1} = yy^{-1}$. Furthermore, $x\sigma = y\sigma$, whence it follows by Theorem 2.2.8(vi) that $x = y$ and that θ is injective.

Next let $((a, E)\rho, k) \in T$. Then $k^{-1} \circ (a, E)\rho \in \mathcal{Y}$. Thus there exists b in E such that $(a, k^{-1})\rho = (b, E)\rho$ and so there exists $x \in k$ such that $a = xx^{-1}$, $b = x^{-1}x$. Hence $((a, E)\rho, k) = ((xx^{-1}, E)\rho, x\sigma) = x\theta$. This shows that θ is surjective.

Finally, let $x, y \in S$. Write $z = xyy^{-1}$. Then $z \in x\sigma$; also $zz^{-1} = xy(xy)^{-1}$ and $z^{-1}z = x^{-1}xyy^{-1}$. Hence

$$(xy(xy)^{-1}, E)\rho = (x^{-1}xyy^{-1}, x\sigma)\rho$$

$$= (x\sigma) \circ (x^{-1}xyy^{-1}, E)\rho$$

$$= (x\sigma) \circ [(x^{-1}x, E)\rho \wedge (yy^{-1}, E)\rho],$$
by Lemma 2.2.16,

$$= (x\sigma) \circ [(xx^{-1}, (x\sigma)^{-1})\rho \wedge (yy^{-1}, E)\rho],$$
by Lemma 2.2.11,

$$= (x\sigma) \circ [(x\sigma)^{-1} \circ (xx^{-1}, E)\rho \wedge (yy^{-1}, E)\rho]$$

$$= (xx^{-1}, E)\rho \wedge (x\sigma) \circ (yy^{-1}, E)\rho.$$

Consequently, θ is a homomorphism, thus completing our proof of the first of our two theorems of McAlister.

Theorem 2.2.19 *An inverse semigroup is E-unitary if and only if it is isomorphic to some P-semigroup.*

A special case of a McAlister triple which naturally comes to mind is that in which $\mathscr{X} = \mathscr{Y}$. In $P(G, \mathscr{Y}, \mathscr{Y})$ we simply have a group acting on the left of a semilattice $\mathscr{Y}$ by order automorphisms so that axioms (P1), (P3), and (P4) are automatic consequences of (P2), and the underlying set of $P(G, \mathscr{Y}, \mathscr{Y})$ is the full cartesian product $\mathscr{Y} \times G$. Such a semigroup is called a *semidirect product* of a semilattice and a group.

Definition 2.2.20 Let G be a group, $\mathscr{Y}$ a semilattice and $\theta: G \to \text{Aut}(\mathscr{Y})$ a homomorphism from G to the automorphism group of $\mathscr{Y}$. Then the *semidirect product* of $\mathscr{Y}$ and G under θ, denoted by $\mathscr{Y} \times_\theta G$, consists of $\mathscr{Y} \times G$ under the multiplication

$$(A, g)(B, h) = (A \wedge gB, gh),$$

where gB denotes $(B)(g\theta)$.

A semidirect product of a semilattice and a group is of course an E-unitary inverse semigroup, because it is a special type of P-semigroup. Let us consider one particular semidirect product. Let X be a non-empty set and put $Y = X \cup X'$, where X' is a set disjoint from X and of the same cardinality. For our semilattice take $P(Y)$, the semilattice of the power set of Y under intersection, and for our group we choose $G(Y)$, the full symmetric group on Y. Let $\theta: G(Y) \to \text{Aut } P(Y)$ be defined by $(g\theta)(A) = g(A)$. Consider $R(X)$, defined as

$$R(X) = \{(A, g) \in P(X) \times_\theta G(Y) : g^{-1}A \in P(X)\}.$$

Lemma 2.2.21 *$R(X)$ is an E-unitary inverse subsemigroup of $P(Y) \times_\theta G(Y)$.*

Proof Let $(A, g), (B, h) \in R(X)$. Then $(A, g)(B, h) = (A \cap gB, gh)$ and $(gh)^{-1}(A \cap gB) = h^{-1}g^{-1}(A \cap gB) = h^{-1}(g^{-1}A) \cap h^{-1}(B) \subseteq h^{-1}(B) \in P(X)$, and so $R(X)$ is closed under multiplication. The inverse of (A, g) is $(g^{-1}A, g^{-1})$, which is also in $R(X)$. Hence $R(X)$ is an inverse subsemigroup of $P(Y) \times_\theta G(Y)$, and the full result now follows as the E-unitary property is clearly inherited by such subsemigroups. ∎

Next we introduce the relation ρ on $R(X)$ defined by $(A, g)\rho(B, h)$ if $A = B$ and $g^{-1}a = h^{-1}a$ for all $a \in A$.

Lemma 2.2.22 *The relation ρ on $R(X)$ is an idempotent-separating congruence.*

Proof From the definition it is clear that ρ is an equivalence on $R(X)$. Suppose that $(A, g)\rho(A, h)$ and take any $(C, k) \in R(X)$. Then

$$(C, k)(A, g) = (C \cap kA, kg), \qquad (C, k)(A, h) = (C \cap kA, kh);$$

now any member of $C \cap kA$ has the form ka ($a \in A$), whence $(kg)^{-1}ka = g^{-1}a = h^{-1}a = (kh)^{-1}ka$, whereupon we see that ρ is a left congruence. Next compare

$$(A, g)(C, k) = (A \cap gC, gk) \quad \text{with} \quad (A, h)(C, k) = (A \cap hC, hk).$$

Let $gc \in A \cap gC$ ($c \in C$). Then $g^{-1}(gc) = h^{-1}(gc)$ yields $hc = gc$. Thus $gc \in A \cap hC$, and we conclude that $A \cap gC \subseteq A \cap hC$; the reverse containment can be similarly shown, and therefore we obtain $A \cap gC = A \cap hC$. Moreover, for any $gc (= hc)$ in $A \cap gC$ we obtain

$$(gk)^{-1}gc = k^{-1}c = k^{-1}h^{-1}hc = (hk)^{-1}gc.$$

Thus ρ is also a right congruence.

Finally, it is obvious that ρ is idempotent-separating for any two ρ-related idempotents $(A, 1)$ and $(B, 1)$ must clearly be equal. ∎

Lemma 2.2.23 $R(X)/\rho$ is isomorphic to $\mathscr{I}_X$.

Proof We shall check that $\theta: R(X)/\rho \to \mathscr{I}_X$, defined by

$$((A, g)\rho)\theta = g^{-1}|A$$

is an isomorphism (that θ is well-defined follows from the definition of ρ and the fact that $g^{-1}A \in P(X)$).

Suppose that $((A, g)\rho)\theta = ((B, h)\rho)\theta$; in other words, $g^{-1}|A = h^{-1}|B$. From this it immediately follows that $A = B$ and that $g^{-1}a = h^{-1}a$ for all $a \in A$; in other words, $(A, g)\rho(B, h)$, whence it follows that θ is injective.

Next let $\alpha: A \to A\alpha$ be any member of $\mathscr{I}_X$. Extend α to a permutation g^{-1} on Y (if X is finite, α can be extended to a permutation on X and so on Y also; if X is infinite $|Y \setminus A\alpha| = |X'| = |Y \setminus A|$, from which the existence of a required bijection follows). We then have

$$((A, g)\rho)\theta = g^{-1}|A = \alpha,$$

which proves that θ is surjective.

Finally, take $(A, g)\rho$ and $(B, h)\rho$ in $R(X)$. Then $((A, g)\rho(B, h)\rho)\theta = (((A, g)(B, h))\rho)\theta = ((A \cap gB, gh)\rho)\theta = (gh)^{-1}|(A \cap gB) = h^{-1}g^{-1}|(A \cap gB)$. On the other hand,

$$((A, g)\rho)\theta((B, h)\rho)\theta = (g^{-1}|A)(h^{-1}|B),$$

and it remains to show equality of domains. By definition of composition in $\mathscr{I}_X$ the domain of this last product is

$$g(g^{-1}A \cap B) = A \cap gB,$$

as required. ∎

Theorem 2.2.24 (McAlister's Second Theorem) *Every inverse semigroup is the image under some idempotent-separating homomorphism of some P-semigroup.*

Proof Let S be an inverse semigroup. By McAlister's First Theorem and Lemma 2.2.23 there exists a P-semigroup P and an idempotent-separating homomorphism $\varphi: P \to \mathscr{I}_S$. Since S is embedded in $\mathscr{I}_S$, we can consider $S\varphi^{-1}$, which is an inverse subsemigroup of P. Hence $S\varphi^{-1}$ is also E-unitary and so there is an isomorphism $\theta: Q \to S\varphi^{-1}$ from some P-semigroup Q onto $S\varphi^{-1}$. Then $\theta\varphi: Q \to S$ is the required idempotent-separating homomorphism onto S. ∎

Remark As a matter of chronology, McAlister's Second Theorem was proved before the first (see McAlister 1974a, b).

Although this ends our brief account of the McAlister Theory, it is by no means the end of the story. There are a number of distinct proofs of these theorems, and many important papers take up the theme from this point. In the language of McAlister and Reilly (1977), the preceding proof constructs an E-unitary 'cover' for S, meaning a pre-image under an idempotent-separating homomorphism. A description of all E-unitary covers of a given inverse semigroup S was given by McAlister and Reilly (1977). The paper of Reilly and Munn (1976) studies E-unitary congruences on P-semigroups, and the P-semigroup structure of such a quotient is given explicitly. Applying these results to the free inverse semigroup allows the authors to recover the McAlister Theorems. The proof of Schein (1975a) adopts a highly intuitive approach, in that the idea of the proof is explained in terms of inverse semigroups of one-to-one mappings; indeed, the above proof by Wilkinson (1983) of McAlister's Second Theorem draws on ideas introduced there. O'Carroll (1976) used the McAlister Theory to show that any E-unitary inverse semigroup can be embedded in the semidirect product of a semilattice and a group, and Wilkinson (1983) showed that a semigroup S is an E-unitary inverse semigroup if and only if S is isomorphic to an 'entire' inverse subsemigroup of some semidirect product P of a semilattice and a group; the adjective 'entire' meaning that S and P share the same maximal group morphic image.

Donald McAlister's engaging survey article (1980) was the leading paper in the Monash Conference on Algebraic Semigroups, and provides an excellent overview of the topic to that time, listing twenty references. In particular, it gives a nice explanation as to how the birooted-tree representation of FI_X can be extracted from that of Scheiblich. A full account of the theory is also in Petrich (1984). A recent appraisal of the significance of McAlister's Second Theorem has been given by Lawson (1991), who suggests that applications will arise in a categorical framework involving so-called inductive groupoids.

A graphical representation of the free product of E-unitary inverse semigroups (which is also E-unitary) has been given by Jones (1982). A graphical solution of the word problem for free combinatorial strict inverse semigroups is due to Reilly (1989) (an inverse semigroup is *strict* if it is a subdirect product of Brandt semigroups and/or groups; Exercise 1.3.4). Recent work by Margolis *et al.* (1987) on word problems for inverse monoids using automata theory is in part inspired by the Munn representation for the free inverse semigroup. Most progress has been made on the E-unitary case, and this looks a promising field. Stephen's diagram method applied to presentations of inverse monoids is exhibited in Stephen (1990).

The theory of E-unitary inverse semigroups has been generalized in separate directions by Margolis and Pin (1987a) and by Szendrei (1987) respectively. The former paper deals with (not necessarily regular) semigroups, the idempotents of which form a semilattice, while Szendrei has formulated a structure theorem for E-unitary regular semigroups. Although the approaches differ, Szendrei shows that they agree in the inverse case. For other directions of generalization of the McAlister Theory, see Fountain (1977), Veeramony (1984), and Lawson (1986). Semidirect products involving inverse semigroups and groups, and semidirect products for arbitrary semigroups, have been introduced by Preston (1986b, c) respectively; also see Nico (1983).

It now seems that the history of the McAlister P-theory may be longer than was first supposed. Schein (1986) claims that the P-theorem can be traced to Golab (1939). Also, the Covering Theorem seems to have been anticipated in differential geometry (Joubert 1966), but to have passed unappreciated, while Munn's Construction given in our proof of the P-theorem is a special case of that in the so-called Maximum Enlargement Theorem of Charles Ehresmann (1984).

The theory of inverse semigroups is not yet well understood or appreciated by the wider mathematical community, but the subject is beginning to manifest itself in surprising ways.

Exercises 2.2

1 (Schein 1975a) Let S be an E-unitary inverse subsemigroup of $\mathscr{I}_X$. Show that any σ-class of S is a compatible set; that is, the union of all members of the class is a member of $\mathscr{I}_X$ (where σ denotes the minimum group congruence on S).

2. Consider FI_X where $X = \{x, y, z\}$, and let
$u = x^2 x^{-3} xyy^{-1} xy^{-1} yzxx^{-1} zz^{-1}$,
$v = zz^{-1} xx^{-1} xyy^{-1} x^{-1} z^{-1} x^{-1} y^2 y^{-1}$.

 (a) Calculate the birooted tree representations of u, v, and uv.
 (b) Express u, v, and uv as corresponding members of the set $\mathscr{L}_X$ of Theorem 2.2.2.

(c) Write u, v, and uv in Schein right canonical form.
(d) Give the Scheiblich representation of u, v and uv.

3. A congruence ρ on a semigroup S is said to *saturate* a subset A of S if A is the union of ρ-classes. A congruence ρ on a semigroup S is *idempotent pure* (or *idempotent determined*) if ρ saturates $E(S)$.

  (a) Show that if ρ is an idempotent pure congruence on an E-unitary inverse semigroup S then S/ρ is also E-unitary.
  Let S be an arbitrary inverse semigroup with generating set X whence there exists a homomorphism $\varphi: FI_X \to S$.
  (b) Let ρ be the congruence FI_X generated by $\beta \circ \beta^{-1}$, where $\beta = \varphi|E(FI_X)$. Show that ρ is idempotent pure.
  (c) Hence deduce McAlister's Second Theorem.

4. Let S be the inverse subsemigroup of I_5 generated by the map
$$\begin{pmatrix} 1 & 2 & 3 & 4 \\ 2 & 3 & 1 & 5 \end{pmatrix}.$$
  (a) Show that S is E-unitary with $|S| = 7$.
  (b) Show that $|S/\sigma| = |E| = 3$, and deduce that S is not a semidirect product of a semilattice and a group.

5. (McAlister 1974a) Let $P(G, \mathcal{X}, \mathcal{Y})$ be a P-semigroup.
   (i) $(A, g) \mathcal{R} (B, h)$ if and only if $A = B$;
   (ii) $(A, g) \mathcal{L} (B, h)$ if and only if $g^{-1}A = h^{-1}B$;
   (iii) $(A, g) \mathcal{H} (A, 1)$ if and only if $gA = A$ (thus the $\mathcal{H}$-class of the idempotent $(A, 1)$ is isomorphic to the stabilizer of A under G).
   (iv) $(A, 1) \mathcal{D} (B, 1)$ if and only if $B = gA$ for some $g \in G$ (thus there is a one-to-one correspondence between the set of orbits of G and the set of $\mathcal{D}$-classes of P).
   (v) $(A, 1) \leqslant_\mathcal{J} (B, 1)$ if and only if there exists $g \in G$ such that $A \leqslant gB$;
   (vi) $(A, g) \sigma (B, h)$ if and only if $g = h$.

6. (McAlister 1974b) Let μ denote the greatest idempotent-separating congruence on a P-semigroup $P(G, \mathcal{X}, \mathcal{Y})$ (see Exercise 1.4.6). For each $A \in \mathcal{Y}$, let $C_A = \{g \in G: gB = B \text{ for all } B \leqslant A\}$. Prove that $(A, g)\mu(B, h)$ if and only if $A = B$ and $gh^{-1} \in C_A$.

7. Call two P-semigroups $P(G, \mathcal{X}, \mathcal{Y})$ and $P(H, \mathcal{U}, \mathcal{V})$ *equivalent* if there is an order isomorphism $\theta: \mathcal{X} \to \mathcal{U}$ such that $\mathcal{Y}\theta = \mathcal{V}$, and an isomorphism φ of G onto H such that the diagram below commutes for each $g \in G$:

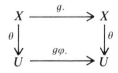

Prove that equivalent P-semigroups are isomorphic.

Remark The converse is also true, (McAlister 1974*b*, Theorem 1.3), thus solving the isomorphism problem for P-semigroups, and so showing that the representation of an E-unitary semigroup as a P-semigroup is unique up to equivalence.

3 Biordered sets

3.1 Introduction

In Section 1.5 the notion of the biordered set of a semigroup S was introduced, along with the associated idea of a sandwich set of an ordered pair of idempotents, $S(e,f)$. That an element may always be drawn from $S(e,f)$ in case S is regular was used implicitly in Theorem 1.4.17, and sandwich sets were also used to describe those regular semigroups in which the natural partial order is compatible with multiplication (Theorem 1.5.10 and Corollary 1.5.11).

In order to build semigroups from biordered sets, it is convenient to be able to consider the concept abstractly without explicit reference to semigroups. In Section 3.2 we define biordered sets using a collection of axioms that are readily seen to be satisfied by the biordered set of idempotents, $E(S)$, of a semigroup. That these axioms are appropriate has been proved by Easdown (1985), who showed conversely that any abstract biordered set is the biordered set of some semigroup. We shall not record that proof here, but we do have reason to demonstrate a weaker result that any biordered set is a biordered subset of some semigroup (Easdown 1984d) as the proof introduces a useful construction which is the analogue of the idempotent-separating representations of semigroups introduced in Section 1.4. As an application of this construction we prove that any fundamental regular semiband is determined by its biordered set, a result proved by Hall (1973) and Nambooripad (1979).

Another aim in our development of the theory of biordered sets is that of characterizing those biordered sets which can occur as the biordered set of some interesting class of semigroup. Known special characteristics of the biordered set of a semigroup can be useful: for instance, we can always select a member from a given sandwich set, $S(e,f)$, when dealing with a regular semigroup. A famous result of Nambooripad (1975) states that a biordered set is the biordered set of some regular semigroup if and only if all its sandwich sets are non-empty. A proof of this theorem, due to Easdown (1984c), will be presented.

On the other hand, in the attempt to construct a semigroup beginning with its biordered set, or part of it, it is clearly helpful to know in advance any special properties that the biordered set must possess. The descriptions of the biordered sets of eventually regular, group-bound, periodic, and finite semigroups (Easdown 1984b) involve the idea of the sandwich set of a given n-tuple of idempotents. This generalization was introduced by Pastijn (1980), and a description of such sandwich sets in case S is regular is given. Another contemporaneous paper on biordered sets of certain types is that of Tret'yakova (1986). A further paper on this topic is Easdown (1988).

3.2 An axiomatic system for biordered sets

Let $(E, \circ)$ be a non-empty set with a partial multiplication and domain $D_E \subseteq E \times E$. Define relations $\rightarrow$ and $\succ\!\!-\, \subseteq E \times E$ by

$$e \rightarrow f \quad \text{if and only if } (f, e) \in D_E \text{ and } fe = e;$$
$$e \succ\!\!- f \quad \text{if and only if } (e, f) \in D_E \text{ and } ef = e.$$

Note that $e \rightarrow\!\!\!\!\!\times f$ or $e \times\!\!\!\!-\!\!\!\!- f$ imply that $e = f$.

Definition 3.2.1 The partial algebra E is called a *biordered set* if it satisfies the following axioms:

(B1) $\rightarrow$ and $\succ\!\!-$ are pre-orders (reflexive and transitive, but not necessarily antisymmetric) and $D_E = \rightarrow \cup \succ\!\!- \cup (\rightarrow \cup \succ\!\!-)^{-1}$

(B21) $e \rightarrow f \Rightarrow$ [diagram with e, f, ef]

(B21)* $e \succ\!\!- f \Rightarrow$ [diagram with e, f, fe]

(B22) [diagram with e, f, g] $\Rightarrow fe \succ\!\!- ge$.

(B22)* [diagram with e, f, g] $\Rightarrow ef \rightarrow eg$.

(B31) $e \rightarrow f \rightarrow g \Rightarrow (eg)f = ef$.

(B31)* $e \succ\!\!- f \succ\!\!- g \Rightarrow f(ge) = fe$.

(B32) [diagram with e, f, g] $\Rightarrow (gf)e = (ge)(fe)$.

(B32)* [diagram with e, f, g] $\Rightarrow e(fg) = (ef)(eg)$.

(B4) [diagram with e, f, g] and $fe \succ\!\!- ge \Rightarrow$ there exists f' such that [diagram with e, f', g] and $f'e = fe$

(B4)* [diagram with e, f, g] and $ef \rightarrow eg \Rightarrow$ there exists f' such that [diagram with e, f', g] and $ef' = ef$.

Note that successive axioms rely on predecessors for their sense: for example, in (B21), if $e \to f$ then by (B1), $(e, f) \in D_E$ so that ef is defined; in (B31), if $e \to f \to g$ then $e \to g$ by transitivity of $\to$, due to (B1), so by (B21) $eg \to f$, whence $(eg)f$ is defined. Each axiom can be made self-contained by requiring that the appropriate products be defined in E, but this is unnecessary for our purposes, as we shall assume that E satisfies all axioms. For an investigation of the independence of the axioms, see Premchand (1984). The axioms as given above are a modified form of those of Nambooripad (1979).

Observe the duality between our axioms obtained by interchanging arrow types and reversing the order of products and bracketing; (B1) is self-dual, but the axioms marked by asterisk are dual to their non-asterisked counterparts.

The biordered set of a semigroup can be used as a source of insight for abstract biordered sets, but care needs to be exercised as the existence of products cannot be taken for granted.

As is proved in Nambooripad (1975), the (B4) axioms can be replaced by an alternative dual pair given in terms of sandwich sets. For $e, f \in E$ define, as in Section 1.5,

$$M(e, f) = \{g \in E \colon g \succ\!\!\!- e \text{ and } g \to f\}.$$

Define a pre-order, $\prec$, on $M(e, f)$ as follows: for $g, h \in M(e, f)$, $g \prec h$ if and only if $eg \to eh$ and $gf \succ\!\!\!- hf$. Define the *sandwich set* of e and f, denoted by $S(e, f)$, by

$$S(e, f) = \{h \in M(e, f) \colon \forall g \in M(e, f), g \prec h\}.$$

The (B4) axioms may be replaced by the following:

(B4′) $\quad \begin{array}{c} e \\ \nearrow \ \nwarrow \\ f \quad\ g \end{array} \Rightarrow S(f, g)e = S(fe, ge)$

and

(B4′)* $\quad \begin{array}{c} e \\ \swarrow \ \searrow \\ f \quad\ g \end{array} \Rightarrow eS(f, g) = S(ef, eg).$

We shall not make use of this alternative form of the (B4) axiom.

If E and F are biordered sets and $\theta \colon E \to F$ is a mapping, then θ is called a *(biordered set) morphism* if θ satisfies the following.

Definition 3.2.2 $(e, f) \in D_E \Rightarrow (e\theta, f\theta) \in D_F$ and $(ef)\theta = e\theta f\theta$.

We call θ a *biordered set isomorphism* if it is an isomorphism of partial algebras; that is, if both θ and θ^{-1} are biordered set morphisms. The requirement that θ^{-1} be a morphism is not automatically satisfied by partial algebras: without it the possibility would remain that F could have products for which there were no counterparts in E. Nambooripad uses the term *bimorphism* for biordered set morphism.

We call $F \subseteq E$ a *biordered subset* of E if F is a *full partial subalgebra* of E, in the sense that $D_F = D_E \cap (F \times F)$ and, furthermore, that F satisfies all the biordered set axioms with respect to the restrictions of $\to$ and $\succ\!\!-$ to F. By inspection of the list of axioms, it is clear that a full partial subalgebra F of E is a biordered subset of E if (and only if) F satisfies (B4) and (B4)*.

Proposition 3.2.3 *The biordered set $(E, \circ)$ of a semigroup S (as defined in Section 1.5) is a biordered set in the above sense.*

Proof The reader may easily check that the biordered set axioms hold for $(E, \circ)$, although we draw attention to two points: if $e \to f$ in $(E, \circ)$ then $(ef)^2 = ef$, so that $(e, f) \in D_E$, with a dual remark applying if $e \succ\!\!- f$; also (B4) is satisfied by $f' = fg$, for then

$$f'^2 = fgfg = f(eg)(ef)g = (fe)(ge)fg = f(efg) = f^2g = f'$$

and

$$f'e = fge = f(eg)e = (fe)(ge) = fe,$$

while, clearly, $f' \succ\!\!- g$ and $f' \to e$; dually (B4)* is satisfied by taking $f' = gf$. ∎

The products D_E of a biordered set $E(S)$ of a semigroup S are referred to as the *basic products* by Nambooripad, as opposed to other idempotent products which may or may not arise in the partial algebra of idempotents $(E(S), \circ)$.

Three basic properties of abstract biordered sets now follow. In fact, the (B4) axioms are not invoked in the next three results.

Lemma 3.2.4 *If $e \to f \to g$, then $(ef)g = e(fg) = (eg)(fg)$.*

Proof By applying (B21) to each of the given arrows we obtain

$$e \leftrightarrow ef \succ\!\!- f \leftrightarrow fg \succ\!\!- g.$$

In particular,

yields $(ef)g \succ\!\!- fg$, by (B22). Thus

$$\begin{align} (ef)g &= ((ef)g)(fg) && \text{(as } (ef)g \succ\!\!- fg) \\ &= (ef)(fg) && \text{(by (B31) as } ef \to fg \to g) \\ &= e(fg) && \text{(by (B31) as } e \to fg \to f) \\ &= (eg)(fg) && \text{(by (B31) as } e \to fg \to g). \end{align}$$ ∎

114 | Techniques of semigroup theory

Lemma 3.2.5

and $fe \succ\!\!\!-\!\!\!\prec ge$ implies that $f \succ\!\!\!-\!\!\!\prec g$.

Proof

$$\begin{aligned}
gf &= (gf)g && \text{(since } gf \twoheadrightarrow g \text{ by (B21)*)} \\
&= ((gf)e)g && \text{(by (B31), since } gf \to g \to e) \\
&= ((ge)(fe))g && \text{(by (B32))} \\
&= (ge)g && \text{(since } ge \succ\!\!-\!\! fe) \\
&= g && \text{(since } ge \leftrightarrow g, \text{ by (B21))}.
\end{aligned}$$
∎

The third of our lemmas is a result for abstract biordered sets that has already been stated for biordered sets of semigroups (Exercise 1.5.7).

Lemma 3.2.6 *If $(e, f) \in D_E$ then $ef \in S(f, e)$.*

Proof If $f \to e$ then $fe \in M(f, e)$. Suppose that $g \in M(f, e)$. Then $fg \twoheadrightarrow f$ by (B21)*, and by axiom (B22) $ge \succ\!\!-\!\! fe$, and so $g \prec f$ in $M(f, e)$. Therefore, in this case, $ef = fe \in S(f, e)$. Similarly, $fe \in M(e, f)$ by (B21), and if $g \in M(e, f)$ then $g \to f \leftrightarrow fe$. Hence, by (B22)*, $eg \to e(fe)$. Since $f \leftrightarrow fe$, $(fe)f = f$ and hence $gf \twoheadrightarrow (fe)f$. Thus $g \prec fe$ in $M(e, f)$ and so $fe \in S(e, f)$.

Dually, if $f \succ\!\!-\!\! e$ then $fe = fe \in S(e, f)$ and $ef \in S(f, e)$, thus completing the proof. ∎

As mentioned before, it is now known that all biordered sets are biordered sets of semigroups, so that in working with abstract biordered sets we are not denying ourselves any additional structure which might have been possessed by biordered sets of arbitrary semigroups. The proof of Easdown (1985) takes the free semigroup on the elements of a given biordered set factored by the congruence generated by all the relations of the biordered set, and then demonstrates that the biordered set of the resultant semigroup coincides with that originally given. The proof breaks into the three distinct problems of showing that:

(i) there is no collapse of elements of the original biordered set under the factoring operation;
(ii) no new arrows are created in the biordered set; and
(iii) no new idempotents are created.

The main result of this section enables (i) and (ii) to be proved quite easily, but the proof of (iii) involves a difficult word argument. It will be enough for our purposes to show that any biordered set may be embedded in a certain semigroup of partial mappings, yielding that biordered sets are biordered subsets of idempotents of semigroups (Easdown 1984d).

To this end, let E be a biordered set. Denote the equivalence relations $\succ\!\!\prec$ and $\leftrightarrow$ on E by $\mathscr{L}'$ and $\mathscr{R}'$ respectively, and denote the set of $\mathscr{L}'$- and of $\mathscr{R}'$-classes of E by $E/\mathscr{L}'$ and $E/\mathscr{R}'$ respectively. An arbitrary member of $E/\mathscr{L}'$ $[E/\mathscr{R}']$ will be denoted by L $[R]$ and the $\mathscr{L}'$- $[\mathscr{R}']$ class containing a particular $e \in E$ will be written as L_e $[R_e]$; the notation is chosen in recognition of the fact that biordered sets model biordered sets of semigroups, in which case the $\mathscr{L}'$ and $\mathscr{R}'$ relations are just the corresponding Green's relations restricted to idempotents.

For convenience, we sometimes write the full transformation semigroup on a set X by $T(X)$, while its dual will be written as $T^*(X)$.

We now introduce a representation of E which is constructed by mimicking the representation of an eventually regular semigroup S introduced in Section 1.4 to describe the maximum idempotent-separating congruence μ on S. Suppose that $\infty \notin E/\mathscr{L}' \cup E/\mathscr{R}'$, and put $X = E/\mathscr{L}' \cup \{\infty\}$ and $Y = E/\mathscr{R}' \cup \{\infty\}$.

Definition 3.2.7

(i) $\rho: E \to T(X)$, where

$$\rho_e: L \mapsto \begin{cases} L_{xe} & \text{if there exists } x \in L \text{ such that } x \to e; \\ \infty & \text{otherwise;} \end{cases}$$

$\infty \mapsto \infty$.

(ii) $\lambda: E \to T(Y)$, where

$$\lambda_e: R \mapsto \begin{cases} R_{ex} & \text{if there exists } x \in R \text{ such that } x \succ\!\!- e; \\ \infty & \text{otherwise;} \end{cases}$$

$\infty \mapsto \infty$.

(iii) $\varphi: E \to T(X) \times T^*(Y)$, where $e \mapsto \varphi_e = (\rho_e, \lambda_e)$.

Note that ρ and λ are single-valued by (B22) and (B22)* respectively, and are idempotents in $T(X)$ and $T^*(Y)$ respectively.

By the following lemma ρ and λ are morphisms of E into $E(T(X))$ $E(T^*(Y))$ respectively.

Lemma 3.2.8 $(e, f) \in D_E \Rightarrow \rho_{ef} = \rho_e \rho_f$ and $\lambda_{ef} = \lambda_e \lambda_f$.

Proof Suppose that $(e, f) \in D_E$. The argument falls into two cases, each with two subcases.

(i) Suppose that $e \to f$. We first show that $\rho_e (= \rho_{fe}) = \rho_f \rho_e$. Let $L \in E/\mathscr{L}'$. Suppose that $L\rho_e \ne \infty$, so there exists $x \in L$ such that $x \to e$. Then $L\rho_e = L_{xe}$ and $L\rho_f \rho_e = L_{(xf)e}$, since by transitivity of $\to$ and by (B21)

But $(xf)e = xe$, by (B31), so $L\rho_e = L\rho_f \rho_e$. Suppose that $L\rho_f \rho_e \ne \infty$, so that there exists $x \in L$ such that $x \to f$ and there exists $y \in L_{xf}$ such that $y \to e$; so that, by (B21) and transitivity of the arrows we obtain

In particular, $y \rightarrowtail f \leftarrow x$ and $y = yf \succ\!\!\prec xf$, so by (B4) and (B21) there exists y' such that

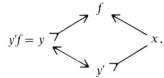

Since $y'f \succ\!\!\prec xf$, $y' \succ\!\!\prec x$, by Lemma 3.2.5. But
$$y' \leftrightarrow y'f = y \to e,$$
so that $y' \to e$ and hence $L\rho_e = L_{y'e} \ne \infty$. Therefore, if $L\rho_e = \infty$ then $L\rho_e \rho_f = \infty$. We deduce that $\rho_e = \rho_f \rho_e$.

Now we show that $\rho_{ef} = \rho_e \rho_f$. Suppose that $L\rho_{ef} \ne \infty$, so that there exists $x \in L$ such that $x \to ef$. Hence, by (B21) and transitivity, we obtain:

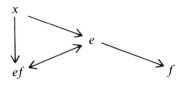

and then again using (B21) and transitivity we obtain:

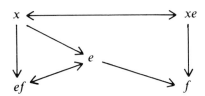

so that

$$L\rho_e\rho_f = L_{xe}\rho_f = L_{(xe)f} = L_{x(ef)} \quad \text{(by Lemma 3.2.4)}$$
$$= L\rho_{ef}.$$

Suppose that $L\rho_e\rho_f \neq \infty$, so there exists $x \in L$ such that $x \to e$, whence

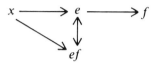

so that $L\rho_{ef} \neq \infty$. Therefore $\rho_{ef} = \rho_e\rho_f$.

(ii) Suppose that $e \succ\!\!\!- f$. We first show that $\rho_e(=\rho_{ef}) = \rho_e\rho_f$. Suppose that $L\rho_e \neq \infty$, so that there exists $x \in L$ such that $x \to e$. Hence

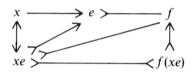

so $L\rho_e\rho_f = L_{xe}\rho_f = L_{f(xe)}\rho_f = L_{(f(xe))f} = L_{f(xe)} = L_{xe} = L\rho_e$. If $L\rho_e\rho_f \neq \infty$ then $L\rho_e \neq \infty$. Thus $\rho_e = \rho_e\rho_f$.

Now we show that $\rho_{fe} = \rho_f\rho_e$. Suppose that $L\rho_{fe} \neq \infty$, so there exists $x \in L$ such that $x \to fe$. Hence

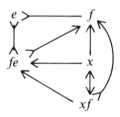

In particular, $xf \succ\!\!\!- f \prec\!\!\!- e$ and $xf = f(xf) \to fe$, so that by (B4)* and (B21)* there exists x' such that

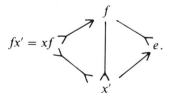

Combining these two diagrams gives

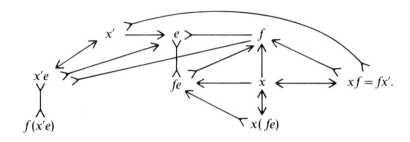

But
$$f(x'e) = (fx')(fe) \quad \text{(by (B32)*)}$$
$$= (xf)(fe)$$
$$= x(fe) \quad \text{(by (B31))}$$

so that $L\rho_f\rho_e = L_{xf}\rho_e = L_{x'e} = L_{f(x'e)} = L_{x(fe)} = L\rho_{fe}$. Suppose that $L\rho_f\rho_e \neq \infty$, so that there exist $x \in L$ such that $x \to f$ and there exists $y \in L_{xf}$ such that $y \to e$. Hence

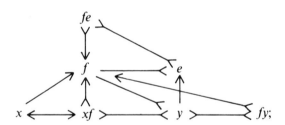

so, in particular,

$$(fy)f = fy \rightarrowtail\longleftarrow xf,$$

so by (B4) applied to $fy \to f \leftarrow x$, there exists y' such that

Biordered sets | 119

and Lemma 3.2.5 we obtain

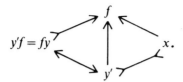

By (B22)*, $fy \to fe$, which yields

so that $L\rho_{fe} \neq \infty$. Therefore $\rho_{fe} = \rho_f \rho_e$. Thus we have shown that $(e, f) \in D_E \Rightarrow \rho_{ef} = \rho_e \rho_f$. By the dual argument we also have $(e, f) \in D_E \Rightarrow \lambda_{ef} = \lambda_e \lambda_f$. ∎

Lemma 3.2.9 *If $\varphi_e \varphi_f = \varphi_e$ then $e \succ\!\!-\!\!\prec f$ and $\varphi_f \varphi_e = \varphi_e$ implies $e \to f$.*

Proof Suppose that $\varphi_e \varphi_f = \varphi_e$, so that $\rho_e \rho_f = \rho_e$. Then $L_e = L_e \rho_e = L_e \rho_e \rho_f = L_e \rho_f$, so that there exists $x \in L_e$ such that $x \to f$, and $L_{xf} = L_e$. Hence

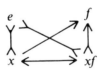

so $e \succ\!\!-\!\! f$. If $\varphi_f \varphi_e = \varphi_e$ then $\lambda_f \lambda_e = \lambda_e$, so that dual reasoning yields $e \to f$.

Theorem 3.2.10 *The mapping φ is an injective biordered set morphism from E into $E(T(X) \times T^*(Y))$. Also $E\varphi$ is a biordered subset of $E(T(X) \times T^*(Y))$ and $E \simeq E\varphi$ as biordered sets.*

Proof By Lemma 3.2.8, φ is a morphism from E into $E(T(X) \times T^*(Y))$. If $\varphi_e = \varphi_f$ then, by Lemma 3.2.9, $e \succ\!\!-\!\!\prec f$ and $e \leftrightarrow f$, so that $e = f$. Hence φ is injective.

We show that $E\varphi$ is a partial subalgebra of $E(T(X) \times T^*(Y))$.

Let $\varphi_e, \varphi_f \in E\varphi$. If $\varphi_e \succ\!\!-\!\! \varphi_f$ then, by Lemma 3.2.9, $e \succ\!\!-\!\! f$, so that by Lemma 3.2.8 $\varphi_f \varphi_e = \varphi_{fe} \in E\varphi$. Similarly, if $\varphi_e \to \varphi_f$ then $\varphi_e \varphi_f \in E\varphi$. Thus $E\varphi$ is a partial subalgebra.

Let $\varphi^{-1}: E\varphi \to E$ denote the map whereby $\varphi_e \mapsto e$. Thus, if $(\varphi_e, \varphi_f) \in D_{E\varphi}$ then, by Lemma 3.2.9, $(e, f) \in D_E$ so, by Lemma 3.2.8,

$$(\varphi_e \varphi_f)\varphi^{-1} = (\varphi_{ef})\varphi^{-1} = ef = (\varphi_e)\varphi^{-1}(\varphi_f)\varphi^{-1},$$

and so φ^{-1} is a morphism of partial algebras; that is, φ is an isomorphism of partial algebras. Hence $E\varphi \simeq E$ as partial algebras, so $E\varphi$ is a biordered set, and hence a biordered subset of $E(T(X) \times T^*(Y))$. ∎

Theorem 3.2.10 is the fundamental result in Easdown's theory of biordered sets. The representation φ is the main tool, and will be used to prove all the major results that follow.

Corollary 3.2.11 *Biordered sets are biordered subsets of biordered sets of semigroups.*

Recently, Easdown has developed an alternative axiomization of biordered sets, based on the original formulation of Nambooripad (1975). A biordered set is characterized as a 'dually ordered set' (a concept that generalizes that of a partially ordered set), together with a partial multiplication defined using certain right and left translation mappings with a series of special properties. The practical importance of this approach is that it gives a clear direction for the construction of skeletons of semigroups, and examples illustrate this point (Easdown 1991).

Exercises 3.2

1. (a) Show that the element f' in axiom (B4) is unique for a biordered set of a semigroup.
   (b) Show that the biordered set of a semigroup satisfies the axioms B4' and (B4')*.
2. Show that if $(ef)g$ and $e(fg)$ are both defined in some biordered set E then they are equal. However, it is possible for one, but not the other, of these products to be defined in E.

Remark It can be shown that the first statement above is true if E is a partial algebra satisfying just (B1), (B2), (B3), and their duals.

3. Let E be a biordered set and F the free semigroup on elements of E. Let $\rho = \{(z, xy): (x, y) \in D_E \text{ and } xy = z\}$. Show that:

   (i) $e, f \in E$ and $e\rho^* f$ then $e = f$ (no collapse);
   (ii) $e \to f \Leftrightarrow e \, \rho^* fe$, and $e \succ\!\!\!— f \Leftrightarrow e \, \rho^* \, ef$ (no new arrows).

Remark Easdown (1985) showed that, in addition, if $w \in F$ and $w^2 \rho^* w$, then $w \rho^* e$ for some $e \in E$ (no new idempotents), thus establishing that biordered sets are biordered sets of semigroups.

3.3 Reconstructing some semibands from their biordered sets

In this section we return to the question posed in Section 1.5 as to the extent to which the nature of a semigroup may be gleaned from knowledge of its biordered set. A path of investigation is revealed by comparing the representation of S defined in Section 1.4, here denoted by φ^0, with the representation φ of the biordered set $E(S)$ introduced in the previous section. We show that the information carried by the biordered set $E(S)$ is equivalent to that carried by the image of the core of S under φ^0. From this we can deduce, for instance, that for any regular semiband S, S/μ is determined by its biordered set, where μ is the maximum idempotent-separating congruence on S (Theorem 1.4.14). The content of this section is from Easdown and Hall (1984).

For convenience we restate our two representations. Let S be any semigroup and E any biordered set. Let $\mathscr{L}^0$ [$\mathscr{R}^0$] denote the set of regular $\mathscr{L}$-[$\mathscr{R}$-] classes of S, and $\mathscr{L}'$ [$\mathscr{R}'$] the set of $\succ\!\!\prec$ [$\leftrightarrow$]-classes of E. Let $L[R]$ denote a member of $\mathscr{L}'$ [$\mathscr{R}'$], and L^0 [R^0] a member of $\mathscr{L}^0$ [$\mathscr{R}^0$]; let L_e [R_e] denote the $\succ\!\!\prec$ [$\leftrightarrow$]-class containing a given member $e \in E$, and L_x^0 [R_x^0] denote the (regular) $\mathscr{L}$- [$\mathscr{R}$-] class containing any $x \in \text{Reg}(S)$.

Definition 3.3.1

(i) $\rho^0: S \to T(\mathscr{L}^0 \cup \{\infty\})$ by, for all $s \in S$ and $x \in \text{Reg}(S)$,

$$\rho_s^0: L_x^0 \mapsto \begin{cases} L_{xs}^0 & \text{if } x \, \mathscr{R}^S xs, \\ \infty & \text{otherwise,} \end{cases}$$

$\infty \mapsto \infty$.

(ii) $\lambda^0: S \to T^*(\mathscr{R}^0 \cup \{\infty\})$ by, for all $s \in S$ and $x \in \text{Reg}(S)$,

$$\lambda_s^0: R_x^0 \mapsto \begin{cases} R_{sx} & \text{if } x \, \mathscr{L}^S sx, \\ \infty & \text{otherwise,} \end{cases}$$

$\infty \mapsto \infty$.

(iii) $\varphi^0: S \to T(\mathscr{L}^0 \cup \{\infty\}) \times T^*(\mathscr{R}^0 \cup \{\infty\})$ by $\varphi_s^0 = (\rho_s^0, \lambda_s^0)$.

(iv) $\rho: E \to T(\mathscr{L} \cup \{\infty\})$ by, for all $e \in E$,

$$\rho_e: L_x \mapsto \begin{cases} L_{xe} & \text{if } x \to e \text{ for some } x \in L, \\ \infty & \text{otherwise,} \end{cases}$$

$\infty \mapsto \infty$.

122 | Techniques of semigroup theory

(v) $\lambda: E \to T^*(\mathcal{R} \cup \{\infty\})$ by, for all $e \in E$,

$$\lambda_e: R_x \mapsto \begin{cases} R_{ex} & \text{if } x \succ\!\!-e \text{ for some } x \in R, \\ \infty & \text{otherwise,} \end{cases}$$

$$\infty \mapsto \infty.$$

(vi) $\varphi: E \to T(\mathcal{L} \cup \{\infty\}) \times T^*(\mathcal{R} \cup \{\infty\})$ by $\varphi_e = (\rho_e, \lambda_e)$.

Recall from Proposition 1.4.11 and remarks preceding it that φ^0 is a well-defined idempotent-separating representation of S, and if S is eventually regular then $\ker \varphi^0$ is also the maximum idempotent-separating congruence on S. By Theorem 2.10, the mapping φ is a biordered set isomorphism onto $E\varphi$.

To achieve our aim of demonstrating the equivalence of the biordered set $E = E(S)$ and $\langle E(S) \rangle \varphi^0$ we show that, for each $e \in E$, the mapping ρ_e (defined in terms of the biordered set $E(S)$ alone) is a reconstruction of the mapping ρ_e^0. Combining with the dual argument yields that $\varphi_e = (\rho_e, \lambda_e)$ reconstructs $\varphi_e^0 = (\rho_e^0, \lambda_e^0)$ and our claim will follow.

The following lemma is a formal statement of the claim that, for each e in $E = E(S)$, ρ_e is a reconstruction of ρ_e^0. Observe that for all x, y in E we have $x \succ\!\!\prec y$ if and only if $x \mathcal{L}^S y$, so that the mapping $\circ: \mathcal{L}' \cup \{\infty\} \to \mathcal{L}^0 \cup \{\infty\}$ defined by $\infty \mapsto \infty$, $L_e \mapsto L_e^0$ (for all $e \in E$) is a bijection. Observe further that for any $x, y, e \in E$, $x \to e$ implies that $x \leftrightarrow xe$ which, in turn, is true if and only if $x \mathcal{R}^S xe$. It follows from the definitions of our representations that $L\rho_e \neq \infty$ implies $L^0\rho_e^0 \neq \infty$ and, moreover, in this case $(L\rho_e)^0 = L_{xe}^0 = L^0\rho_e^0$. We also wish to show that this holds if $L\rho_e = \infty$. In summary, our statement is as follows.

Lemma 3.3.2 *For any semigroup S and idempotent e in S, the following diagram commutes:*

$$\begin{array}{ccc} \mathcal{L} \cup \{\infty\} & \xrightarrow{\rho_e} & \mathcal{L} \cup \{\infty\} \\ \Big\downarrow {\circ} & & \Big\downarrow {\circ} \\ \mathcal{L}^0 \cup \{\infty\} & \xrightarrow{\rho_e^0} & \mathcal{L}^0 \cup \{\infty\}. \end{array}$$

From the foregoing it suffices to prove that if $L^0\rho_e^0 \neq \infty$ then $L\rho_e \neq \infty$. This follows from the next lemma concerning regular $\mathcal{D}$-classes.

Lemma 3.3.3 *For any semigroup S, any $x \in \text{Reg}(S)$ and any $e \in E(S)$, if $x \mathcal{R}^S xe$ then there exists an idempotent g of S such that $g \to e$ and $x \mathcal{L}^S g \mathcal{R}^S ge \mathcal{L}^S xe$.*

Proof Since $x \in \text{Reg}(S)$ there exists $f \in E(S)$ such that $f \mathcal{L}^S x$. Also, since $x \mathcal{R}^S xe$ there exists $a \in S$ such that $xea = x$. Put $g = eaf$.

From Green's Lemma we obtain $f\mathcal{R}^S fe$ and $fea = f$. Hence $fg = f(eaf) = (fea)f = f^2 = f$; clearly, $gf = g$ so that $g \mathcal{L}^S f \mathcal{L}^S x$ and by another application of Green's Lemma we obtain, $g \mathcal{R}^S ge \mathcal{L}^S xe$. Finally, it is clear that $eg = g$ and $g^2 = (eaf)(eaf) = ea(fea)f = eaf^2 = eaf = g$. ∎

As a consequence of Lemma 3.3.3 and its dual for the λ_e and λ_e^0 we obtain the next theorem.

Theorem 3.3.4 *If E is the biordered set of any semigroup S, then $\langle E\varphi \rangle$ is isomorphic to $\langle E \rangle \varphi^0 = \langle E\varphi^0 \rangle$; in fact, the map*

$$\Phi\colon \langle E\varphi \rangle \to \langle E \rangle \varphi^0 = \langle E\varphi^0 \rangle, \qquad \text{whereby } \varphi_{e_1}\varphi_{e_2}\cdots \varphi_{e_n} \mapsto \varphi^0_{e_1\ldots e_n}$$

(for all $e_1, \ldots, e_n \in E$)

is well-defined, and is an isomorphism of semigroups.

Corollary 3.3.5 *If S is a semiband and $E = E(S)$ is its biordered set of all idempotents, then there exists an idempotent-separating morphism from S onto $\langle E\varphi \rangle$, namely $\varphi^0 \circ \Phi^{-1}$, induced by the map $\varphi\colon E \to E\varphi$.*

Corollary 3.3.6 *If S is an eventually regular semiband then S/μ is isomorphic to $\langle E\varphi \rangle$, where $E = E(S)$ is the biordered set of S. In particular, a fundamental idempotent-generated regular semigroup is determined by its biordered set.*

Proof This follows from Corollary 3.3.5, and the fact that $\mu = \ker \varphi^0$. ∎

Recall that a semigroup is idempotent-consistent with respect to a congruence σ if each idempotent σ-class contains an idempotent, and S is idempotent-consistent if idempotent-consistent with respect to all of its congruences. Corollary 1.4.9 says that all eventually regular semigroups are idempotent-consistent.

Theorem 3.3.7 *Take any biordered set E. Then E is the set of all idempotents of some semigroup S which is idempotent-consistent with respect to the congruence $\ker \varphi^0$ if and only if*

$$E\varphi = E(\langle E\varphi \rangle).$$

Proof Suppose that E is the biordered set of all idempotents of some semigroup S which is idempotent-consistent with respect to $\ker \varphi^0$; that is, $E(S\varphi^0) = E\varphi^0$. Immediately we have $E(\langle E\varphi^0 \rangle) = E\varphi^0$, but by considering the isomorphism $\Phi\colon \langle E\varphi\rangle \to \langle E\varphi^0\rangle$ of Theorem 3.3.4 we see that

$$E\varphi = (E\varphi^0)\Phi^{-1} = (E(\langle E\varphi^0\rangle))\Phi^{-1} = E(\langle E\varphi\rangle),$$

as required.

Conversely, suppose that $E\varphi = E(\langle E\varphi\rangle)$ and put $S = \langle E\varphi\rangle$. By Theorem 3.2.10, E is isomorphic to $E\varphi$, the biordered set of S, and so we can finish the proof by showing that S is idempotent-consistent with respect ker φ^0.

Since the biordered set of S, namely $E(\langle E\varphi\rangle)$, is isomorphic to the biordered set E there is no loss of generality in assuming that the biordered set of S is E; this enables us to apply Theorem 3.3.4 which, together with $S = \langle E\rangle$, yields

$$E\varphi^0 = (E\varphi)\Phi = (E(\langle E\varphi\rangle))\Phi = E(\langle E\varphi\rangle\Phi) = E(\langle E\rangle\varphi^0) = E(S\varphi^0),$$

thus completing the proof. ∎

The following theorem, which follows easily from Theorems 3.3.4 and 3.3.7, is a prototype of several theorems, each characterizing the biordered sets of a particular class of idempotent-consistent semigroups.

Theorem 3.3.8 *Let $\mathscr{C}$ be a class of idempotent-consistent semigroups closed under the taking of homomorphic images and cores. Let E be any biordered set. Then $E = E(S)$ for some $S \in \mathscr{C}$ if and only if $E\varphi = E(\langle E\varphi\rangle)$ and $\langle E\varphi\rangle \in \mathscr{C}$.*

This is a powerful result, as it reduces the question of whether a biordered set E comes from a certain class of semigroup to a study of the idempotents of one particular semigroup which is constructed from E. Moreover, the theorem is applicable to many interesting classes of semigroup. For instance, $\mathscr{C}$ could be any of the following classes: finite semigroups, periodic semigroups, bands, regular semigroups, completely regular semigroups, group-bound semigroups, and eventually regular semigroups (that the last four classes are closed under the taking of cores follows from the argument of Theorem 1.4.18; also see Exercise 1.4.15). For example, to see if a finite biordered set E comes from some finite semigroup, simply construct $\langle E\varphi\rangle$ and check whether its biordered set is E (this can be done effectively; Easdown 1984b). If not, then E comes from no finite semigroup. In particular, let us re-examine Example 1.5.6.

Example 3.3.9

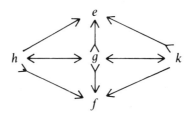

Since $h \to e$ and $k \to f$, both he and kf are idempotents. Thus, if this diagram is to represent the biordered set of some semigroup we must resolve the products he and kf. Up to isomorphism there are three possible biordered sets:

(i) E_1: $he = kf = g$;
(ii) E_2: $he = k$, $kf = h$;
(iii) E_3: $he = g$, $kf = h$.

The alternative where $he = k$ and $kf = g$ is isomorphic as a biordered set to E_3; all other candidates for the products either collapse the biordered set or introduce new arrows: for example, if $he = f$ then $ehe = he = f$, whence $f \rightarrowtail e$.

Consider $E_1 \varphi$. The only unknown products in $E_1 \varphi$ are $e\varphi f\varphi$ and $f\varphi e\varphi$. All $\mathscr{L}'$- and $\mathscr{R}'$-classes of E are singletons except the $\mathscr{R}'$-class $R = \{g, h, k\}$. The mappings $\rho_e, \rho_f, \rho_g, \lambda_e, \lambda_f, \lambda_g$ are given below, it being understood that ∞ is a fixed point of all mappings:

$$\rho_e = \begin{pmatrix} e & f & g & h & k \\ e & \infty & g & g & k \end{pmatrix}, \quad \lambda_e = \begin{pmatrix} e & f & R \\ e & \infty & R \end{pmatrix},$$

$$\rho_f = \begin{pmatrix} e & f & g & h & k \\ \infty & f & g & h & g \end{pmatrix}, \quad \lambda_f = \begin{pmatrix} e & f & R \\ \infty & f & R \end{pmatrix},$$

$$\rho_g = \begin{pmatrix} e & f & g & h & k \\ \infty & \infty & g & g & g \end{pmatrix}, \quad \lambda_g = \begin{pmatrix} e & f & R \\ \infty & \infty & R \end{pmatrix}.$$

Thus we see that $e\varphi f\varphi = g\varphi = f\varphi e\varphi$, and so E_1 is the biordered set of a band and we have produced its multiplication table.

The biordered set E_2, however, comes from no eventually regular semigroup, in fact no idempotent-consistent semigroup because $E_4 = E\langle E_2\varphi \rangle$ is strictly larger than E_2; λ_e, λ_f are as before, but now we have

$$\rho_e = \begin{pmatrix} e & f & g & h & k \\ e & \infty & g & k & k \end{pmatrix}, \quad \rho_f = \begin{pmatrix} e & f & g & h & k \\ \infty & f & g & h & h \end{pmatrix},$$

$$\rho_e \rho_f = \begin{pmatrix} e & f & g & h & k \\ \infty & \infty & g & h & h \end{pmatrix}, \quad \rho_f \rho_e = \begin{pmatrix} e & f & g & h & k \\ \infty & \infty & g & k & k \end{pmatrix},$$

and thus $e\varphi f\varphi$ and $f\varphi e\varphi$ are two new idempotents. If we identify $e\varphi f\varphi$ and $f\varphi e\varphi$ with ef and fe respectively, then the biordered set E_4 is given below:

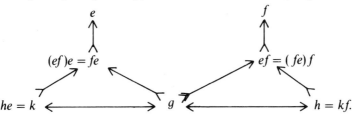

Not all arrows are drawn but can be deduced by transitivity; the set E_4 is now closed under multiplication and so forms a band.

The previous examples lead to the following question: When is a given biordered set embeddable in a band? For example, E_3 can never be embedded in a band, for suppose that E_3 were a biordered subset of a band E, so in particular ef is idempotent. Hence

$$g = h(ef) = (kf)(ef) = k(ef)^2 = k(ef) = h,$$

a contradiction.

However, E_3 is the biordered set of some finite semigroup. It is left to the reader to show that $S = \langle E_3 \varphi \rangle$ has six elements and that $E(S) \simeq E$; the 'new' member is ef, which is not an idempotent. Note that S is not regular as ef has no inverse; indeed, E_3 is the smallest biordered set which comes from a finite semigroup, but from no regular semigroup.

We next turn to the problem of characterizing the biordered sets of special classes of semigroup.

Exercises 3.3

1. By calculating $\langle E_3 \varphi \rangle$ in Example 3.3.9, show that E_3 comes from a six-element semigroup in which the new element ef is not regular. Deduce that E_3 comes from no regular semigroup.
2. Recall the biordered set E defined by the diagram

   introduced in Exercise 1.5.2. Show that $\langle E\varphi \rangle$ is a completely 0-simple semigroup.

3.4 Biordered sets of certain classes of semigroup

A natural class with which to begin our categorization is those semigroups which themselves consist of only idempotents; that is, bands. In this we follow Easdown (1984c). The effectiveness of the arrow notation is fully realized in this section, as the diagrams allow many basic products of a biordered set to be displayed simultaneously. The reader can then remind himself of any of a large number of relationships as required throughout the course of an argument.

Theorem 3.4.1 *Let E be a biordered set. Then $E = E(S)$ for some band S if and only if $(\forall e, f \in E)(\exists h \in M(f, e))$ such that $(\forall x \in M(e, f))$ the following diagram holds:*

Condition 3.4.2

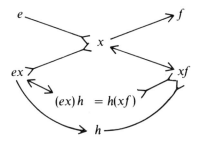

Remark The double arrows in Condition 3.4.2 follow by axioms (B21) and (B21)*, and are included merely to illustrate the relationships between the elements.

Proof Suppose that $E = E(S)$ for some band S and let $e, f \in E$. Put $h = ef$, so that $h \in M(f, e)$. If $x \in M(e, f)$ then $h(ex) = efefx = efx = ex$, and dually $(xf)h = xf$. Furthermore, $(ex)h = exef = exf = efxf = h(xf)$, thereby showing that Condition 3.4.2 holds.

Conversely, suppose that the conditon of the theorem holds. Recall our representation $\varphi: E \to E(T(X) \times T^*(Y))$ and Theorem 3.2.10. Let $e, f \in E$. We shall show that $\varphi_e \varphi_f = \varphi_h$, where h is as in Condition 3.4.2, for we shall have then shown that $E\varphi = \langle E\varphi \rangle$; that is, $\langle E\varphi \rangle$ is a band with biordered set $E\varphi \simeq E$, by Theorem 3.2.10.

We show that $\rho_e \rho_f \doteq \rho_h$. Suppose that $L\rho_e\rho_f \neq \infty$, so that $x \to e$ for some $x \in L$, $y \to f$ for some $y \in L_{xe}$, and $L\rho_e\rho_f = L_{yf}$. Observing that $y \succ\!\!\prec xe \succ\!\!\!-\, e$, we see that $y \in M(e, f)$, and we obtain the next diagram:

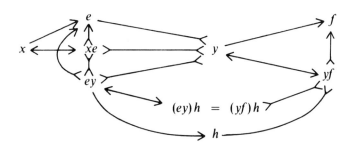

Thus $ey \rightarrowtail e \leftarrow x$ and $(ey)e = ey \succ\!\!\prec xe$, so by (B4) and Lemma 3.2.5 there exists some $x' \in E$ such that

$$x'e = ey \quad \text{[diagram]} \quad x.$$

Observe that $x' \in L = L_x$ and $x' \leftrightarrow ey \rightarrow h$, whence $x' \rightarrow h$. Hence

$$L\rho_h = L_{x'h} = L_{(x'e)k} \quad \text{(by axiom (B31), since } x' \rightarrow h \rightarrow e\text{)}$$
$$= L_{(ey)h}$$
$$= L_{(yf)h} = L_{yf} = L\rho_e\rho_f.$$

Now suppose that $L\rho_h \neq \infty$, so that $x \rightarrow h$ for some $x \in L$. Since h is a member of $M(f, e)$ we can apply Condition 3.4.2 to $f \mathrel{-\!\!\prec} h \rightarrow e$, to obtain $h' \in M(e, f)$ such that

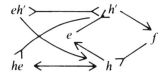

so that $eh' \rightarrow\!\!\times he$, whence $eh' = he$. Hence

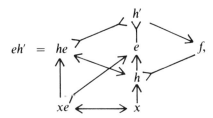

so that $xe \rightarrowtail e \mathrel{-\!\!\prec} h'$ and $e(xe) = xe \rightarrow eh'$. Thus, by axiom (B4)*, there exists some $x' \in E$ such that

$$ex' = xe \quad \text{[diagram]} \quad h'.$$

Observe $x' \rightarrow h' \rightarrow f$, so that $x' \rightarrow f$, and we have

$$L\rho_e\rho_f = L_{xe}\rho_f = L_{x'f} \neq \infty.$$

Thus, if $L\rho_e\rho_f = \infty$ then $L\rho_h = \infty$.

Therefore we have shown that $\rho_e \rho_f = \rho_h$. Dually $\lambda_f \lambda_e = \lambda_h$, so that $\varphi_e \varphi_f = \varphi_h$. This completes the proof of the theorem. ∎

The proof has in fact furnished another natural characterization of bi-ordered sets of bands.

Corollary 3.4.3 *A biordered set E is the biordered set of some band S if and only if $\langle E\varphi \rangle = E\varphi$.*

A third interpretation of biordered sets of bands can be framed in terms of the notion of 'solidity', which is closely related to Condition 3.4.2.

Definition 3.4.4 Call $M(e,f)$ *solid* if there exists some $y \in E$ such that for all $x \in M(e,f)$ the pair (ex, xf) is *complete at y*, meaning that

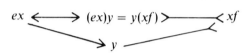

Theorem 3.4.5 *Let E be a biordered set. Then E is the biordered set $E(S)$ of some band S if and only if for all $e, f \in E$ the sandwich set $S(e,f)$ is non-empty and $M(e,f)$ is solid.*

Proof First suppose that $E = E(S)$ for some band S and let $e, f \in E$. Then Condition 3.4.2 holds for E, which says in particular that $M(e,f)$ is solid, while $S(e,f)$ is non-empty because S is regular (Theorem 1.5.3).

Conversely, suppose that for all $e, f \in E$, $S(e,f)$ is non-empty, and $M(e,f)$ is solid, so there exists some $y \in E$ such that, for all $x \in M(e,f)$,

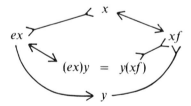

Let $z \in S(e,f)$ and let $h = (ez)y$. We show that $h \in M(f,e)$ and that y can be replaced by h in the preceding diagram (that is, the pair (ex, xf) is complete at h), whereupon the result will follow by Theorem 3.4.1. Now

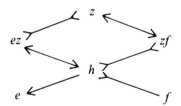

130 | Techniques of semigroup theory

so that $h \in M(f, e)$. Take any $x \in M(e, f)$. Since $z \in S(e, f)$ we have

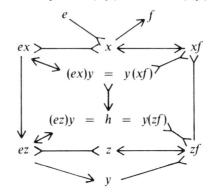

so that

$$(ex)h = [(ex)y]h \quad \text{by (B31)}$$
$$= (ex)y$$
$$= y(xf)$$
$$= h[y(xf)]$$
$$= h(xf) \quad \text{by (B31)*}.$$

This completes the proof. ∎

We now move on to biordered sets of regular semigroups. Recall from Section 1.5 that a biordered set E is regular if none of its sandwich sets are empty. We shall prove the theorem of Nambooripad that if E is regular then $E = E(S)$ for some regular semigroup S. Again, we call upon Theorem 3.2.10 and observe that it will therefore be sufficient to show that $\langle E\varphi \rangle$ is both regular and $E\langle E\varphi \rangle = E\varphi$.

Lemma 3.4.6 If $e, f \in E$ and $h \in S(e, f)$ then $\varphi_e \varphi_h \varphi_f = \varphi_e \varphi_f$ in $\langle E\varphi \rangle$.

Proof Suppose that $h \in S(e, f)$. We shall show that $\rho_e \rho_h \rho_f = \rho_e \rho_f$.

Suppose that $L\rho_e \rho_f \neq \infty$, so $x \to e$ for some $x \in L$, and $y \to f$ for some $y \in L_{xe}$. Hence $y \in M(e, f)$, so $ey \to eh$ and $yf \to hf$, since $h \in S(e, f)$. The next diagram summarizes our position:

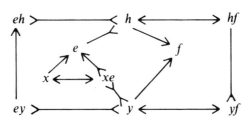

By (B4)* there exists some $y' \in M(e, h)$ such that $ey' = ey$. Hence we may augment our diagram:

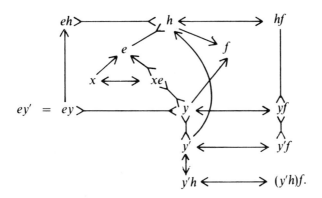

That $yf \succ\!\!\prec y'f$ follows from $f \leftarrow y \succ\!\!\prec y' \rightarrow f$ and (B22). But

$$(y'h)f = y'(hf) \quad \text{(by Lemma 3.2.4)}$$
$$= (y'f)(hf) \quad \text{(by (B31))}$$
$$= y'f \quad (\text{since } y'f \succ\!\!- hf).$$

Hence

$$L\rho_e\rho_h\rho_f = L_{xe}\rho_h\rho_f = L_{y'}\rho_h\rho_f = L_{(y'h)}\rho_f = L_{(y'h)f} = L_{y'f} = L_{yf} = L\rho_e\rho_f.$$

If $L\rho_e\rho_h\rho_f \neq \infty$ then $x \rightarrow e$ for some $x \in L$, and $y' \rightarrow h \rightarrow f$ for some $y' \in L_{xe}$, so that $L\rho_e\rho_f \neq \infty$. Hence if $L\rho_e\rho_f = \infty$ then also $L\rho_e\rho_h\rho_f = \infty$.

Thus we have shown that $\rho_e\rho_f = \rho_e\rho_h\rho_f$. Dually, $\lambda_f\lambda_e = \lambda_f\lambda_h\lambda_e$; so that $\varphi_e\varphi_f = \varphi_e\varphi_h\varphi_f$, as required. ∎

Let $x_1, \ldots, x_n \in E$, where E is a regular biordered set. Define $y_{n-1}, \ldots, y_1 \in E$ inductively as follows: choose $y_{n-1} \in S(x_{n-1}, x_n)$ as this set is not empty; suppose that y_{n-i} is defined, for $i \leq n - 2$, satisfying $y_{n-i} \succ\!\!- x_{n-i}$, and choose $y_{n-i-1} \in S(x_{n-i-1}, x_{n-i}y_{n-i})$ as this set is nonempty. Thus we have

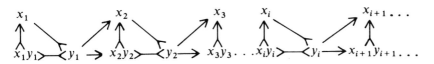

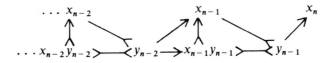

Lemma 3.4.7 *Let $x_1, \ldots, x_n, y_1, \ldots, y_{n-1} \in E$, as above. Then*

$$\varphi_{x_1} \cdots \varphi_{x_n} \varphi_{y_{n-1}} \cdots \varphi_{y_1} = \varphi_{x_1 y_1}.$$

Proof We show by induction that, for $i = 1$ through to $n - 1$,

$$\varphi_{x_1} \cdots \varphi_{x_i} \varphi_{y_i} \cdots \varphi_{y_1} = \varphi_{x_1 y_1}. \tag{1}$$

Now (1) is true for $i = 1$ since φ is a morphism, by Theorem 3.2.10. Suppose that (1) is true for i. Then

$$\varphi_{x_1} \cdots \varphi_{x_{i+1}} \varphi_{y_{i+1}} \cdots \varphi_{y_1} = (\varphi_{x_1} \cdots \varphi_{x_i}) \varphi_{x_{i+1} y_{i+1}} (\varphi_{y_i} \cdots \varphi_{y_1})$$
(since φ is a morphism)

$$= \varphi_{x_1} \cdots \varphi_{x_i} \varphi_{y_i} \cdots \varphi_{y_1}$$
(since $y_i \to x_{i+1} y_{i+1}$ and φ is a morphism)

$$= \varphi_{x_1 y_1},$$

by the inductive hypothesis. Hence

$$\varphi_{x_1} \cdots \varphi_{x_n} \varphi_{y_{n-1}} \cdots \varphi_{y_1} = \varphi_{x_1} \cdots \varphi_{x_{n-1}} \varphi_{y_{n-1}} \cdots \varphi_{y_1}$$
(since $y_{n-1} \to x_n$ and φ is a morphism)

$$= \varphi_{x_1 y_1},$$

by (1). ∎

Lemma 3.4.8 *Let $x_1, \ldots, x_n, y_1, \ldots, y_{n-1} \in E$, as above. Then the product $\varphi_{y_{n-1}} \cdots \varphi_{y_1}$ is an inverse of the product $\varphi_{x_1} \cdots \varphi_{x_n}$.*

Proof We show, for values of i from 1 through to $n - 1$, that

$$\varphi_{y_{n-1}} \cdots \varphi_{y_{n-i}} \text{ is an inverse of } \varphi_{x_{n-i}} \cdots \varphi_{x_n}. \tag{2}$$

Now

$$\varphi_{y_{n-1}} \cdots \varphi_{y_{n-i}} \varphi_{x_{n-i}} \cdots \varphi_{x_n} \varphi_{y_{n-1}} \cdots \varphi_{y_{n-i}}$$

$$= \varphi_{y_{n-1}} \cdots \varphi_{y_{n-i}} \varphi_{x_{n-i} y_{n-i}}$$
(by Lemma 3.4.7 applied to the sequence $x_{n-i}, \ldots, x_n, y_{n-i}, \ldots, y_{n-1}$)

$$= \varphi_{y_{n-1}} \cdots \varphi_{y_{n-i}} \quad \text{(since } y_{n-i} \succ\!\!\!-\, x_{n-i} \text{ and } \varphi \text{ is a morphism)}.$$

We show by induction, for $i = 1$ through to $n - 1$, that

$$(\varphi_{x_{n-i}} \cdots \varphi_{x_n})(\varphi_{y_{n-1}} \cdots \varphi_{y_{n-i}})(\varphi_{x_{n-i}} \cdots \varphi_{x_n}) = \varphi_{x_{n-i}} \cdots \varphi_{x_n}. \tag{3}$$

Now

$$(\varphi_{x_{n-1}} \varphi_{x_n}) \varphi_{y_{n-1}} (\varphi_{x_{n-1}} \varphi_{x_n}) = \varphi_{x_{n-1}} \varphi_{y_{n-1}} \varphi_{x_n}$$
(since $x_n \leftarrow y_{n-1} \succ\!\!\!-\, x_{n-1}$ and φ is a morphism)

$$= \varphi_{x_{n-1}} \varphi_{x_n} \quad \text{(by Lemma 3.4.6, since } y_{n-1} \in S(x_{n-1}, x_n),$$

so that (3) holds for $i = 1$. Suppose that it holds for i. Then

$$(\varphi_{x_{n-i-1}} \cdots \varphi_{x_n})(\varphi_{y_{n-1}} \cdots \varphi_{y_{n-i-1}})(\varphi_{x_{n-i-1}} \cdots \varphi_{x_n})$$
$$= \varphi_{x_{n-i-1}y_{n-i-1}}\varphi_{x_{n-i-1}}(\varphi_{x_{n-i}} \cdots \varphi_{x_n})(\varphi_{y_{n-1}} \cdots \varphi_{y_{n-i}})(\varphi_{x_{n-i}} \cdots \varphi_{x_n})$$

(by (1) and the inductive hypothesis)

$$= \varphi_{x_{n-i-1}}\varphi_{y_{n-i-1}}\varphi_{x_{n-i}}\varphi_{y_{n-i}}(\varphi_{x_{n-i}} \cdots \varphi_{x_n})$$

(since φ is a morphism, $y_{n-i-1} \succ\!\!\!- x_{n-i-1}$, and by Lemma 3.4.7)

$$= \varphi_{x_{n-i-1}}\varphi_{x_{n-i}y_{n-i}}(\varphi_{x_{n-i}} \cdots \varphi_{x_n})$$

(by Lemma 3.4.6, since $y_{n-i-1} \in S(x_{n-i-1}, x_{n-i}y_{n-i})$)

$$= \varphi_{x_{n-i-1}}(\varphi_{x_{n-i}} \cdots \varphi_{x_n})(\varphi_{y_{n-1}} \cdots \varphi_{y_{n-i}})(\varphi_{x_{n-i}} \cdots \varphi_{x_n})$$
(by Lemma 3.4.7)

$$= \varphi_{x_{n-i-1}} \cdots \varphi_{x_n} \quad \text{(by the inductive hypothesis).}$$

Thus (3) holds, and therefore (2) also holds. ∎

To complete the proof of Nambooripad's Theorem, Easdown uses the following lemma, which is accredited to Hall.

Lemma 3.4.9 *Let E be a biordered set. If $\alpha \in \langle E\varphi \rangle$ is idempotent and $\varphi_e \mathscr{L} \alpha \mathscr{R} \varphi_f$ for some $e, f \in E$, where $\mathscr{L}$ and $\mathscr{R}$ denote Green's relations on $\langle E\varphi \rangle$, then $\alpha \in E\varphi$.*

Proof Suppose that $\alpha = \alpha^2 \in \langle E\varphi \rangle$ and $\varphi_e \mathscr{L} \alpha \mathscr{R} \varphi_f$ for some $e, f \in E$. We then have the following egg-box picture:

φ_e		$\varphi_e \varphi_f$
α		φ_f

Now $L_e\rho_e = L_e \neq \infty$, so $L_e\rho_f = L_e\rho_e\rho_f \neq \infty$, since $\rho_e \mathcal{R} \rho_e\rho_f$ (with respect to the semigroup $\langle E\rho \rangle$). Hence $x \to f$ for some $x \in L_e$, so

so by Green's Lemma, and since φ is a morphism

φ_e		$\varphi_e\varphi_f$
φ_x		$\varphi_{xf} = \varphi_x\varphi_f$
α		φ_f

Hence $\varphi_{xf} \succ\!\!\!\!-\!\!\!\!\times \varphi_f$, so that $\varphi_{xf} = \varphi_f$, whence $\varphi_x \mathcal{H} \alpha$; therefore $\varphi_x = \alpha$, because both are idempotents.

Theorem 3.4.10 *If E is a regular biordered set then $\langle E\varphi \rangle$ is a regular semigroup with biordered set $E\varphi \simeq E$.*

Proof Suppose that $\alpha = \varphi_{x_1} \ldots \varphi_{x_n} \in \langle E\varphi \rangle$ for some $x_1, \ldots, x_n \in E$. Form a sequence $y_1, \ldots, y_{n-1} \in E$ using the method given preceding Lemma 3.4.7. By Lemma 3.4.8, $\varphi_{y_{n-1}} \ldots \varphi_{y_1}$ is an inverse of α, so that α is regular in $\langle E\varphi \rangle$. This shows that $\langle E\varphi \rangle$ is a regular semigroup.

Now, since $\varphi_{y_{n-1}} \ldots \varphi_{y_1}$ is inverse to α, it follows from Lemma 3.4.7 that

$$\alpha \mathcal{R} \alpha\varphi_{y_{n-1}} \ldots \varphi_{y_1} = \varphi_{x_1y_1}$$

and so $\alpha \mathcal{R} \varphi_f$, where f denotes x_1y_1. Dually, there exists some $e \in E$ such that $\varphi_e \mathcal{L} \alpha$. If α is idempotent then, by Lemma 3.4.9, $\alpha \in E\varphi$. This proves that $E(\langle E\varphi \rangle) = E\varphi$ and, by Theorem 3.2.10, $E \simeq E\varphi$. This completes the proof. ∎

Sandwich sets

The sandwich set of a pair of idempotents, $S(e, f)$ was first defined by Nambooripad (1975), and the idea was extended to a finite sequence of

Biordered sets | 135

idempotents by Pastijn (1980). As motivation for this generalization we shall run through a proof of Easdown (1986) of a theorem of Hall (1973) characterizing completely regular semibands (for an alternative proof, see Exercise 1.4.14).

Theorem 3.4.11 *Let $E = E(S)$ for some regular semiband $S = \langle E \rangle$. Then S is completely regular if and only if E is solid, in the sense that, for all $e, f, g \in E$,*

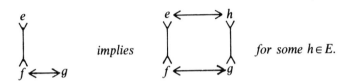

for some $h \in E$.

Proof The necessity is clear because if $e \mathcal{L} f \mathcal{R} g$ then $R_e \cap L_g$ is non-empty, and so contains an idempotent since S is completely regular.

Suppose that E is solid, and let $e_1 \ldots e_n \in S$, where $e_1, \ldots, e_n \in E$. It is sufficient to locate an idempotent x for which $e_1 \ldots e_n \mathcal{H} x$.

Let $a \in V(e_1 \ldots e_n)$ and, for $i = 1$ to $n-1$, put

$$x_i = e_{i+1} \ldots e_n a e_1 \ldots e_i.$$

Then each $x_i \in E$, and we have the following arrows and products:

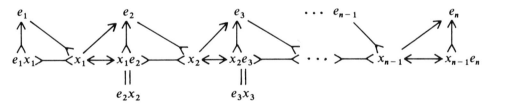

Moreover, $e_1 \ldots e_n = e_1 x_1 e_2 x_2 e_3 \ldots e_{n-1} x_{n-1} e_n = e_1 x_1 e_2 \ldots e_n$.

The reader will recall that the above construction of idempotents forming a chain of alternating double left and right arrows uses the technique of Fitzgerald that was introduced in our Theorem 1.4.18.

It is now an easy matter to locate an idempotent in the $\mathcal{H}$-class of $e_1 \ldots e_n$. We have the following egg-box picture in S, which follows by successive applications of Green's Lemma;

e_1x_1	$e_1x_1e_2$	$e_1x_1e_2e_3$	...	$e_1x_1e_2\ldots e_{n-1}$	$e_1x_1e_2\ldots e_n = e_1\ldots e_n$
$x_1 \leftrightarrow x_1e_2$					
	$x_2 \leftrightarrow x_2e_3$				
		$\vdots$			
				$x_{n-1} \leftrightarrow x_{n-1}e_n$	

Since E is solid, there must be an idempotent in the $\mathscr{H}$-class of $e_1x_1e_2\ldots e_i$ for $i = 2$ through to $i = n$, which completes the proof of the theorem.

Suppose that the sequence $(x_1, \ldots x_{n-1})$ has been constructed as in the above proof. Let b be an element of S such that $b = b(e_1 \ldots e_n)b$, which is weaker than requiring that $b \in V(e_1 \ldots e_n)$; $e_1 \ldots e_n$ is called a *pre-inverse of b*. Define, for $i = 1$ to $n-1$,

$$f_i = e_{i+1} \ldots e_n b e_1 \ldots e_i.$$

Then f_i is idempotent for each i, and we obtain the picture shown in Fig. 3.1.

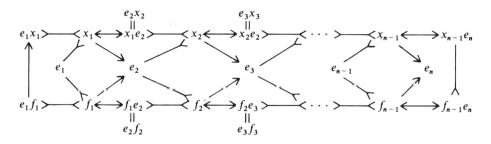

Fig. 3.1.

Notice that the x_i and the f_i are defined using semigroup-theoretic properties, and that these imply biordered set-theoretic properties. We use the latter to define sandwich sets of finite sequences of idempotents.

For elements $e_1, \ldots, e_n$ of a biordered set E, call a sequence $(f_1, \ldots, f_{n-1})$ *complete* if $e_i \mathrel{-\!\!\!<} f_i \to e_{i+1}$ and $f_i e_{i+1} = e_{i+1} f_{i+1}$ for each $i = 1$ to $n - 1$. Following Pastijn (1980), define

$$M(e_1, \ldots, e_n) = \{(f_1, \ldots, f_{n-1}) \mid (f_1, \ldots, f_{n-1}) \text{ is complete}\}.$$

Note that if $i = 2$ this is just the definition of $M(e_1, e_2)$ as introduced in Section 1.5, as the condition $f_i e_{i+1} = e_{i+1} f_{i+1}$ is satisfied vacuously.

Definition 3.4.12 The *sandwich set* $S(e_1, \ldots, e_n)$ is defined as

$$S(e_1, \ldots, e_n) = \{(x_1, \ldots, x_{n-1}) \in M(e_1, \ldots, e_n) \mid \text{for all}$$
$$(f_1, \ldots, f_{n-1}) \in M(e_1, \ldots, e_n)$$
$$\text{both } e_1 f_1 \rightarrow e_1 x_1 \text{ and } f_{n-1} e_n \succ\!\!\!- x_{n-1} e_n\}.$$

Figure 3.1 defines a pre-order on members of $M(e_1, \ldots, e_n)$, induced by the right and left arrows, and the sandwich set is simply the maximal layer with respect to this pre-order.

Theorem 3.4.13 *A biordered set E comes from a completely regular semigroup if and only if E is both regular and solid.*

Proof If E comes from a completely regular semigroup S, then E is also the biordered set of the core, $\langle E \rangle$ of S, which is itself completely regular by Theorem 1.4.18. It then follows from Theorem 3.4.11 that E is regular and solid.

Conversely, since E is regular it is, by Theorem 3.4.10, the biordered set of some regular semigroup S. It is then also the biordered set of the (regular) semiband $\langle E \rangle$, and since E is also solid the result follows from Theorem 3.4.11. ∎

If $E = E(S)$, where S is a regular semigroup, then a given sandwich set $S(e_1, \ldots, e_n)$ consists of all sequences $(x_1, \ldots, x_{n-1})$ defined as in the proof of Theorem 3.4.11 as a ranges over $V(e_1 \ldots e_n)$, as will now be shown (Pastijn 1980). Henceforth let E be the biordered set of some semiband S.

First we make several observations concerning a complete sequence $(f_1, \ldots, f_{n-1}) \in M(e_1, \ldots, e_n)$:

$$f_i e_{i+1} = e_{i+1} f_{i+1} \leqslant e_{i+1} \quad i = 1, 2, \ldots, n-2,$$
$$e_1 f_1 \leqslant e_1, \text{ and } f_{n-1} e_n \leqslant e_n.$$

Moreover,

$$e_1 f_1 \succ\!\!\!\!-\!\!\!\prec f_1 \leftrightarrow f_1 e_2 = e_2 f_2 \succ\!\!\!\!-\!\!\!\prec f_2 \ldots e_i f_i \succ\!\!\!\!-\!\!\!\prec f_i \leftrightarrow f_i e_{i+1}$$
$$= e_{i+1} f_{i+1} \succ\!\!\!\!-\!\!\!\prec f_{i+1} \leftrightarrow \ldots \succ\!\!\!\!-\!\!\!\prec f_{n-1} \leftrightarrow f_{n-1} e_n. \quad (4)$$

Lemma 3.4.14 *Let $(f_1, \ldots, f_{n-1}) \in M(e_1, \ldots, e_n)$. Then $f_i = f_i f_{i-1} \ldots f_1 e_1 \ldots e_i = e_{i+1} \ldots e_n f_{n-1} \ldots f_i = e_{i+1} \ldots e_n f_{n-1} \ldots f_1 e_1 \ldots e_i$ for all $i = 1, \ldots, n-1$.*

Proof We show that $f_i = f_i f_{i-1} \ldots f_1 e_1 \ldots e_i$ inductively. For $i = 1$ the statement says that $f_1 \succ\!\!-\! e_1$, which is true by definition. Let us suppose that

$$f_{i-1} = f_{i-1} \ldots f_1 e_1 \ldots e_{i-1}$$

with $2 \leq i \leq n-1$. Then

$$f_{i-1} e_i = e_i f_i \succ\!\!-\!\!\prec f_i$$

and so

$$f_i = f_i(e_i f_i) = f_i f_{i-1} e_i = f_i f_{i-1} \ldots f_1 e_1 \ldots e_{i-1} e_i.$$

Let us next verify that $f_i = e_{i+1} \ldots e_n f_{n-1} \ldots f_i$ by induction; the statement certainly being true for $i = n-1$ since $e_n f_{n-1} = f_{n-1}$. Let us suppose that

$$f_{i+1} = e_{i+2} \ldots e_n f_{n-1} \ldots f_{i+1}$$

with $1 \leq i \leq n-2$. In consequence of (4) we obtain

$$f_i \leftrightarrow f_i e_{i+1} = e_{i+1} f_{i+1},$$

whence

$$f_i = (f_i e_{i+1}) f_i = e_{i+1} f_{i+1} f_i = e_{i+1} e_{i+2} \ldots e_n f_{n-1} \ldots f_{i+1} f_i,$$

which gives the required equality.

From the foregoing we have

$$f_i = f_i^2 = (e_{i+1} \ldots e_n f_{n-1} \ldots f_i)(f_i \ldots f_1 e_1 \ldots e_i)$$
$$= e_{i+1} \ldots e_n f_{n-1} \ldots f_1 e_1 \ldots e_i,$$

as required. ∎

Lemma 3.4.15 *Let* $(f_1, \ldots, f_{n-1}) \in M(e_1, \ldots, e_n)$. *If* $a = e_1 \ldots e_n$ *and* $a' = f_{n-1} \ldots f_1$, *then* $a'aa' = a'$. *In fact*, $aa'a = (e_1 f_1)(e_2 f_2) \ldots (e_{n-1} f_{n-1}) e_n$ *and* a' *are mutually inverse. Moreover*, $aa' = e_1 f_1$ *and* $a'a = f_{n-1} e_n$.

Proof By Lemma 3.4.14, $f_1 = e_2 \ldots e_n f_{n-1} \ldots f_1$, and thus

$$e_1 f_1 = (e_1 \ldots e_n)(f_{n-1} \ldots f_1) = aa'.$$

Again by Lemma 3.4.14, $f_{n-1} = f_{n-1} \ldots f_1 e_1 \ldots e_{n-1}$ and so

$$f_{n-1} e_n = (f_{n-1} \ldots f_1)(e_1 \ldots e_n) = a'a.$$

We show inductively that $f_{n-1} \ldots f_i \in L_{f_i} \cap R_{f_{n-1}}$; the statement certainly being true for $i = n-1$. Suppose that $f_{n-1} \ldots f_{i+1} \in L_{f_{i+1}} \cap R_{f_{n-1}}$ with $1 \leq i \leq n-2$; then by (4)

$$f_{n-1} \ldots f_{i+1} \mathscr{L} f_{i+1} \mathscr{L} e_{i+1} f_{i+1} \mathscr{R} f_i,$$

and so, since $\mathscr{L}$ and $\mathscr{R}$ are right and left congruences respectively, $(f_{n-1} \ldots f_{i+1}) f_i \in L_{f_i} \cap R_{f_{n-1} \ldots f_{i+1}}$. Using the inductive hypothesis we conclude that:

$$f_{n-1} \ldots f_i \in L_{f_i} \cap R_{f_{n-1}} \quad \text{for all } i = 1, \ldots, n-1.$$

Hence, in particular,

$$f_{n-1} \ldots f_1 \in L_{f_1} \cap R_{f_{n-1}} = L_{e_1 f_1} \cap R_{f_{n-1} e_n} = L_{aa'} \cap R_{a'a}.$$

Since $aa' \in E(L_{a'}) = E(L_{f_{n-1} \ldots f_1})$, we have $a'(aa') = a'$. It follows immediately that $aa'a \in V(a')$ and $aa'a$ is the unique inverse of a' in the $\mathscr{H}$-class $L_{a'a} \cap R_{aa'}$.

It remains to verify the stated expression for $aa'a$. By Lemma 3.4.14,

$$f_i = e_{i+1} \ldots e_n f_{n-1} \ldots f_1 e_1 \ldots e_i = e_{i+1} \ldots e_n a' e_1 \ldots e_i$$

for all $i = 1, 2, \ldots, n-1$; hence

$$(e_1 f_1)(e_2 f_2) \ldots (e_{n-1} f_{n-1}) e_n = e_1(e_2 \ldots e_n a' e_1) e_2(e_3 \ldots e_n a' e_1 e_2) \ldots$$
$$e_{n-1}(e_n a' e_1 \ldots e_{n-1}) e_n$$
$$= (aa')^{n-1} a = aa'a,$$

as required. ∎

Theorem 3.4.16 *Let $a = e_1 \ldots e_n$ be any element in the core of some semigroup S. Then $M(e_1, \ldots, e_n)$ consists of all sequences of the form*

$$(e_2 \ldots e_n a' e_1, \ldots, e_{i+1} \ldots e_n a' e_1 \ldots e_i, \ldots, e_n a' e_1 \ldots e_{n-1}),$$

in which $a' = a'aa'$.

Proof If $(f_1, \ldots, f_{n-1}) \in M(e_1, \ldots, e_n)$, then $a' = f_{n-1} \ldots f_1$ is a core element of S such that $a'aa' = a'$, and for any $i = 1, \ldots, n-1$ we obtain $f_i = e_{i+1} \ldots e_n a' e_1 \ldots e_i$: all this follows from Lemmas 3.4.14 and 3.4.15.

Conversely, suppose that a is a pre-inverse of a'; that is, $a'aa' = a'$. Then for any $i = 1, \ldots, n-1$, $f_i = e_{i+1} \ldots e_n a' e_1 \ldots e_i \in E$ and $e_i \prec f_i \to e_{i+1}$ since

$$f_i^2 = (e_{i+1} \ldots e_n a' e_1 \ldots e_i)(e_{i+1} \ldots e_n a' e_1 \ldots e_i)$$
$$= e_{i+1} \ldots e_n a' a a' e_1 \ldots e_i = e_{i+1} \ldots e_n a' e_1 \ldots e_i = f_i.$$

Also it is immediate that $f_i e_{i+1} = e_{i+1} f_{i+1}$ for all $i = 1, \ldots, n-2$. We conclude that $(f_1, \ldots, f_{n-1}) \in M(e_1, \ldots, e_n)$. ∎

Having found a description for $M(e_1, \ldots, e_n)$, we seek a formulation for $S(e_1, \ldots, e_n)$ along similar lines.

Lemma 3.4.17 *Let $(f_1, \ldots, f_{n-1}) \in S(e_1, \ldots, e_n)$ and put $e_1 \ldots e_n = a$. Then $e_1 f_1 e_2 f_2 \ldots f_{n-1} e_n = b$ is regular and $H_b \leq H_a$. If c is regular and $H_c \leq H_a$, then $H_c \leq H_b$. In particular, if $(g_1, \ldots, g_{n-1}) \in M(e_1, \ldots, e_n)$ then $c = e_1 g_1 \ldots g_{n-1} e_n$ is regular and $H_c \leq H_a$ and $H_c \leq H_b$. The sandwich set $S(e_1, \ldots e_n)$ consists of those $(g_1, \ldots, g_{n-1}) \in M(e_1, \ldots, e_n)$ for which $e_1 g_1 e_2 g_2 \ldots g_{n-1} e_n = c \mathscr{H} b$.*

140 | Techniques of semigroup theory

Proof Put $a' = f_{n-1} \ldots f_1$; then, by Lemma 3.4.15, $b = aa'a$ and a' are inverses of each other. Hence $H_b \leqslant H_a$ and b is regular. Let c be any regular member of $aS \cap Sa$ and take $c' \in V(c)$. Since $c \in Sa$ there exist idempotents $h_1, \ldots, h_k$ such that $c = h_1 \ldots h_k e_1 \ldots e_n$. Let us put

$$t_1 = h_2 \ldots h_k e_1 \ldots e_n c' h_1$$
$$\vdots$$
$$t_p = h_{p+1} \ldots h_k e_1 \ldots e_n c' h_1 \ldots h_p \quad (1 \leqslant p \leqslant k-1)$$
$$\vdots$$
$$t_k = e_1 \ldots e_n c' h_1 \ldots h_k$$
$$t_{k+1} = e_2 \ldots e_n c' h_1 \ldots h_k e_1$$
$$\vdots$$
$$t_{k+q} = e_{q+1} \ldots e_n c' h_1 \ldots h_k e_1 \ldots e_q \quad (1 \leqslant q \leqslant n-1)$$
$$\vdots$$
$$t_{k+n-1} = e_n c' h_1 \ldots h_k e_1 \ldots e_{n-1}.$$

Then $c = h_1 t_1 h_2 t_2 \ldots h_k t_k e_1 t_{k+1} \ldots e_{n-1} t_{k+n-1} e_n$ and

$$(t_1, \ldots, t_{k+n-1}) \in M(h_1, \ldots, h_k, e_1, \ldots, e_n).$$

Evidently, $c \in L(e_1 t_{k+1} e_2 t_{k+2} \ldots e_{n-1} t_{k+n-1} e_n)$ and

$$(t_{k+1}, \ldots, t_{k+n-1}) \in M(e_1, \ldots, e_n).$$

By Lemma 3.4.15 we obtain $e_1 t_{k+1} \ldots e_{n-1} t_{k+n-1} e_n \mathscr{L} t_{k+n-1} e_n$; hence c is a member of $L(t_{k+n-1} e_n)$. Since $t_{k+n-1} e_n \succ\!\!\!-\, f_{n-1} e_n$ we then have $c \in L(f_{n-1} e_n)$. Once more by Lemma 3.4.15, we have $b = aa'a \mathscr{L} a'a = f_{n-1} e_n$, and consequently $c \in Sb$. Dually we can show that $c \in bS$ also.

If $(g_1, \ldots, g_{n-1}) \in S(e_1 \ldots, e_n)$ the above argument shows that

$$e_1 g_1 e_2 g_2 \ldots e_{n-1} g_{n-1} e_n = c \in bS \cap Sb.$$

By the same argument, $H_b \leqslant H_c$, whence $b \mathscr{H} c$.

Conversely, if $(g_1, \ldots, g_{n-1}) \in M(e_1, \ldots, e_n)$ and $e_1 g_1 e_2 g_2 \ldots e_{n-1} g_{n-1} e_n = c \mathscr{H} b$, then $e_1 f_1 \mathscr{R} b \mathscr{R} c \mathscr{R} e_1 g_1$, and $f_{n-1} e_n \mathscr{L} b \mathscr{L} c \mathscr{L} g_{n-1} e_n$, by Lemma 3.4.15, and thus $(g_1, \ldots, g_{n-1}) \in S(e_1, \ldots, e_n)$. ∎

Theorem 3.4.18 *If $a = e_1 \ldots e_n \in \text{Reg}(S)$, then $S(e_1, \ldots, e_n)$ is non-empty and $S(e_1, \ldots, e_n)$ consists of all elements $(f_1, \ldots, f_{n-1}) = (e_2 \ldots e_n a' e_1, \ldots, e_{i+1} \ldots e_n a' e_1 \ldots e_i, \ldots, e_n a' e_1 \ldots e_{n-1})$ for which $a' \in V(a)$.*

Proof By the previous lemma, $S(e_1, \ldots, e_n)$ consists of all $(g_1, \ldots, g_{n-1})$ of $M(e_1, \ldots, e_n)$ for which $e_1 g_1 e_2 \ldots g_{n-1} e_n \mathscr{H} a$. If

$$(f_1, \ldots, f_{n-1}) = (e_2 \ldots e_n a' e_1, \ldots, e_{i+1} \ldots e_n a' e_1 \ldots e_i, \ldots,$$
$$e_n a' e_1 \ldots e_{n-1})$$

with $a' \in V(a)$ then, by Theorem 3.4.16, $(f_1, \ldots, f_{n-1}) \in M(e_1, \ldots, e_n)$ and

$e_1 f_1 e_2 f_2 \ldots f_{n-1} e_n = a$; consequently we see that $(f_1, \ldots, f_{n-1})$ is a member of $S(e_1, \ldots, e_n)$.

Conversely, let $(g_1, \ldots, g_{n-1}) \in S(e_1, \ldots, e_n)$. By the previous lemma $e_1 g_1 e_2 \ldots g_{n-1} e_n \mathcal{H} a$. Put $a'' = g_{n-1} g_{n-2} \ldots g_1$; by Lemma 3.4.15 a'' and $aa''a = e_1 g_1 e_2 \ldots g_{n-1} e_n$ are mutual inverses and

$$aa''a \mathcal{R} aa'' = e_1 g_1 \in E(S); \qquad aa''a \mathcal{L} a''a = g_{n-1} e_n \in E(S).$$

We conclude that $a \mathcal{R} aa''$ and $a \mathcal{L} a''a$, and thus $aa''a = a$; this implies that $a'' \in V(a)$. By Lemma 3.4.14 we obtain

$$(g_1, \ldots, g_{n-1}) = (e_2 \ldots e_n a'' e_1, \ldots, e_n a'' e_1 \ldots e_{n-1}),$$

thus completing the proof. ∎

Corollary 3.4.19 *A biordered set E is regular if and only if $S(e_1, \ldots, e_n)$ is non-empty for all $e_1, \ldots, e_n \in E$.*

Proof Only the forward direction requires explanation. If E is regular then E is the biordered set of some regular semigroup, and hence of some regular semiband; the result then follows from the previous theorem.

Exercises 3.4

1. A biordered set E can be that of at most one band.
2. Show that a biordered set E is solid in the sense used in Theorem 3.4.11 if and only if $S(e, f)$ is solid for all $e, f \in E$.
3. Consider the biordered set E defined by the diagram:

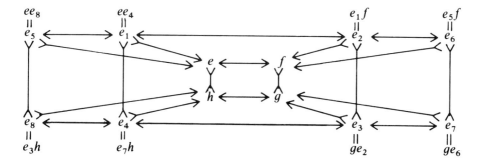

(a) Show that $\langle E\varphi \rangle$ is an ideal extension of a rectangular band by a completely 0-simple semigroup, the non-zero $\mathcal{H}$-classes of which are groups of order two, and deduce that E is the biordered set of a union of groups, but not the biordered set of any band.

(b) Show that $M(e, g)$ is not solid, and from this deduce the final conclusion of part (a).

4. Show that if S is eventually regular then for any $e_1, \ldots, e_n \in E(S)$, $S(e_1, \ldots, e_n)$ is *eventually non-empty*, meaning that there exists a positive integer k such that $S(e_1, \ldots, e_n)^k$ is non-empty, where $(e_1, \ldots, e_n)^k$ denotes

$$\underbrace{(e_1, \ldots, e_n, \ e_1, \ldots, \ e_n, \ldots, \ e_1, \ldots, e_n)}_{k \text{ times}}$$

Remark The converse is also true. For this and similar characterizations of the biordered sets of finite, periodic, and group-bound semigroups, see Easdown (1984b).

4 Zigzags and their applications

4.1 The Zigzag Theorem

We begin this chapter with a new and short proof of the Zigzag Theorem of John Isbell, which characterizes semigroup dominions by means of certain sequences of equations known as zigzags. A variety of applications concerning epimorphisms and amalgams will illustrate the usefulness of this result.

First we look at the connection between epimorphisms and dominions. Observe that to say that a homomorphism $\alpha: S \to T$ is epi (pre-cancellative) is the same as saying that whenever homomorphisms $\beta, \gamma: T \to V$ are such that $\beta|S\alpha = \gamma|S\alpha$, then $\beta = \gamma$. If this is the case, we say that $S\alpha$ is *dense in* T, or that $S\alpha$ is *epimorphically embedded in* T via the inclusion $i: S\alpha \to T$. Therefore the study of epimorphisms is equivalent to the study of epimorphic embeddings.

Suppose that the semigroup inclusion $i: U \to S$ is epi, that is, U is dense in S. We may think of U as a 'large' or 'dominating' part of S, in the sense that the action of any morphism from S is determined by its action on U. In general, however, the domination of U may only be partial. To make this precise, Isbell (1966) defined the *dominion of* U *in* S, denoted by $\mathrm{Dom}(U, S)$, to consist of all elements $d \in S$ which are *dominated by* U in the sense that whenever $\alpha, \beta: S \to T$ are morphisms such that $\alpha|U = \beta|U$ then $d\alpha = d\beta$. It is easy to see that $\mathrm{Dom}(U, S)$ is a subsemigroup of S containing U; indeed, it is routine to check that $\mathrm{Dom}(U, S)$ is a closure operator on the subsemigroups of S in that $U \subseteq \mathrm{Dom}(U, S)$ and if $U \subseteq V \subseteq S$ then

$$\mathrm{Dom}(U, S) \subseteq \mathrm{Dom}(V, S) \quad \text{and} \quad \mathrm{Dom}(\mathrm{Dom}(U, S), S) = \mathrm{Dom}(U, S).$$

We call a semigroup U *closed in* S if $\mathrm{Dom}(U, S) = U$, and U is *absolutely closed* if it is closed in every containing semigroup S. At the other extreme, U is dense in S or epimorphically embedded in S if $\mathrm{Dom}(U, S) = S$. A weaker condition on U than that of being absolutely closed is that of being *saturated*, which means that U cannot be properly epimorphically embedded in another semigroup. All these definitions can of course be applied equally well to other algebras.

There are several points to note. First, it follows from these definitions that to say that every epi from a semigroup U is onto, is equivalent to saying that every morphic image of U is saturated (there exists a saturated semigroup with a morphic image which is not; Higgins 1985b). Also it is not always true that a subsemigroup U is dense in its own dominion (Exercise 4.1.5). Finally, the notion of dominion could be introduced more generally for subsets U of S, but then we see that $\text{Dom}(U, S) = \text{Dom}(\bar{U}, S)$, where $\bar{U}$ is the subsemigroup of S generated by U.

An example of a non-surjective epimorphism was given in Exercise 1.1.9, but to illustrate the foregoing ideas and the relationships between them more fully we require the Zigzag Theorem. The first proofs of Isbell (1966) and Philip (1974) were topological in flavour. The algebraic proofs of Howie (1976) and Storrer (1976) are based on work by Stenstrom (1971) on tensor products of monoids. Yet another proof, using the geometric method of semigroup diagrams, is due to David Jackson. This latter approach also employs HNN extensions of semigroups to solve the problem. Here we shall follow Jackson's lead in using what is essentially an HNN extension for our embedding (instead of the more intractable free product with amalgamation; see Howie 1976) to derive a short and direct proof of the Zigzag Theorem. However, our proof is from first principles, and no explicit reference is made to the concept of an HNN extension of a semigroup (Higgins 1990b).

Theorem 4.1.1 (the Zigzag Theorem) *Let U be a subsemigroup of S. Then $d \in \text{Dom}(U, S)$ if and only if $d \in U$ or there exists a series of factorizations of d as follows:*

$$d = u_0 y_1 = x_1 u_1 y_1 = x_1 u_2 y_2 = x_2 u_3 y_2 = \ldots = x_m u_{2m-1} y_m = x_m u_{2m},$$

where $u_i \in U$, $x_i, y_i \in S$, $u_0 = x_1 u_1$, $u_{2i-1} y_i = u_{2i} y_{i+1}$, $x_i u_{2i} = x_{i+1} u_{2i+1}$ $(1 \leq i \leq m-1)$ and $u_{2m-1} y_m = u_{2m}$.

Such equations are known as a *zigzag in S over U with value d, length m, and spine $u_0, u_1, \ldots, u_{2m}$* (in that order).

At first sight the theorem may seem to give a complicated list of conditions that would be difficult to apply in practice. It is more appealing (and helps to explain the term 'zigzag') if the zigzag factorizations are listed one under the other in the order given:

$$
\begin{aligned}
d = \; & u_0 y_1 \\
& \overbrace{x_1 u_1}\, y_1 \\
& x_1\, \overbrace{u_2 y_2} \\
& \overbrace{x_2 u_3}\, y_2 \\
& \vdots
\end{aligned}
$$

Zigzags and their applications | 145

$$x_i u_{2i-1} y_i$$
$$\overbrace{x_i u_{2i} y_{i+1}}$$
$$\overbrace{x_{i+1} u_{2i+1} y_{i+1}}$$
$$\vdots$$
$$\overbrace{x_m u_{2m-1} y_m}$$
$$\overbrace{x_m u_{2m}},$$

where the braces indicate the zigzag equations, one of which is obtained from each pair of consecutive lines by cancelling the common term, x_i or y_i, as the case may be; the pattern of these equalities then forms a zigzag path down the list of equations. It is of course possible, and sometimes convenient, to regard the list of equations in reverse order, so that our first equation is written $d = x_1 u_0$, while the last factorization of d is written $u_{2m} y_m$. To familiarize himself with such zigzag equations, the reader might pause to write down appropriate (length one) zigzags for the elements of $S \backslash U$ from Exercise 1.1.9. Other examples are given in Exercise 4.1.1.

Proof The reverse implication follows from a straightforward manipulation of the zigzag. Suppose that d is the value of the length m zigzag Z, as given in the statement of the theorem. Let α, β be two homomorphisms from S to a semigroup T such that $\alpha|U = \beta|U$ and extend α and β to homomorphisms from S^1 to T^1 by defining $1\alpha = 1 = 1\beta$ if $1 \notin S$. Taking $x_0 = y_{m+1} = 1$, we show inductively that $d\alpha = (x_i u_{2i}) \beta y_{i+1} \alpha$ for $i = 0, 1, \ldots, m$. In particular, this gives the required $d\alpha = (x_m u_{2m})\beta = d\beta$. Now.

$$d\alpha = (u_0 y_1)\alpha = u_0 \alpha y_1 \alpha = u_0 \beta y_1 \alpha = (x_0 u_0)\beta y_1 \alpha.$$

Suppose inductively that $d\alpha = (x_i u_{2i})\beta y_{i+1} \alpha$ $(0 \leq i < m)$. Then using the zigzag equations we obtain

$$d\alpha = (x_{i+1} u_{2i+1})\beta y_{i+1}\alpha = x_{i+1}\beta u_{2i+1}\beta y_{i+1}\alpha = x_{i+1}\beta u_{2i+1}\alpha y_{i+1}\alpha$$
$$= x_{i+1}\beta (u_{2i+1} y_{i+1})\alpha = x_{i+1}\beta (u_{2i+2} y_{i+2})\alpha = x_{i+1}\beta u_{2i+2}\alpha y_{i+2}\alpha$$
$$= x_{i+1}\beta u_{2i+2}\beta y_{i+2}\alpha = (x_{i+1} u_{2i+2})\beta y_{i+2}\alpha,$$

thus completing the proof in this direction.

Conversely, suppose that $d \in \text{Dom}(U, S) \backslash U$. Form a semigroup H by adjoining a new element t to S subject to the relations $t^2 = 1$, $tu = ut$, and $tut = u$ for all $u \in U$ (the latter relation is a consequence of the others, but it is convenient in the word argument to follow to consider it an elementary relation). Consider the homomorphisms $\varphi_1, \varphi_2 \colon S \to H$ whereby $s\varphi_1 = s$ and $s\varphi_2 = tst$ (in fact both these maps are embeddings, but we do not use this). Clearly, $\varphi_1 | U = \varphi_2 | U$ so that $tdt = d$, or what is the same, $td = dt$ in H. We prove that this latter equation implies that d is the value of some zigzag in S over U.

Since $td = dt$, there is a sequence of transitions of minimal length $I \colon td \to \ldots \to dt$, where each transition $pwq \to pw'q (p, w, w', q \in H)$ is either a *t-transition*, i.e. involves a relation in which t occurs, or is a *refactorization*, i.e. $w = w'$ in S. We claim that no transition in I involves any of the relations $t^2 = 1$ or $tut = u (u \in U)$. Suppose to the contrary that I contains a transition $\alpha \colon pq \to pt^2q (p, q \in H)$. Clearly α is not the final transition of I, so consider the next transition $\beta \colon pt^2q \to$. The right-hand side of β has one of the forms (i) pq; (ii) $p't^2q$; (iii) pt^2q'; (iv) the product p has the form $p'u$ or $p'tu$ ($u \in U$); or (v) a similar remark applies to q. In case (iv) β could then have the form $\beta \colon p'ut^2q \to p'tutq$ or $\beta \colon p'tut^2q \to p'utq$, as the case may be. But then the pair α, β could be replaced by the single transition $p'uq \to p'tutq$ or $p'tuq \to p'utq$, as the case may be. Dual remarks apply to case (v). Therefore cases (i), (iv), and (v) contradict our minimum length assumption on I. In cases (ii) and (iii), however, the two transitions α and β can be performed in the opposite order without changing the net effect. It follows that all transitions of the form $pq \to pt^2q$ can be taken to appear together at the end of I, and thus there are none.

Next suppose that α has the form $puq \to ptutq$ and once again consider the next transition β. If p has the form $p'v$, $p't$, or $p'tv$ ($v \in U$) then β could have the form $p'vtutq \to p'tvutq$, $p'ttutq \to p'utq$, or $p'tvtutq \to p'vutq$ respectively. In each case the pair α, β could be replaced by the single transitions: $p'vuq \to p'tvutq$, $p'tuq \to p'utq$, or $p'tvuq \to p'vutq$ respectively. Again, dual remarks apply if q has one of the corresponding forms vq', tq', or vtq'. Another possibility for β is $ptutq \to put^2q$ or $ptutq \to pt^2uq$, but here again α and β could be replaced by just one transition. Again, for α to cancel β would contradict the minimality of the length of I, leaving as the remaining possibilities that β involves only p, or only q, or has the form $ptutq \to pt^2ut^2q$. Arguing similarly to above, we conclude that all the transitions of the form $puq \to ptutq$ can be taken to occur together at the end of I, and so there are none.

Call a t-transition of the form $putq \to ptuq$ [$ptuq \to putq$] a *left* [*right*] *transition*, so that our sequence I consists entirely of refactorizations and of left and right transitions, with exactly one occurrence of the letter t in each word of I. Suppose that ptq is a product occurring in I, and that the next t-transition in the sequence is a left transition. We may assume that this transition occurs immediately or is preceded by just one refactorization of the form $ptq \to p'utq$, for it is clear that any refactorization of p can be performed in one step, while any refactorization of q can be postponed until after the left transition. Next suppose that I contains two left transitions with no intervening right transition, which we may assume have the form

$$putq \to ptuq \to p'vtuq \to p'tvuq, \quad \text{or simply} \quad p'vutq \to p'vtuq \to p'tvuq$$

($u, v \in U$). In the latter case the pair of transitions can be replaced by a single left transition, while the three transitions of the first case can be replaced by two:

$$putq \to p'vutq \to p'tvuq.$$

Coupling all this with the dual arguments for right transitions allows us to conclude that I consists of alternate left and right transitions, separated by a single refactorization; furthermore, both the first and last t-transitions are right. The sequence I therefore implies equalities in H of the form:

$$td = tu_0 y_1 = u_0 t y_1 = x_1 u_1 t y_1 = x_1 t u_1 y_1 = x_1 t u_2 y_2 = x_1 u_2 t y_2$$
$$= x_2 u_3 t y_2 = \ldots = \ldots = x_{m-1} u_{2m-2} t y_m = x_m u_{2m-1} t y_m$$
$$= x_m t u_{2m-1} y_m = x_m t u_{2m} = x_m u_{2m} t = dt,$$

for some $m \geq 1$, $u_i \in U$ ($0 \leq i \leq 2m$), $x_i, y_i \in S^1$, and $u_0 = x_1 u_1$, $u_{2i-1} y_i = u_{2i} y_{i+1}$, $x_i u_{2i} = x_{i+1} u_{2i+1}$ ($1 \leq i \leq m-1$), and $u_{2m-1} y_m = u_{2m}$. In fact, $x_i, y_i \in S$, for if $x_i = 1$ then in S we have

$$d = u_0 y_1 = x_1 u_1 y_1 = x_1 u_2 y_2 = \ldots = u_{2i} y_{i+1} = \ldots = x_m u_{2m};$$

and so that I could be shortened by beginning with $td \to tu_{2i} y_{i+1}$, with a dual remark applying if some $y_i = 1$. In other words, d is the value of the zigzag defined by deleting all instances of the letter t in the above equations, thus completing the proof. ∎

Remark To see that φ_1 and φ_2 are monos, consider H/ρ, where ρ is the congruence generated by the relation $t = 1$; H/ρ can be identified with S^1 and $\varphi_i \rho^\natural$ is essentially the inclusion of S into S^1 ($i = 1, 2$). Hence $\varphi_i \rho^\natural$ is injective and so is φ_i. Note also that our proof actually shows that if $td = rt$ ($d, r \in S$) then $d \in \text{Dom}(U, S)$, whence $d = tdt = r$; thus it follows that $S\varphi_1 \cap S\varphi_2 = \text{Dom}(U, S)$.

The Zigzag Theorem is a tool that will allow us to solve many problems concerning epimorphisms for various interesting classes of semigroup. For example, we shall shortly prove that all left simple semigroups are absolutely closed. The proof relies on the general observation that two zigzags, Z and Z' in S over U, with the same spine, are *equivalent* in that they share a common value. For the moment call their respective values d and d'; let Z be of the form given in the Zigzag Theorem and Z' be defined by $d' = u_0 z_1 = w_1 u_1 z_1 = w_1 u_2 z_2 = \ldots = w_m u_{2m-1} z_m = w_m u_{2m}$. We obtain

$$d = u_0 y_1 = w_1 u_1 y_1 = w_1 u_2 y_2 = \ldots = w_m u_{2m-1} y_m = w_m u_{2m} = d',$$

as claimed.

A zigzag in S over U is called *left [right] inner* if $x_1, x_2, \ldots, x_m \in U$ [$y_1, y_2, \ldots, y_m \in U$]; the value of any such zigzag is of course $x_m u_{2m} \in U$ [$u_0 y_1 \in U$]. We say that U is *left [right] inner in S* if every zigzag in S over U is equivalent to a left [right] inner zigzag, and U is *left [right] inner* if this is so for every containing semigroup S (the term *right-isolated* for a right-inner semigroup is also in use).

Henceforth the phrase 'let Z be a zigzag in S over U' shall be taken to mean that Z has the general form given in the statement of the Zigzag Theorem.

Theorem 4.1.2 (Howie and Isbell 1967) *A left [right] simple semigroup U is left [right] inner. In particular, left [right] simple semigroups are absolutely closed.*

Proof We shall deal with the case in which U is left simple, so let Z be a zigzag in S over U. We require an equivalent left inner zigzag in S over U: $u_0 y_1 = v_1 u_1 y_1 = v_1 u_2 y_2 = \ldots = v_m u_{2m}$, with $v_1, v_2, \ldots, v_m \in U$. The left simplicity of U means precisely that every equation of the form $b = xa$ ($a, b \in U$) has a solution in U. In particular, there exists $v_1 \in U$ such that $u_0 = v_1 u_1$; next we find $v_2 \in U$ such that $v_1 u_2 = v_2 u_3$, and so on, producing the required list $v_1, v_2, \ldots, v_m$, where v_i is a solution to the equation $v_{i-1} u_{2i-2} = v_i u_{2i-1}$ ($2 \leq i \leq m$).

The case in which U is right simple is dealt with dually; in this instance the zigzag equations are more conveniently written in reverse order. ∎

The fact that $\mathcal{T}_X$ is absolutely closed was first correctly proved by Scheiblich and Moore (1973). Shorter proofs of this fact have been found, however, by Shoji (1980), and by Hall (1982a). We include the latter argument here.

Denote $\mathcal{T}_X$ by U, and suppose that U is contained in S. For each x in X we denote by $[x]$ the constant function on X with range $\{x\}$, and for each $s \in S$ we define $\varphi(s) \in U$ as follows: for all $x \in X$,

$$x\varphi(s) = \begin{cases} \operatorname{ran}([x]s) & \text{if } [x]s \in U, \\ p & \text{otherwise,} \end{cases}$$

where p is a fixed member of X.

Note that if $[x]s \in U$ then $[x]s$ is a constant map on X since $[x]s = [x][x]s$. The map $\varphi: S \to U$ fixes U pointwise but is not a morphism (if it were, we could deduce immediately that U was absolutely closed by comparing the action of φ (considered as a mapping into S) to that of the identity mapping on S). However, we do have the following.

Lemma 4.1.3 *For any $u, v \in U = \mathcal{T}_X$, and $s, t \in S$:*

(i) $u = vt$ *implies* $u = v\varphi(t)$;
(ii) $us = vt$ *implies* $u\varphi(s) = v\varphi(t)$.

Proof. We prove statement (ii): statement (i) then follows as a special case, for we may assume without loss of generality that S is a monoid.

From $us = vt$ we have, for any $x \in X$, $[x]us = [x]vt$, whence $[xu]s = [xv]t$. Thus

$$x(u\varphi(s)) = (xu)\varphi(s) = \begin{cases} \operatorname{ran}[xu]s & \text{if } [xu]s \in U, \\ p & \text{otherwise,} \end{cases}$$

$$= \begin{cases} \operatorname{ran}[xv]t & \text{if } [xv]t \in U, \\ p & \text{otherwise,} \end{cases}$$

$$= (xv)\varphi(t) = x(v\varphi(t)),$$

and so $u\varphi(s) = v\varphi(t)$, as required. ∎

Corollary 4.1.4 *The full transformation semigroup $\mathcal{T}_X$ is right inner, and in particular it is absolutely closed.*

Remark It follows of course from this corollary that any semigroup may be embedded in an absolutely closed semigroup, and that the embedding can preserve finiteness. It can also be deduced that the partial transformation semigroup on X is absolutely closed (see Exercise 4.1.10).

Inverse semigroups are also right inner (and equally left inner); a proof of this is in Howie (1976). We shall prove that inverse semigroups are absolutely closed in another way, using Proposition 4.1.6 below, a result the primary purpose of which is to prove that generalized inverse semigroups are saturated.

Lemma 4.1.5 *Let U be a generalized inverse semigroup. Let $a \in U$, $a' \in V(a)$, $e \in E(U)$, and $\varphi: U \to S$ a homomorphism. Then:*

(i) *aea' and $a'ea$ are idempotents;*
(ii) *$U\varphi$ is a generalized inverse semigroup.*

Proof The statement of (i) is in fact true for an orthodox semigroup and was proved in Theorem 1.1.9, while (ii) is an easy consequence of Lallement's Lemma (1.1.7). ∎

In the proof of the next result, a', u', etc. denote arbitrary inverses of a, u, etc.

Proposition 4.1.6 *Let U be a generalized inverse subsemigroup of a semigroup S. For all $s, t \in S$ and for all $e, f \in E(U)$ suppose that $seft = sfet$. Then U is closed in S.*

150 | Techniques of semigroup theory

Proof Suppose that $d \in \text{Dom}(U, S) \setminus U$. Then there is a zigzag Z in S over U with value d; suppose that the length of Z is one, so that Z has the form

$$d = u_0 y$$
$$= x u_1 y$$
$$= x u_2.$$

Then
$$d = x u_1 y = (x u_1) u_1'(u_1 y) = u_0 u_1' u_2 \in U.$$

Therefore to complete the proof it is sufficient to show that any zigzag in S over U with value d and length $m > 1$ can be replaced by a zigzag of length $m - 1$ with the same value. Hence suppose now that Z has length m.

Now
$$d = x_1 u_1 y_1 = (x_1 u_1) u_1'(u_1 y_1) = u_0 u_1' u_2 y_2.$$

Also
$$(x_2 u_3) u_2' u_1 u_1' u_2 = x_1 (u_2 u_2')(u_1 u_1') u_2 = x_1(u_1 u_1')(u_2 u_2' u_2).$$

Therefore
$$x_2 u_3 u_2' u_1 u_1' u_2 = (x_1 u_1) u_1' u_2 = u_0 u_1' u_2.$$

This calculation justifies the first two lines of the following zigzag of length $m - 1$:

$$d = u_0 u_1' u_2 y_2$$
$$= x_2 (u_3 u_2' u_1 u_1' u_2) y_2$$
$$= x_2 (u_3 u_2' u_1 u_1' u_2 u_3' u_4) y_3$$
$$\vdots$$
$$= x_{m-1}(u_{2m-3} u_{2m-4}' \ldots u_1 u_1' \ldots u_{2m-3}' u_{2m-2}) y_m$$
$$= x_m(u_{2m-1} u_{2m-2}' \ldots u_1 u_1' \ldots u_{2m-3}' u_{2m-2}) y_m$$
$$= x_m(u_{2m-1} u_{2m-2}' \ldots u_1 u_1' \ldots u_{2m-1}' u_{2m}).$$

The argument at the ith stage is as follows. Given

$$d = x_i(u_{2i-1} u_{2i-2}' \ldots u_1 u_1' \ldots u_{2i-2}) y_i$$

then we also have

$$u_{2i-1} u_{2i-2}' \ldots u_1 u_1' \ldots u_{2i-2} y_i =$$
$$u_{2i-1}[u_{2i-1}' u_{2i-1}] [u_{2i-2}' \ldots u_1 u_1' \ldots u_{2i-2}] y_i.$$

The bracketed terms are idempotents, the latter from Lemma 4.1.5(i) and induction. Since idempotents commute within products this equals

$$u_{2i-1} u_{2i-2}' \ldots u_1 u_1' \ldots u_{2i-2} u_{2i-1}' u_{2i-1} y_i$$
$$= u_{2i-1} u_{2i-2}' \ldots u_1 u_1' \ldots u_{2i-1}' u_{2i} y_{i+1}.$$

Therefore, $d = x_i(u_{2i-1}u'_{2i-2} \cdots u_1u'_1 \cdots u_{2i})y_{i+1}$, which is the next line of the zigzag. The argument when $i = m$ is the same except that y_{m+1} is absent. An argument for two consecutive lines involving the same y_i is dual to that above, thus completing the proof. ∎

Corollary 4.1.7 *Inverse semigroups are absolutely closed.*

Theorem 4.1.8 (Higgins 1981) *A generalized inverse semigroup U is saturated and any epimorphism from U is onto.*

Proof Suppose that U is contained in S and that $\text{Dom}(U, S) = S$. Let $s, t \in S$. If $s \in U$ then $s = s(s's)$, with $s's \in U$. If $s \in S \setminus U$ then by The Zigzag Theorem there exists $u \in U$, $s_1 \in S$ such that $s = s_1 u$. Using a similar argument for t we conclude that there exist $u, v \in U$ such that $s = s_1 u$, $t = vt_1$ for some $s_1, t_1 \in S$. Then for $e, f \in E(U)$ we obtain:

$$seft = (s_1 u)ef(vt_1) = s_1 u(u'uefvv')vt_1 = s_1 u(u'ufevv')vt_1$$
$$= (s_1 u)fe(vt_1) = sfet.$$

By Proposition 4.1.6 we deduce that U is closed in S, whence it follows that $U = S$, and so we conclude that U is saturated. The second part of the statement follows from Lemma 4.1.5(ii). ∎

Another corollary of Proposition 4.1.6 is that any subband of a normal band is closed, a fact proved by zigzag manipulation by Scheiblich (1976), but a stronger result, that the variety of normal bands has the so-called strong amalgamation property (defined in Section 4.1.2), was established by Imaoka (1976). Other results along these lines are found in Scheiblich and Hsieh (1982). Generalized inverse semigroups are not in general absolutely closed; indeed, the 2×2 rectangular band lacks this property (Howie and Isbell 1967; see also Example 13 of Higgins 1986b).

A wide class of saturated semigroups was identified by Hall and Jones (1983) in the class of all semisimple semigroups with only finitely many $\mathscr{J}$-classes: in particular, this includes all finite regular semigroups and completely [0-]simple semigroups. In Higgins (1986a) it was shown that the condition on the $\mathscr{J}$-classes can be relaxed to that of having no infinite chains.

Before we prove this theorem we make a couple of technical observations concerning zigzags. If Z is a zigzag in S over U with value d of minimum length m, then $x_i, y_i \in S \setminus U$ for all $i = 1, 2, \ldots, m$, for if some x_i, say, were in U then we would have a zigzag in S over U with value d and length $m - i$, beginning with $d = u_{2i}y_{i+1}$. Next suppose that U is properly epimorphically embedded in S and that two successive lines of a zigzag Z in S over U are $x_i u_{2i} y_{i+1} = x_{i+1} u_{2i+1} y_{i+1}$ ($1 \leq i \leq m - 1$). Then since $y_{i+1} \in S \setminus U$ we have, by the Zigzag Theorem, $y_{i+1} = ay'_{i+1}$ for some $a \in U$, $y'_{i+1} \in S \setminus U$. We may then construct a modified zigzag Z', with value d with the two given lines replaced by

$$x_i u_{2i} a y'_{i+1} = x_{i+1} u_{2i+1} a y'_{i+1}$$

because the necessary equalities are provided by

$$u_{2i-1} y_i = u_{2i} a y'_{i+1}, \quad x_i u_{2i} a = x_{i+1} u_{2i+1} a, \quad \text{and}$$

$$u_{2i+1} a y'_{i+1} = u_{2i+2} y_{i+2}.$$

A similar remark applies to the two initial lines of the zigzag: $u_0 y_1 = x_1 u_1 y_1$. We call the process of passing from Z to Z' 'expansion of Z at y_{i+1} via the factorization $y_{i+1} = a y'_{i+1}$'. Of course, we can perform the dual operation of expansion at some x_i. The relevant section of the zigzags Z and Z' are given below:

$$\begin{array}{cc}
\vdots & \vdots \\
x_i u_{2i-1} y_i & x_i u_{2i-1} y_i \\
x_i u_{2i} y_{i+1} & x_i u_{2i} a y'_{i+1} \\
x_{i+1} u_{2i+1} y_{i+1} & x_{i+1} u_{2i+1} a y'_{i+1} \\
x_{i+1} u_{2i+2} y_{i+2} & x_{i+1} u_{2i+2} y_{i+2} \\
\vdots & \vdots
\end{array}$$

Theorem 4.1.9 *A completely semisimple semigroup U is saturated if it has no infinite chain of $\mathscr{J}$-classes.*

Proof Throughout the proof, Green's relations are understood to be the relations on U. Suppose that U is a completely semisimple semigroup with no infinite chain of $\mathscr{J}$-classes properly epimorphically embedded in S. Take $d \in S \setminus U$. Let F_d be the collection of all $\mathscr{J}$-classes J, such that some $u \in J$ is the first spine member of some zigzag of minimum length in S over U with value d. Denote the collection of all minimal members of F_d by $\bar{F}_d$ and put

$$J_R = \bigcup_{d \in S \setminus U} \bar{F}_d.$$

Replacing the word 'first' by 'last' in the definition of F_d gives us the dual collection of $\mathscr{J}$-classes which we denote by G_d, and similarly let $\bar{G}_d$ denote the collection of all minimal members of G_d and put

$$J_L = \bigcup_{d \in S \setminus U} \bar{G}_d.$$

Take J to be a fixed maximal member of $J_R \cup J_L$; without loss we assume that $J \in J_R$ so there exists $d \in S \setminus U$ such that $d = u_0 y_1$, say, is the first line of a zigzag Z of minimum length m in S over U with value d and $u_0 \in J$. The minimality condition on J guarantees that the $\mathscr{J}$-class of the first spine member of Z is invariant under any expansion. We may also assume that the $\mathcal{J}$-class corresponding to each particular spine member is invariant under expansion at x_i or y_i ($1 \leq i \leq m$), because we can expand Z at each x_i and y_i until the $\mathcal{J}$-class of each spine member is fixed under any further expansion, which must occur as U satisfies the descending chain condition on $\mathcal{J}$-classes. We shall assume that this process has been carried out.

Now $u_0 = x_1 u_1$ implies that $u_0 = u_0 u_1' u_1$ ($u_1' \in V(u_1)$) so that $J \leq J_{u_1}$. Expand Z at x_1 by $x_1 = x_1' a$, say, with $J_a \in J_L$. Then

$$J \leq J_{u_1} = J_{au_1} \leq J_a,$$

where the equality is guaranteed by invariance of $\mathcal{J}$-class under expansion. By the maximality of J in $J_R \cup J_L$ we know that $J < J_a$ is impossible, thus yielding $J = J_a = J_{au_1}$. Since U is completely semisimple, each $\mathcal{J}$-class is a $\mathcal{D}$-class and each principal factor is completely [0-] simple, whence it follows that each $\mathcal{R}$-class is minimal within its own $\mathcal{J}$-class by Theorem 1.3.13. Now $R_{au_1} \leq R_a$ and $au_1 \mathcal{J} a$, so that $au_1 \mathcal{R} a$ and there exists t in U^1 such that $a = au_1 t$; but then

$$x_1 = x_1' a = x_1' au_1 t = x_1 u_1 t = u_0 t \in U,$$

a contradiction, thus completing the proof. ∎

Not all completely semisimple semigroups are saturated; the first example of a non-saturated regular semigroup is that of the band in Higgins (1983c). Trotter (1986) constructed a band with a properly epimorphically embedded subband, thus showing that, in general, epimorphisms are not necessarily surjective in the category of bands and band homomorphisms. Scheiblich (1976) proved that epis are onto in the category of all bands satisfying the ascending chain condition on $\mathcal{D}$-classes, and as a corollary (since finitely generated bands are finite) it follows that epis are also necessarily onto in the category of finite bands.

Other notable absolutely closed classes of semigroups are: finite monogenic semigroups (Howie and Isbell 1967); the full and partial transformation semigroups on a set X (see Hall 1982a); the semigroup of all endomorphisms of a vector space over a division ring (Isbell 1974); and self-injective semigroups (Shoji 1980). The semigroup varieties consisting entirely of absolutely closed semigroups were determined by the author (1983a), and are precisely those varieties consisting entirely of left groups, entirely of right groups, or entirely of semilattices of groups.

Other classes of saturated semigroups include commutative semigroups with the descending chain condition on principal ideals (Howie and Isbell 1967) finite semigroups the idempotent-generated ideal of which satisfies some permutation identity (Higgins 1984b), and nilpotent semigroups (Higgins 1984a). The determination of all saturated semigroup varieties is an open problem, although the question has been settled for commutative varieties (Higgins 1983b; Khan 1982) and for heterotypical varieties, which are those varieties not containing the variety of all semilattices (Higgins 1984a). For a survey on the status of this question, see Higgins (1985a); see also Khan (1983, 1985a).

On the other hand, there are examples of finite null semigroups, finite right normal bands, and finite rectangular bands which are not absolutely closed (Hall 1982a; Higgins 1983a; Howie and Isbell 1967; see also Scheiblich 1976). In general, the following types of semigroup are not saturated: commutative cancellative semigroups (just consider the inclusion $i: \mathbb{N}(+) \to \mathbb{Z}(+)$), commutative periodic semigroups (Higgins 1983b), bands (Higgins 1983c), and finite idempotent-generated semigroups. Some of these examples are in Exercises 1, 2, 5 and 8 of this section. The author's paper (1986b) studies the dominating subsemigroups of inverse subsemigroups, completely regular semigroups, together with $\mathcal{T}_X$, $\mathcal{PT}_X$, and $\mathcal{I}_X$, and features a useful description of the dense subsemigroups of semilattices of groups (Exercise 4.1.4) and of 0-rectangular bands. Other examples concerning dominions are found in Isbell (1968, 1969, 1974). The preservation of epimorphism-related properties of semigroups under the taking of morphic images, of ideals, and of adjunction of an identity, is studied in the author's paper (1985b) (Exercise 4.1.6), but here a number of basic questions remain open:

(1) Is the class of absolutely closed semigroups closed under the taking of
 (a) morphic images; or (b) ideals?
(2) Is the class of saturated semigroups closed under the taking of
 (a) direct products; (b) ideals; and (c) adjunction of an identity?

Since epimorphisms are akin to surmorphisms, they might be expected to have structure-preserving qualities, and indeed the epimorphic image of a commutative semigroup is commutative. We show a stronger result, the proof of which is curiously reminiscent of the technique used in our proof of the Zigzag Theorem.

Theorem 4.1.10 (Isbell 1966) *Let U be a commutative subsemigroup of S. Then $\mathrm{Dom}(U, S)$ is commutative.*

Proof Let Z be a zigzag for some $d \in \mathrm{Dom}(U, S)$ and $u \in U$. Then

$$ud = uu_0 y_1 = u_0 u y_1 = x_1 u_1 u y_1 = x_1 u u_1 y_1 = x_1 u u_2 y_2 = \ldots$$
$$\ldots = x_m u_{2m-1} u y_m$$
$$= x_m u u_{2m-1} y_m = x_m u u_{2m} = x_m u_{2m} u = du,$$

from which it follows that U is central in its dominion. It remains to show that any two members $d, t \in \mathrm{Dom}(U, S) \setminus U$ also commute. Since t commutes with all members of the spine of the zigzag for d, (by the previous argument), we can repeat that argument with u replaced by t to conclude that $td = dt$. ∎

N. M. Khan has generalized this result in two directions by showing that every permutation identity is preserved by epimorphisms (Khan 1985b), and that any equational class (variety) of commutative semigroups is closed under the taking of epimorphisms (Khan 1982). The former result involves an exceedingly difficult zigzag manipulation, but the latter result affords an elegant proof, which uses a compact vector notation to handle the arbitrary identities involved.

Theorem 4.1.11 (Khan 1982) *All subvarieties of the variety of commutative semigroups are closed under the taking of epimorphisms.*

Proof Suppose that U is a commutative semigroup properly epimorphically embedded in S (necessarily commutative by Theorem 4.1.10). We are required to show that any identity satisfied by U is also satisfied by S. Take any identity φ satisfied by U, say,

$$\varphi: u(x_1, x_2, \ldots, x_n) = v(x_1, x_2, \ldots, x_n),$$

where u and v are words in the variables $x_1, x_2, \ldots, x_n$. Take any n elements $d_1, d_2, \ldots, d_n \in S$. If $d \in U$, then there is a zigzag in S^1 over U with value d in $d = d1 = 1d1 = 1d$. Next let Z be a zigzag in S over U of length m, say. Then Z can be replaced by a zigzag Z' of length $m + 1$ with the same value defined by the equations

$$u_0 y_1 = x_1 u_1 y_1 = x_1 u_1 y_1 = x_1 u_1 y_1 = x_1 u_2 y_2 = \ldots = x_m u_{2m}.$$

The point of these two observations is to yield the fact that any given $d_1, d_2, \ldots, d_n \in S$ all have zigzags of some common length m in S^1 over U, say,

$$d_j = u_0^j y_1^j = x_1^j u_1^j y_1^j = x_1^j u_2^j y_2^j = \ldots = x_m^j u_{2m}^j;$$

where

$$u_0^j = x_1^j u_1^j, \quad u_{2i-1}^j y_i^j = u_{2i}^j y_i^j, \quad x_i^j u_{2i}^j = x_{i+1}^j u_{2i+1}^j \quad (1 \leq i \leq m-1)$$

and $u_{2m-1}^j y_m^j = u_{2m}^j$, where $u_i^j \in U$ ($j = 1, \ldots, n$, $i = 0, \ldots, 2m$) and x_i^j, $y_i^j \in S^1$ ($j = 1, \ldots, n$, $i = 1, \ldots, m$). (The indices here are of course superscripts, not powers.)

In the following we shall make free use of the commutativity of S^1 and shall use the following notation:

$$\tilde{x} = (x_1, x_2, \ldots, x_n);$$

in this notation, our identity φ is simply $u(\tilde{x}) = v(\tilde{x})$. Define

$$\tilde{d} = (d_1, d_2, \ldots, d_n), \quad \tilde{u}_i = (u_i^1, u_i^2, \ldots, u_i^n) \quad (0 \leq i \leq 2m),$$
$$\tilde{x}_i = (x_i^1, x_i^2, \ldots, x_i^n), \quad \tilde{y}_i = (y_i^1, y_i^2, \ldots, y_i^n) \quad (1 \leq i \leq m).$$

We wish to prove that $u(\tilde{d}) = v(\tilde{d})$. In the following lemma we denote the cartesian product of n copies of a semigroup S by S^n.

Lemma 4.1.12 *The element $\tilde{d} \in S^n$ is in $\mathrm{Dom}(U^n, (S^1)^n)$.*

Proof We have a zigzag with value $\tilde{d}$ of length m in $(S^1)^n$ over U as follows:
$$\tilde{d} = \tilde{u}_0 \tilde{y}_1, \qquad \tilde{u}_0 = \tilde{x}_1 \tilde{u}_1,$$
$$\tilde{u}_{2i-1}\tilde{y}_i = \tilde{u}_{2i}\tilde{y}_{i+1}, \quad \tilde{x}_i\tilde{u}_{2i} = \tilde{x}_{i+1}\tilde{u}_{2i+1} \qquad (1 \leq i \leq m-1),$$
$$\tilde{u}_{2m-1}\tilde{y}_m = \tilde{u}_{2m}, \qquad \tilde{x}_m\tilde{u}_{2m} = \tilde{d},$$
where $\tilde{x}_i, \tilde{y}_i \in (S^1)^n$ $(1 \leq i \leq m)$ and $\tilde{u}_i \in U^n$ $(0 \leq i \leq 2m)$. $\square$

Using the above zigzag we can now complete the proof of Khan's Theorem. We have
$$u(\tilde{d}) = u(\tilde{u}_0\tilde{y}_1) = u(\tilde{u}_0)u(\tilde{y}_1) = v(\tilde{u}_0)u(\tilde{y}_1) \qquad \text{(since } U \text{ satisfies } \varphi)$$
$$= v(\tilde{x}_1\tilde{u}_1)u(\tilde{y}_1) = v(\tilde{x}_1)v(\tilde{u}_1)u(\tilde{y}_1) = v(\tilde{x}_1)u(\tilde{u}_1)u(\tilde{y}_1)$$
$$= v(\tilde{x}_1)u(\tilde{u}_2\tilde{y}_1) = v(\tilde{x}_1)u(\tilde{u}_3\tilde{y}_2)$$
$$= \ldots = v(\tilde{x}_m)u(\tilde{u}_{2m-1}\tilde{y}_m) = v(\tilde{x}_m)u(\tilde{u}_{2m}) = v(\tilde{x}_m)v(\tilde{u}_{2m}) = v(d),$$
as required. ■

It was proved by N. M. Khan jointly with the author that commutativity can be replaced by another permutation identity π: $x_1 x_2 \ldots x_n = x_{1\pi} x_{2\pi} \ldots x_{n\pi}$, in the preceding theorem if and only if either $1\pi \neq 1$ or $n\pi \neq n$ (see Higgins 1984*b*). The status of the open problem of determining all such epimorphically closed varieties of semigroups is given in Higgins (1985*a*).

Exercises 4.1

1. Show that the following inclusions are epimorphisms;
    (i) $(\mathbb{N}, +) \to (\mathbb{Z}, +)$;
    (ii) $((0, 1], .) \to (\mathbb{R}^+, .)$;
    (iii) $(\mathbb{N}, .) \to (\mathbb{Q}^+, .)$.
2. (a) Find a properly epimorphically embedded subsemigroup of the five-element combinatorial Brandt semigroup (Exercise 1.3.4) $S = M^0[1; I, I, \Delta]$, where $|I| = 2$, 1 denotes the trivial group, and Δ denotes the 2×2 identity matrix.
   (b) Consider the semigroup S of (a), but replace the group by $\langle a \rangle$, an infinite cyclic group. Show that S is dominated by the finite subsemigroup $U = \{x, y, e, f, 0\}$, where $x = (a; 1, 2)$, $y = (1; 1, 2)$, $e = (1; 1, 1)$, and $f = (1; 2, 2)$.
   (c) In general, show that an inverse semigroup S is dominated by a subsemigroup U if U is full, i.e. $E(U) = E(S)$, and $\bar{U} = S$, where $\bar{U}$ is the inverse subsemigroup of S generated by U. Conversely, show that $\mathrm{Dom}(U, S) = S$ implies that $\bar{U} = S$, but not necessarily that U is full.

Finally, show that neither of the conditions U is full nor $\bar{U} = S$ is sufficient to imply that $\text{Dom}(U, S) = S$.

3. Let U be a subsemigroup of a completely regular semigroup S and let $\hat{U}$ denote the completely regular subsemigroup of S generated by U. Prove that:
   (i) $\text{Dom}(U, S) = \text{Dom}(\hat{U}, S)$ and is itself completely regular;
   (ii) a finite union of groups has no proper dense subsemigroup;
   (iii) a subsemigroup of a completely simple semigroup is dense if and only if $\hat{U} = S$.

4. (Continuation of Exercise 4.1.3) Let U be a subsemigroup of a semilattice of groups $S = \bigcup_{e \in E} H_e$, where $\{H_e\}_{e \in E}$ are the maximal subgroups of S indexed by $E = E(S)$. Show that:
   (i) $\bar{U}$ is the union of the subgroups generated by each of the $U \cap H_e$;
   (ii) U is dense in S if and only if each maximal subgroup H_e is generated as a group by its intersection with U.

5. (a) Show that the dominion of a countable semigroup is countable.
   (b) (Hall 1982a) Let D be the null semigroup $\{0, u_1, u_2, \ldots\}$, where 0 denotes the zero of D, and let $U = \{0, u_1, u_2\}$. Let S be the commutative semigroup $D \cup \langle a \rangle$, where $\langle a \rangle$ is an infinite cyclic semigroup, 0 is the zero of S, and $u_i a^j = u_{i+j}$ ($i, j = 1, 2, \ldots$).
   Show that (i) $\text{Dom}(U, S) = D$ and (ii) $\text{Dom}(U, D) = U$.

6. Let $S = \mathcal{T}_X$, where X is denumerable, let U be the Baer–Levi subsemigroup of $\mathcal{T}_X$ (Exercise 1.1.11), and let V be the subsemigroup of all injections of S. Identify U^1 with $U \cup \{1\}$, where 1 is the identity map on S. Show that although U is absolutely closed, $\text{Dom}(U^1, S) = V$.

7. Each of the transformation semigroups $\mathcal{T}_X, \mathcal{PT}_X$, and $\mathcal{I}_X$ have no proper dense subsemigroup if X is finite. (The converse is also true; Higgins 1986b.)

8. Let T_6 denote the full transformation semigroup on $\{1, 2, 3, 4, 5, 6\}$. Let S be the subsemigroup of T_6 generated by the maps $a_1 = (111444)$, $a_2 = (113446)$, $a_3 = (223556)$, $x = (333666)$, and $y = (323656)$. Let U be the subsemigroup of S generated by $\{a_1, a_2, a_3\}$. Then $|U| = 4$ and U is a right normal band (i.e. satisfies the identities $x = x^2$ and $xyz = yxz$), S is a regular subsemigroup of order 7, and U is not closed in S.

Remark For an example of fourteen-element band with a nine-element subband which is not closed, see Scheiblich (1976).

9. The following example shows that Theorem 4.1.10 cannot be generalized to weaker permutation identities that commutativity. Let F_X be the free semigroup on $X = \{x_1, x_2, \ldots\}$ and let $S = F_X/\rho$, where ρ is the congruence generated by the relation ρ_0 which consists of pairs

$(w_1 w_2 \ldots w_n, w_{1\pi} w_{2\pi} \ldots w_{n\pi})$, where π is any permutation on $n \geqslant 3$ symbols and w_i is any member of

$$U = \left\langle \bigcup_{n=0}^{\infty} \{x_{3n+1} x_{3n+2}, x_{3n+2}, x_{3n+2} x_{3n+3}\} \right\rangle.$$

Show that $x_{3n+1} x_{3n+2} x_{3n+3} \in \text{Dom}(U\rho^\natural, S)$ for all $n = 0, 1, 2, \ldots$ and that $\text{Dom}(U\rho^\natural, S)$ satisfies no permutation identity, although $U\rho^\natural$ satisfies all permutation identities but commutativity.

10. Use the fact that $\mathscr{PT}_X$ is isomorphic to the subsemigroup of $\mathscr{T}_{X \cup \{\infty\}}$ of all mappings which fix ∞ to modify the proof of Lemma 4.1.3 to show that $\mathscr{PT}_X$ is absolutely closed.

11. (a) Suppose that U is a semigroup with identity 1 and zero 0, with U embedded in S. Prove that 1 and 0 are the respective identity and zero of $\text{Dom}(U, S)$.

    (b) (Isbell 1974) Let S be the semigroup with generators and relations given by

    $$S = \langle a, x; a^3 = a^2, x^3 = 0, xax = x, a^2 xa^2 = a^2, a^2 x^2 a^2 = a^2 \rangle$$

    and U the subsemigroup of S generated by $\{a, ax, xa\}$. Prove that S is finite, that U is dense in S, but that $0 \notin U$.

    (c) By way of contrast, show that if $\text{Dom}(U, S)$ has a right identity e and that U is finite, then $e \in U$.

12. A semigroup U with zero 0 is called *right T-nilpotent* if for every sequence of elements $u_1, u_2, \ldots$ of U the product $u_1 u_2 \ldots u_n = 0$ for some positive integer n. Prove that every right T-nilpotent semigroup is saturated.

Remark The concept of left T-nilpotence is defined dually (compare with Exercise 1.2.20) and a semigroup is *T-nilpotent* if it is both right and left T-nilpotent. In general, any T-nilpotent ideal extension of a saturated semigroup is saturated, and any right T-nilpotent ideal extension of a saturated regular semigroup or a saturated finite semigroup is saturated (Higgins 1984a).

4.2 Zigzags and amalgams

The study of epimorphisms and zigzags is not as isolated a topic as it may at first appear, since it is equivalent to the study of so-called special amalgams. Indeed, the study of dominions provides a doorway to the study of general semigroup amalgams, a broad and fundamental area of research.

The connection can be glimpsed upon examination of our proof of the Zigzag Theorem. By the remark following the proof of the Zigzag Theorem we can regard φ_1 and φ_2 as embeddings which provide two copies of S, with a common subsemigroup U, embedded in H; their overlap in H is, however, not just U but $\text{Dom}(U, S)$. This affords an example of a weak embedding of the amalgam $[S, S'; U]$, where S' is the image of S under an isomorphism which fixes the points of U; the embedding would be called strong if $S\varphi_1 \cap S\varphi_2 = U$, that is if U were closed in S.

To explain this fully we introduce the appropriate definitions.

Definition 4.2.1 A *semigroup amalgam* $[S_i; U]_{i \in I}$ is an indexed family $\{S_i; i \in I\}$ of semigroups, the pairwise intersection of which is a common *core* subsemigroup U. A *weak embedding* of $\mathscr{A} = [S_i; U]_{i \in I}$ is a set of monomorphisms $\lambda_i: S_i \to W$ such that $\lambda_i | U = \lambda_j | U = \lambda$, say, for all $i, j \in I$. The embedding is *strong* if $S_i \lambda_i \cap S_j \lambda_j = U\lambda$ for all i, j.

If no ambiguity results, and if $|I| = 2$, we often denote an amalgam $\mathscr{A}$ by $[S, T; U]$. This amalgam is a partial semigroup: some products are meaningful, and $x(yz) = (xy)z$ provided that both sides are defined. The central question concerning amalgams is whether or not $[S, T; U]$ can be strongly or at least weakly embedded in some semigroup W. In general, even weak embedding is sometimes impossible (Exercise 4.2.1).

Definition 4.2.2 Let $\mathscr{C}$ be a class of semigroups. An amalgam $\mathscr{A} = [S_i; U]_{i \in I}$ is in $\mathscr{C}$ if each S_i and U is a member of $\mathscr{C}$; it is *weakly [strongly] embeddable* in $\mathscr{C}$ if $\mathscr{A}$ can be weakly [strongly] embedded in some semigroup in $\mathscr{C}$. We say that $\mathscr{C}$ has the *weak [strong] amalgamation property* if every amalgam in $\mathscr{C}$ is weakly [strongly] embeddable in C.

The properties introduced in Definition 4.2.2 are written as WAP and SAP respectively, and we sometimes simply say that an amalgam can be *embedded*, meaning that it can be strongly embedded. When determining whether a class $\mathscr{C}$ has the WAP or SAP it is sufficient to consider amalgams involving two semigroups S, T with a common core U, provided that $\mathscr{C}$ is closed under the taking of unions of directed subfamilies of $\mathscr{C}$ (see, for example, Petrich 1984). The proof of this is a straightforward argument in transfinite induction. We shall assume any class $\mathscr{C}$ of semigroups under consideration to be closed under the taking of isomorphisms.

Definition 4.2.3 A *special amalgam* $\mathscr{A} = [S, S'; U]$ is an amalgam together with an isomorphism $': S \to S'$ which fixes U pointwise. A class of semigroups $\mathscr{C}$ has the *special amalgamation property*, SpAP, if every special amalgam $[S, S'; U]$, the members of which come from $\mathscr{C}$, can be embedded in a member of $\mathscr{C}$.

Remark A special amalgam $[S, S'; U]$ can always be weakly embedded in S.

160 | Techniques of semigroup theory

Dominions can be characterized in terms of special amalgams.

Theorem 4.2.4 *Let U be a subsemigroup of S. Then $d \in \mathrm{Dom}(U, S)$ if and only if for every weak embedding of the special amalgam $[S, S'; U]$ into some semigroup W by monomorphisms $\lambda: S \to W$, $\lambda': S' \to W$, we have $d\lambda = d'\lambda'$.*

Proof Suppose that $[S, S'; U]$ is weakly embedded in W by monos λ and λ'. Then $\lambda|U = \lambda'|U$, whence $\lambda|U = {}'\circ \lambda'|U$, since the amalgam is special, and thus λ and ${}'\circ \lambda'$ are homomorphisms from S to W which agree on U. Hence if $d \in \mathrm{Dom}(U, S)$ then $d\lambda = d'\lambda'$.

Conversely, consider a special amalgam $[S, S'; U]$, where without loss we suppose that $S'\setminus U$ is disjoint from our semigroup H introduced in the proof of the Zigzag Theorem. Consider the weak embedding of $[S, S'; U]$ into H whereby $\lambda: S \to H$ is the natural inclusion while $\lambda': S' \to H$ is defined by $s'\lambda' = tst$ $(s \in S)$. Now supposing that $d\lambda = d'\lambda'$ for some d in S we obtain $d = tdt$, whence $d \in \mathrm{Dom}(U, S)$. ∎

Since the core of an amalgam is the unifying object in its structure, it is natural to ask if there are embeddability implications due to the character of the core of the amalgam. With this in mind we say that a semigroup U is a *weak [strong] amalgamation base* if every amalgam with core U is weakly [strongly] embeddable in a semigroup. Analogously, we can say that U is a *special amalgamation base* if every special amalgam with U as core is embeddable in a semigroup. From Theorem 4.2.4 we see that a semigroup U is closed in S if and only if the special amalgam $[S, S'; U]$ is embeddable, whereupon it follows that U is a special amalgamation base if and only if U is absolutely closed.

Our proof of Theorem 4.2.4 shows that a special amalgam $[S, S'; U]$ is embeddable in a semigroup only if it is embeddable in the semigroup H used in the proof. In general, the amalgam $\mathscr{A} = [S, T; U]$ can be embedded in some semigroup only if $\mathscr{A}$ can be embedded in the so-called free product with amalgamation of S and T over U, denoted by $S *_U T$. This is the most free semigroup that can be generated by $S \cup T$ which respects all the relations implicit in the amalgam $[S, T; U]$; any pair of members of $[S, T; U]$ which are identified in $S *_U T$ must also be identified in any attempt to embed $[S, T; U]$ in any semigroup. Therefore $S *_U T$ can be viewed as the test object for the study of any collapse of the amalgam which must necessarily occur when S and T are mapped by monomorphisms into some semigroup. In defining $S *_U T$ let $(.)$ denote the multiplication operations of both S and of T, and let juxtaposition indicate the multiplication operation in the free semigroup $F(S \cup T)$.

Definition 4.2.5 $S *_U T = F(S \cup T)/\rho$, where ρ is the congruence generated by
$$\rho_0 = \{(s_1 \cdot s_2, s_1 s_2), (t_1 \cdot t_2, t_1 t_2): s_i \in S, t_i \in T, i = 1, 2\}.$$

Remark If $S \cap T = \emptyset$, Definition 4.2.5 can be read as defining the *free product*, $S * T$, of S and T. The existence of $S *_U T$ is theoretically useful, but free products with amalgamation can be intractable; for instance, grappling directly with $S *_U S'$ has yet to yield a proof of the Zigzag Theorem for semigroups. This approach has worked, however, in the category of commutative semigroups, where $S *_U S'$ affords a relatively simple description (Exercise 4.2.2). Indeed, the Zigzag Theorem for commutative semigroups was the result which suggested, and eventually led to, the corresponding result for semigroups. (Neither is a consequence of the other; conceivably there could exist commutative semigroups $U \subseteq S$ and d in $S \backslash U$ such that d is dominated by U with respect to all homomorphisms from S to any *commutative* semigroup T, yet d has distinct images under two homomorphisms that coincide on U and map to some (non-commutative) semigroup V, in which case we would have d an element of $\text{Dom}(U, S)$ in the category of commutative semigroups, yet there could be no zigzag with value d in S over U. This behaviour can occur, however, in some other semigroup categories: see Exercise 4.2.3.)

There is a basic connection between the weak, strong, and special amalgamation properties of a universal algebraic nature.

Lemma 4.2.6 *A class of semigroups $\mathscr{C}$ has the SAP if (and clearly only if) it has both the WAP and the SpAP.*

Proof Suppose that $\mathscr{A} = [S, T; U]$ is an amalgam in $\mathscr{C}$. Using the WAP we take a semigroup P of $\mathscr{C}$ such that $S, T \subseteq P$ and $U \subseteq S \cap T$. Consider a special amalgam $[P, P'; U]$ and use the SpAP to obtain a semigroup $V \in \mathscr{C}$ such that $P, P' \subseteq V$ and $P \cap P' = U$. We then have S in P and a copy T' of T in P' in such a way that $S \cap T' = U$. It follows from this that $\mathscr{C}$ has the SAP. ∎

The connection between zigzags and general amalgams can be realized through the representation extension property of Hall (1978).

Definition 4.2.7 If U is a subsemigroup of S then U has the *representation extension property in S* if for any set X and any representation $\rho: U \to \mathscr{T}_X$, there is a set Y, disjoint from X, and a representation $\alpha: S \to \mathscr{T}_{X \cup Y}$ such that $\alpha_u | X = \rho_u$ for all $u \in U$. If U has this property in every containing semigroup S we say that U has the *representation extension property* (REP).

Definition 4.2.8 For U a subsemigroup of S we call (S, U) a *strong [weak] amalgamation pair* if every amalgam $[S, T; U]$ is embeddable [weakly embeddable].

Theorem 4.2.9 *A necessary condition for (S, U) to be a weak amalgamation pair is that U has the representation extension property in S.*

Proof Take any representation $\rho: U \to \mathcal{T}_X$ where, without loss, we assume that U and X are disjoint. We extend the binary operation on U to one on $T = U \cup X$ by defining all members of X to act as right zeros on T and putting $xu = x\rho_u$ ($x \in X$, $u \in U$). Then T is a semigroup containing U and by hypothesis the amalgam $[S, T; U]$ can be weakly embedded in a semigroup W, say, by monomorphisms $\varphi: S \to W$, and $\psi: T \to W$ which coincide on U. Identifying S with $S\varphi$ and $X\psi$ with X under the injections φ and $\psi|X$ respectively, we need only put $Y = W \setminus X$ and choose for α the restriction of the right regular representation of W to S. ∎

Corollary 4.2.10 *If U is a weak amalgamation base then U has the representation extension property.*

We next show that all weak amalgamation bases are also special amalgamation bases by proving that the latter is a consequence of the REP.

Theorem 4.2.11 *If U has the representation property in S then U is closed in S.*

Proof Suppose to the contrary that U is not closed in S, so that there exists $d \in S \setminus U$ and a zigzag Z in S over U with value d, which for convenience on this occasion we take to have the form:

$$d = x_1 u_0 = x_1 u_1 y_1 = \ldots = x_i u_{2i-1} y_i = x_{i+1} u_{2i} y_i$$
$$= x_{i+1} u_{2i+1} y_{i+1} = \ldots = x_m u_{2m-1} y_m = u_{2m} y_m.$$

To construct our set X take for each $s \in S \setminus U$, two new elements s', s'' and for each $u \in U$ we take a single new element u', which we also denote by u''. Put $S' = \{s' : s \in S\}$, $U' = U'' = \{u' : u \in U\}$, $S'' = \{s'' : s \in S\}$ and $X = S' \cup S''$ (we assume that $S' \cap S'' = U'$, and that $S \to S'$, $s \mapsto s'$ and $S \to S''$, $s \mapsto s''$ are bijections). Define $\rho: U \to \mathcal{T}_X$ by $s'\rho_u = (su)'$, $s''\rho_u = (su)''$ for all $s \in S$ and $u \in U$ (easily seen to be a representation).

By hypothesis there is a set Y and a representation $\alpha: S \to \mathcal{T}_{X \cup Y}$ such that $\alpha_u | X = \rho_u$ for all $u \in U$. Writing the mappings ρ and α on the right of their arguments and using the zigzag equations we obtain:

$$d' = x_1' u_0' = x_1' u_0 \rho = x_1' u_0 \alpha = x_1'(u_1 y_1)\alpha = x_1' u_1 \alpha y_1 \alpha = x_1' u_1 \rho y_1 \alpha$$
$$= x_1' u_1' y_1 \alpha = x_2' u_2' y_1 \alpha = x_2' u_2 \rho y_1 \alpha = x_2' u_2 \alpha y_1 \alpha$$
$$= x_2'(u_2 y_1)\alpha = x_2'(u_3 y_2)\alpha = \ldots;$$

continuing in this way, the pattern of eight manipulations beginning with the first appearance of x_{i+1}' ($1 \leq i \leq m-1$) is,

Zigzags and their applications | 163

$$x'_{i+1}u'_{2i}y_i\alpha = x'_{i+1}u_{2i}\rho y_i\alpha = x'_{i+1}u_{2i}\alpha y_i\alpha = x'_{i+1}(u_{2i}y_i)\alpha$$
$$= x'_{i+1}(u_{2i+1}y_{i+1})\alpha = x'_{i+1}u_{2i+1}\alpha y_{i+1}\alpha = x'_{i+1}u_{2i+1}\rho y_{i+1}\alpha$$
$$= x'_{i+1}u'_{2i+1}y_{i+1}\alpha = x'_{i+2}u'_{2i+2}y_{i+1}\alpha \qquad \text{(taking } x_{m+1} = 1\text{)}.$$

In particular, if $i = m - 1$, this yields $d' = u'_{2m}y_m\alpha = u''_{2m}y_m\alpha = d''$ (by symmetry); however, this is a contradiction, thus completing the proof. ∎

The results 4.2.10 and 4.2.11 together imply that any weak amalgamation base U is a special amalgamation base; whereupon, by the argument in the proof of Lemma 4.2.6, we obtain the following surprising result.

Theorem 4.2.12 *Any weak amalgamation base is a strong amalgamation base.*

This contrasts with the variety of distributive lattices where each member is a weak amalgamation base, but not all are strong amalgamation bases (Gratzer and Lasker 1971).

All inverse semigroups are known to be amalgamation bases (Howie 1975). This is also proved in Hall (1978), where it is shown that any inverse semigroup U has the so-called strong representation property in any semigroup S, a condition which, as proved by Hall, implies that U is an amalgamation base. These and related results can be read in Petrich (1984). Two further surveys are Howie (1981b) and Higgins (1988c)

A characterization of the representation extension property in terms of the absence of certain types of zigzags is given in the above-cited paper of Hall. Specifically, three types of zigzag equations are introduced for an amalgam $[S, T; U]$. A *length n zigzag of type I over U from $s \in S$ to $t \in T$* with spine $u_1, v_1, \ldots, u_n, v_n, u_{n+1}$ is a sequence of factorizations:

$$s = u_1 s_1, \qquad t_n u_{n+1} = t,$$
$$u_1 = t_1 v_1, \qquad t_1 u_2 = t_2 v_2, \qquad t_2 u_3 = t_3 v_3, \qquad \ldots, \qquad t_{n-1} u_n = t_n v_n,$$
$$v_1 s_1 = u_2 s_2, \qquad v_2 s_2 = u_3 s_3, \qquad \ldots, \qquad v_{n-1} s_{n-1} = u_n s_n, \qquad v_n s_n = u_{n+1},$$

where

$$s_1, s_2, \ldots, s_n \in S, \qquad u_1, v_1, \ldots, u_n, v_n, u_{n+1} \in U, \qquad t_1, t_2, \ldots, t_n \in T.$$

In the case in which $S = T$ these equations match those of an ordinary zigzag as described above.

A *zigzag of length n of type II(a) over U from $t \in T$ to $t' \in T$* with spine $u_1, v_1, \ldots, u_n, v_n$ is a sequence of factorizations:

$$t = t_1 u_1, \qquad t_n v_n = t',$$
$$u_1 = v_1 s_1, \qquad u_2 s_1 = v_2 s_2, \qquad u_3 s_2 = v_3 s_3, \qquad \ldots, \qquad u_n s_{n-1} = v_n,$$
$$t_1 v_1 = t_2 u_2, \qquad t_2 v_2 = t_3 u_3, \qquad \ldots, \qquad t_{n-1} v_{n-1} = t_n u_n,$$

164 | Techniques of semigroup theory

where

$$s_1, \ldots, s_{n-1} \in S, \qquad t_1, \ldots, t_n \in T, \qquad u_1, v_1, \ldots, u_n, v_n \in U.$$

The left–right dual of a type II(a) zigzag is called a *type II(b) zigzag*. Clearly, if the amalgam [S, T; U] is strongly embeddable there is no zigzag of type I from any $s \in S \backslash U$ to any $t \in T \backslash U$ over U, and if [S, T; U] is weakly embeddable there is no zigzag of type II(a) or type II(b) from any $t \in T$ to a different member $t' \in T$.

Hall (1978) then proves the following theorem.

Theorem 4.2.13 *Let U be any subsemigroup of any semigroup S. Then U has the representation extension property in S if and only if for every semigroup T containing U as a subsemigroup, there is no zigzag of type II(a) over U from an element $t \in T$ to a distinct element $t' \in T$. Hence U has the representation extension property if and only if there are no zigzags of type II(a) over U between two distinct elements.*

A new proof that inverse semigroups are amalgamation bases (we may now drop the qualification of weak or strong) is due to Renshaw (1986a), and depends on the fact that inverse semigroups are 'absolutely flat', an idea of Bulman-Fleming and McDowell (1983, 1984) related to tensor products of monoids. The 'homological' approach of Renshaw (1986a, b) to amalgamation questions unifies the concepts of the above-mentioned authors, and many of the principal results on amalgamation appear as corollaries of his theorem that if a semigroup U has the so-called extension property then it is an amalgamation base. The fundamental idea of Renshaw, developed from the corresponding ideas in ring theory (Cohn 1959) is to describe the amalgamated free product $S *_U T$ as a direct limit of certain recursively defined tensor products factored by certain equivalence relations. The embedding of an amalgam $[T_i; U]$ into the amalgamated free product of $[S_i; U]$, where each T_i is a subsemigroup of S_i, is the topic of a pair of the forthcoming papers of Renshaw (1990a, b). Other recent relevant papers are those of Fleischer (1990) and Shoji (1988, 1990).

Exercises 4.2

1. (Kimura 1957) Let $U = \{u, v, 0\}$ be a three-element null semigroup and extend the multiplication of U to one of $S = U \cup \{s\}$ by defining $su = us = v$ and setting all other products equal to 0. Similarly, extend the multiplication of U to one of $T = U \cup \{t\}$ by defining $tv = vt = u$, with all other products equal to 0.

(i) the amalgam $\mathscr{A} = [S, T; U]$ cannot be weakly embedded in any semigroup;
(ii) although S is isomorphic to T, $\mathscr{A}$ is not a special amalgam.

2. (Howie and Isbell 1967) The Zigzag Theorem for the category of commutative semigroups:

   (i) The free commutative product, $S * T$, of two disjoint commutative semigroups can be regarded as the direct product $S^{(1)} \times T^{(1)}$ with the identity $(1, 1)$ removed ($S^{(1)}$ denotes S with an identity 1 adjoined, whether or not S is already a monoid).
   (ii) Show that if $d \in \text{Dom}(U, S)$ in the category of commutative semigroups then there is a sequence of transitions $(1, d) \to \ldots \to (d, 1)$ in $S^{(1)} \times S^{(1)}$, where a typical transition $(x, y) \to (z, t)$ is either

   an r-step, $(x, y) = (p, q)(u, 1)(r, s) = (p, q)(1, u)(r, s)$

   or

   an l-step, $(x, y) = (p, q)(1, u)(r, s) = (p, q)(u, 1)(r, s)$

   where $u \in U$ and $p, q, r, s \in S^{(1)}$.
   (iii) Conclude that $d \in \text{Dom}(U, S)$ in the category of commutative semigroups if and only if there is a zigzag in S over U with value d.

3. (Hall 1982a) A category of semigroups in which zigzags do not fully describe dominions: consider the category V of all semigroups satisfying the identity $x^2 y^2 = y^2 x^2$. Let $S = \{0, e, f, a, a^{-1}\}$ be the five-element combinatorial Brandt semigroup (Exercise 1.3.4) and $U = \{0, a\}$ a two-element null semigroup.

   (i) in the category of semigroups, U is closed in S;
   (ii) if $S, T \in V$ and $\alpha, \beta: S \to T$ are homomorphisms which agree on U then $\alpha = \beta$, so that the V-dominion of U in S is all of S;
   (iii) generally, if $S \in V$ is an inverse semigroup and $U \subseteq S$ generates S as an inverse semigroup then U is epimorphically embedded in S in the category V.

Remark Zigzags fully describe dominions in the categories of semigroups, and commutative semigroups, and the proof of the Zigzag Theorem in Howie (1976) adapts to show that the same is true in the category of finite semigroups. Since every (finite) semigroup can be embedded in a (finite) regular semigroup, it easily follows that the Zigzag Theorem also holds in the corresponding categories of regular semigroups and finite regular semigroups. A long-standing open question is whether or not the Zigzag Theorem holds in the category of bands.

4. (Hall 1978) A subsemigroup U of a semigroup S has the *right* [*left*] *congruence extension property in S* if for every right [left] congruence θ on U there exists a right [left] congruence Θ on S such that $\Theta \cap (U \times U) = \theta$.

(a) If U has the REP in S then U has the right congruence extension property in S. [Hint: extend a right congruence θ on U to U^1, put $X = U^1/\theta$, and consider the representation $\rho\colon U \to \mathcal{T}_X$ whereby $(v\theta)\rho_u = (vu)\theta$, for all $v \in U$, $u \in U$.]

(b) A three-member right zero semigroup U does not have the right congruence extension property. [Hint: if $U = \{u, v_1, v_2\}$ consider $S = U \cup \{e\}$, where $e = e^2$, $ex = x$ ($x \in U$), $ue = v_1$, $v_i e = v_i$ ($i = 1,2$).]

5. (a) From Theorem 4.2.13 and Lemma 4.1.3 deduce that the full transformation semigroup $\mathcal{T}_X$ has the representation extension property.

(b) Use the two cited results together with Exercise 4.1.10 to deduce that the same is true of the partial transformation semigroup on X.

5 Semigroup diagrams and word problems

5.1 Introduction

One of the few outstanding problems in algebraic semigroup theory that can be explained to a layman in a few minutes is the one-relator word problem. You are given two words, w_1 and w_2, together with another pair of words, u and v, all of whose letters come from some alphabet X. Is it possible to transform the word w_1 into the word w_2, given that u and v are treated as 'equal', meaning that u and v can be interchanged, one for the other, as often as you please? The answer will of course depend on the given words. For example, let $w_1 = ab^2$, $w_2 = ab^2a^2$, with our 'relation' given by $ab = aba$. Then

$$w_1 = \underline{ab}b = ab\underline{ab} = ab\underline{ab}a = \underline{ab}abaa = abbaa = w_2.$$

On the other hand, it is easy to see that w_1 cannot be transformed into a^2b^3, because our relation $ab = aba$ can never alter the total number of occurrences of the letter b in a given word.

The one-relator word problem asks: Is it possible to find some procedure, some algorithm (that could in principle be programmed onto a computer) which decides whether or not two given words, w_1 and w_2, can be transformed into one another using some given relation, $u = v$? More generally, we might relax the restriction that we have just one relation, $u = v$, and allow any number of them, $u_i = v_i$, $i \in I$, to be used. This more general problem has been solved in the negative by Post (1947) and, indeed, this was the first example of a so-called unsolvable word problem discovered in algebra. It is now known that even finitely presented semigroups can have an undecidable word problem. An example of such a presentation involving just seven relations on a five-letter alphabet appears in the interesting popular account by Roger Penrose on artificial intelligence (Penrose 1989). However, this still leaves open the possibility of an algorithm which decides the one-relator problem, and at the time of writing this problem remains open (for a survey, see Lallement 1986).

The purpose of this chapter is not to present a detailed account of the work done on the one-relator word problem, but rather to show how diagram methods can be used to solve some special semigroup word problems. In the next section we give a proof, due to John Remmers, that there is an algorithm to solve the word problem for finitely presented semigroups, the defining relations of which satisfy a certain so-called small overlap condition. This parallels a well-developed 'small cancellation' theory in group theory, although it only involves purely semigroup arguments. Results akin to these have also been obtained by Kasincev (1970). In the third section we give a proof, again due to Remmers, of the theorem of Adjan (see Adjan 1966) that the one-relator problem is solvable if u and v have differing initial and terminal letters (a result strengthened by Oganesyan 1984). In the course of doing so, we shall relate semigroup diagrams to the cancellation diagrams of group theory as introduced by Lyndon (see Lyndon and Schupp 1977).

5.2 The word problem for small overlap semigroups

We shall work throughout this section with a semigroup $\langle X; R \rangle$ on a set X of generators with relations R. A word r which occurs in some ordered pair of R is called an *R-word*. For convenience we shall assume that $R = R^c$, the equivalence relation generated by R on the set of all R words; there is no loss of generality in this as passing from R to R^c introduces no new words, and $\langle X; R \rangle = \langle X; R^c \rangle$, with both presentations $(X; R)$ and $(X; R^c)$ being finite or infinite together. A word $p \in F_X$ is a *piece* (relative to R) if there exist R-words r, r' and factorizations $r = upv$, $r' = u'pv'$ ($u, v, u', v' \in F_X^1$) such that either $u \neq u'$ or $v \neq v'$. Note that the possibility that $r = r'$ is not excluded. A piece is a word which is a common segment of R-words in at least two distinct ways.

Definition 5.2.1 A semigroup S with presentation $(X; R)$ is a $C(n)$ *semigroup* ($n \geq 1$) if no R-word is a product of fewer than n pieces.

Note that an R-word need not be a product of pieces at all. Also, a $C(n)$ semigroup is a $C(m)$ semigroup whenever $1 \leq m \leq n$, and any semigroup $\langle X; R \rangle$ is $C(1)$, while $\langle X; R \rangle$ is $C(2)$ if and only if no R-word is a proper segment of another R-word. If $X \subseteq X'$, $R \subseteq R'$, and $(X'; R')$ satisfies $C(n)$, then $(X; R)$ satisfies $C(n)$. Although whether or not a semigroup $\langle X; R \rangle$ is $C(n)$ depends on the given presentation $(X; R)$ (see Exercise 5.2.2) a presentation $(X; R)$ defines a $C(n)$ semigroup if and only if $(X; R^c)$ does also.

In what follows we shall consider a derivation diagram M over $(X; R)$ using the notation of Section 1.7. We write $\bar{\pi}$ for the label of a positive walk π in M, and we assume that M has no *superfluous vertices*, which are vertices for which both the indegree and outdegree are one. The respective lengths of π and $\bar{\pi}$ are denoted by $|\pi|$ and $|\bar{\pi}|$.

We now show that the word problem is solvable for finitely presented $C(3)$ semigroups. We start with a pair of simple observations.

Lemma 5.2.2 *Let D and E be regions of M.*

(i) *If the edge e of M is on ∂D and ∂E, and is either an initial edge of D but not E, or a terminal edge of D but not E, then $\bar{e}$ is a piece.*
(ii) *If β_D and α_E have a common edge e such that $\bar{e}$ is not a piece then $\beta_D = \rho e \sigma$, $\alpha_E = \rho' e \sigma'$ with $\bar{\rho} \equiv \bar{\rho}'$ and $\bar{\sigma} \equiv \bar{\sigma}'$ ($\rho, \sigma, \rho', \sigma'$ positive or empty paths). Furthermore, $(\bar{\alpha}_D, \bar{\beta}_E) \in R$.*

Proof (i) We may suppose that e is a common to β_D and α_E. In the first case $\bar{\beta}_D = \bar{e}v$ and $\bar{\alpha}_E = u\bar{e}v'$, with $u \in F_X$, showing that $\bar{e}$ is a piece; and the second case is similar.

(ii) Note that β_D and $\bar{\alpha}_E$ must be R-words. Since $\bar{e}$ is a segment of $\bar{\beta}_D$ and $\bar{\alpha}_E$ but is not a piece, $\bar{\beta}_D$ and $\bar{\alpha}_E$ are identical words and have factorizations as stated. Since $(\bar{\alpha}_D, \bar{\beta}_D) \in R$, $(\bar{\alpha}_E, \bar{\beta}_E) \in R$, and R is transitive, it follows that $(\bar{\alpha}_D, \bar{\beta}_E) \in R$. ∎

Regions D and E, which have a common edge e not labelled by a piece, will be called a *symmetrically labelled pair of regions* with *interface e*. Interior edges not labelled by pieces are sometimes unavoidable (see Exercise 5.2.1).

The following fact is an immediate consequence of the definition of a $C(2)$ semigroup.

Lemma 5.2.3 *Suppose that $\langle X; R \rangle$ satisfies $C(2)$. Then no side of any region is a proper segment of a side of another region.*

Before proceeding we require a result concerning diagrams with minimal numbers of regions of a certain kind.

If $(X; R)$ is a semigroup presentation, T is a subset of R, and M is a derivation diagram over $(X; R)$, then we use $\|M\|_T$ to denote the number of regions of M the labels of which are in T. If M is a (w, w')-diagram over $(X; R)$, then M is T-*minimal* if $\|M\|_T \leqslant \|M'\|_T$ for any (w, w')-diagram M'.

Theorem 5.2.4 *Let T be a subset of R. Then a subdiagram N of a T-minimal diagram M is T-minimal.*

Proof Suppose that N' is a derivation diagram over $(X; R)$ for $(\bar{\alpha}_N, \bar{\beta}_N)$ with $\|N'\|_T < \|N\|_T$. The idea of the proof is to obtain a contradiction by replacing the submap N of M by N'.

Let σ_1 be a positive or empty walk from the initial vertex of M to the initial vertex of N, and let σ_2 be a positive or empty walk from the terminal vertex of N to the terminal vertex of M (which exist by Lemma 1.7.3). Let M_l be a copy of the subdiagram of M with boundary walk $\alpha_M (\sigma_1 \alpha_N \sigma_2)^{-1}$ and let M_r be a copy of the subdiagram of M with boundary walk $\sigma_1 \beta_N \sigma_2 \beta_M^{-1}$ as shown in Fig. 5.1. By inserting superfluous vertices if necessary into

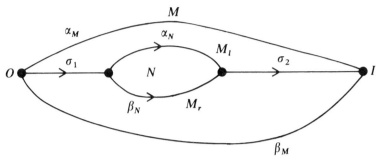

Fig. 5.1

the boundaries of N and N' we may assume that $|\alpha_N| = |\alpha_{N'}|$ and that $|\beta_N| = |\beta_{N'}|$.

We now form a diagram M' by identifying the path $\alpha_{N'}$ of N' with the path α_N of M_l, and identifying the path $\sigma_1 \beta_{N'} \sigma_2$ with the path $\sigma_1 \beta_N \sigma_2$ of M_r. With the obvious labels, M' is a diagram over $(X; R)$ for $(\bar{\alpha}_M, \bar{\beta}_M)$ with $\|M'\|_T < \|M\|_T$, as required. ∎

Lemma 5.2.5 *Let γ be a positive path from O_M to I_M is some diagram M with at least one region. Then there is a least one region D of M such that an entire side α_D or β_D on D is a segment of γ.*

Proof Without loss assume that M is free from superfluous vertices. We proceed by induction on $|\gamma|$. If $|\gamma| = 1$, then γ itself is the side of a region. Otherwise write γ as $\gamma_1 \gamma_2$, with γ_1 and γ_2 non-empty; let P be the common vertex of γ_1 and γ_2. Since M has no superfluous vertices, there is an edge e incident with P not contained in γ; without loss assume that e ends at P. Let δ be a positive path from O to P containing e. The subdiagram enclosed by $\gamma_1 \delta^{-1}$ contains at least on region, and $|\gamma_1| < |\gamma|$. By induction we obtain a region, D, one side of which is a segment of γ_1, and thus of γ. ∎

Corollary 5.2.6 *Let M be a diagram with at least one region. Then there is at least one region D_l (respectively D_r) such that α_{D_l} (respectively β_{D_r}) is a segment of α_M (respectively β_M).*

A (u, v)-diagram is called *short* if no other (u, v)-diagram has less regions (or, in other words, the diagram is R-minimal). By Theorem 5.2.4 a subdiagram of a short diagram is itself short.

Lemma 5.2.7 *Suppose that $(X; R)$ satisfies $C(2)$ and M is short. Then no interior edge of M is an entire side of a region.*

Semigroup diagrams and word problems | 171

Proof Suppose that $e = \beta_D$ for some region D. Since e is interior, e lies on α_E for some region; by Lemma 5.2.3, $e = \alpha_E$. Hence $(\bar{\alpha}_D, \bar{\beta}_E)$ is in R by transitivity of R, and the subdiagram bounded by α_D and β_E is not short, contradicting Theorem 5.2.4. ∎

We now present a series of results on the structure of derivation diagrams over $C(3)$ presentations.

Lemma 5.2.8 *Suppose that $(X; R)$ satisfies $C(3)$ and that e is an edge on the boundaries of distinct regions D, E of M. Then e is an initial edge of D but not of E if and only if e is a terminal edge of E but not of D.*

Proof By symmetry, it suffices to consider just the forward implication. The hypothesis implies that $\bar{e}$ is a piece; hence, by Lemma 5.2.3, e is not a terminal edge of D. It remains to show that e is a terminal edge of E.

To this end, let B denote the set of all interior edges d of M such that d is an initial edge of one of the regions on which it borders but not of the other, and such that d is a terminal edge of neither region. We complete the proof by showing that B is empty.

If this is not the case, we may take e to be maximal with respect to the partial ordering on the edges defined in Exercise 1.7.5. Let f and g be the edges on D and E that immediately follow e (see Fig. 5.2). Since M has no superfluous vertices, f and g are distinct. Therefore f is an initial edge of some region F; by Lemma 5.2.2(i) $\bar{f}$ is a piece, and by choice of e as maximal in B, f is a terminal edge of D. Then ef is an entire side of D, which contradicts $C(3)$ because both $\bar{e}$ and $\bar{f}$ are pieces. ∎

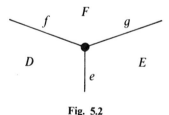

Fig. 5.2

Lemma 5.2.9 *Suppose that $(X; R)$ satisfies $C(3)$ and $e, f,$ and g are consecutive edges on a side of a region D of M. Then e and g are interior edges of M. There are regions E and G such that e is a terminal edge of E and g is an initial edge of G (see Fig. 5.3).*

Proof If e is on ∂M, then since M has no superfluous vertices, f is an initial edge of some region but not a terminal edge of D, contradicting the foregoing lemma. Hence e, and likewise g, are interior edges of M and are a terminal edge of a region E and an initial edge of a region G respectively. ∎

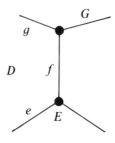

Fig. 5.3

Lemma 5.2.10 *Suppose that $(X; R)$ satisfies $C(3)$. Then:*

(i) *No side of any region M has more than three edges.*
(ii) *If two consecutive edges e, f on a side of a region are interior edges of M, then the entire side of the region containing those edges is interior to M.*

Proof (i) Let e, f, and g be consecutive edges on the side of some region D. By Lemma 5.2.9, e and g are terminal and initial edges respectively of certain regions. By Lemma 5.2.8, e is an initial edge of D and g is a terminal edge of D, so that e, f, and g are the only edges on one side of D.

(ii) By symmetry and part (i), we can take e to be an initial edge of D. If there is an edge g on D following f, then by Lemma 5.2.9, g is interior, and by part (i), efg forms a side of D. ∎

Lemma 5.2.11 *Suppose that $(X; R)$ satisfies $C(3)$ and that P is a vertex of M which is neither the initial nor terminal vertex of any block of M. Then either the indegree or the outdegree of P is one.*

Proof The vertex P is the common vertex of two consecutive edges e and f on a side of some region D. Either e is not a terminal edge of any region, or f is not an initial edge of any region; for otherwise $\bar{e}$ and $\bar{f}$ would be pieces, and by Lemma 5.2.8 ef would be an entire side of D, contradicting $C(3)$; the conclusion now follows. ∎

Lemma 5.2.11 forbids the configuration shown in Fig. 5.4 from arising in a diagram over a $C(3)$ presentation, unless the vertex is a terminal vertex of some block.

Lemma 5.2.12 *Suppose that $(X; R)$ satisfies $C(3)$. Let γ be an interior side of a region D consisting of exactly two edges. Then exactly one of the edges on γ is not labelled by a piece, and D is part of a symmetrically labelled pair of regions, the interface of which is that edge.*

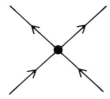

Fig. 5.4

Proof Since $\gamma = ef$, say, then by C(3), one of e and f is not a piece; by symmetry suppose that it is e. Since M has no superfluous vertices, f is an initial edge of some region, so that $\bar{f}$ is a piece by Lemma 5.2.2(i). The rest follows from Lemma 5.2.2(ii). ∎

We can now solve the word problem for finitely presented C(3) semigroups.

Theorem 5.2.13 *Suppose that $(X; R)$ satisfies C(3). Then, for every derivation diagram M over $(X; R)$, $|\alpha_M| = |\beta_M|$.*

Proof We proceed by induction on $\|M\|$, the result being clear if $\|M\| = 0$. Suppose that $\|M\| > 0$ and that all derivation diagrams with fewer than $\|M\|$ regions satisfy the conclusion. Let us assume that M consists of a single block (application of the following argument to each block establishes the general result).

By Corollary 5.2.6, there is a region D of M such that β_D is a segment of β_M, so that $\beta_M = \rho\beta_D\sigma$, say. Denote the respective initial and terminal vertices of D by P and Q respectively. By Lemma 5.2.11, both the indegree of P and the outdegree of Q are no more than 1. By Lemma 5.4.10, $|\alpha_D| \leq 3$. We consider the three possibilities for $|\alpha_D|$.

Case (1): $|\alpha_D| = 1$. Then $\alpha_D = e$ is an edge of M. If e is a boundary edge of M, then M consists entirely of the vertices P, Q and edges α_D, β_D, and the conclusion follows trivially. If e is interior, then by Lemma 5.2.3, $e = \beta_E$ for some region E, and is the interface of a symmetrically labelled pair. By Lemma 5.2.2, the diagram $M' = M - e$ is a derivation diagram; we have $|\alpha_M| = |\alpha_{M'}|$ and $|\beta_M| = |\beta_{M'}|$, whereupon the conclusion follows by induction.

Case (2): $|\alpha_D| = 2$. Then $\alpha_D = ef$ is a path of length 2. Denote the common vertex of e and f by S. By Lemma 5.2.11, either the indegree or the outdegree of S is 1; by symmetry assume that it is the indegree. Then the outdegree of S is at least 2, whence f is an initial edge of some region F. From Lemma 5.2.3 it follows that f is not an entire side of F; whence Q has outdegree 1, and that the edge g following f on F lies on β_M (see Fig. 5.5). Form a new diagram M' by removing β_D and replacing the path fg by a single edge f'. Then $\|M'\| = \|M\| - 1$, and clearly $|\beta_M| = |\beta_{M'}|$. Hence by the inductive hypothesis,

174 | Techniques of semigroup theory

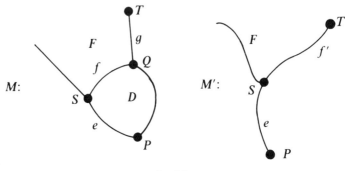

Fig. 5.5

$|\alpha_{M'}| = |\beta_{M'}|$. Since the left side of M has not been altered, $|\alpha_{M'}| = |\alpha_M|$. It follows that $|\alpha_M| = |\beta_M|$.

Case (3): $|\alpha_D| = 3$. We write $\alpha_D = efg$. By Lemma 5.2.9, e is a terminal edge of a region E and g is an initial edge of a region G. By Lemma 5.2.3, e and g are not entire sides of E and of G. Hence there is an edge e_1 on E preceding e and an edge g_1 following g on G (see Fig. 5.6). Since the indegree of P and the outdegree of Q must both be 1, e_1 and g_1 lie on β_M. Form a new diagram M' by removing β_D and replacing the paths $e_1 e$ and $g g_1$ with single edges e' and g' respectively. The argument is now completed just as in Case (2). ∎

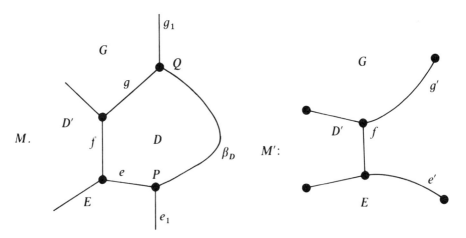

Fig. 5.6

Theorem 5.2.14 Suppose that $S = \langle X; R \rangle$ satisfies $C(3)$ and that there is an upper bound δ for the lengths of R-words. If w, $w' \in F_X$ and $w = w'$ in S then $|w'| \leq \delta |w|$.

Proof Suppose that M is a (w, w')-diagram. Then α_M and β_M have factorizations

$$\alpha_M = \alpha_1 \alpha_2 \ldots \alpha_m \quad \text{and} \quad \beta_M = \beta_1 \beta_2 \ldots \beta_m,$$

where α_i and β_i are the left and right sides respectively of the ith block, M_i, of M. We have $|\alpha_i| = |\beta_i|$, by Theorem 5.2.13. Let $w_i = \bar{\alpha}_i$ and $w'_i = \bar{\beta}_i$. For each block M_i which has at least one region, every edge in that block is labelled by a non-empty segment of some R-word. Hence

$$|w'_i| = |\bar{\beta}_i| \leqslant \delta|\beta_i| = \delta|\alpha_i| \leqslant \delta|\bar{\alpha}_i| = \delta|w_i|.$$

This inequality also holds for blocks with no regions. Hence

$$|w'| = \sum_{i=1}^{m} |w'_i| \leqslant \delta \sum_{i=1}^{m} |w_i| = \delta|w|. \qquad \blacksquare$$

Theorem 5.2.15 *Every finitely presented $C(3)$ semigroup has a solvable word problem.*

Proof Let $\langle X; R \rangle$ be a $C(3)$ semigroup with X, R finite sets. Invoke Theorem 5.4.14 with δ the length of a longest R-word. Let $t = |X|$, and for a given word w let $K = \delta|w|$. Then the number of words length not exceeding K is

$$L = t^K + t^{K-1} + \ldots + t = \frac{t}{t-1}(t^K - 1).$$

If $w = w'$ in $\langle X; R \rangle$, then there is a derivation for the pair (w, w') of length not exceeding L. Since there are only finitely many such derivations, they can be systematically enumerated, implying a solution to the word problem. $\blacksquare$

Corollary 5.2.16 *If $\langle X; R \rangle$ is a finitely presented $C(3)$ semigroup, every word is congruent to only finitely many distinct words.*

The application of semigroup diagrams to word problems is to be found in Kasincev (1970). Howie and Pride (1986) use semigroup diagrams to reduce the one-relator word problem for semigroups to a number of outstanding cases, one of which has been disposed of by Jackson (1986), again using diagram theory. In a forthcoming paper (Jackson 1992) the one-relator problem is reduced to the case in which the relator is 'achievable', which roughly means that the letters of the relator can actively participate in sequences of transitions for the corresponding semigroup presentation. Another feature of interest in this paper is a recursive construction of the class of semigroup diagrams which are there called 'feathers' or 'feathery maps'. The recent paper by Power (1990) on so-called 2-categories employs the same class of diagram, but there they are known as 'pasting-schemes'.

Exercises 5.2

1. Consider the semigroup S with generating set $X = \{s, t, x, y, u\}$ and presentation $(X; xy = t, sx = ux)$.

   (a) Show that S satisfies $C(n)$ for all $n \geq 1$.
   (b) Show that any derivation of the equation $st = ut$ involves the relation $xy = t$ in such a way that y, which is not a piece, labels an interior edge of the corresponding derivation diagram.

In exercises 5.2.2, 5.2.3, and 5.2.4 let n denote a fixed natural number and X the ordered alphabet $\{x_1, x_2, \ldots, x_n\}$. For $k \leq n$, the *up-word*, u_k, is $x_1 x_2 \ldots x_k$, while the *down-word*, d_k, is $x_k x_{k-1} \ldots x_2 x_1$. The *shuffle-words*, s_k, are $s_k = x_1 x_{h+1} x_2 x_{h+2} \ldots x_h x_{2h}$ if $k = 2h$ is even, and $s_k = x_1 x_{h+2} x_2 x_{h+3} \ldots x_{2h+1} x_{h+1}$ if $k = 2h + 1$ is odd. Let Q_n be the semigroup with presentation $(X; u_n = d_n)$.

2. (a) Show that $(X; u_n = d_n)$ satisfies $C(n)$.
   (b) Show that $Q_n = \langle X; u_n = d_n, x_1 u_n = x_1 d_n \rangle$, but that the corresponding presentation fails to satisfy $C(2)$.

A relator of the form (w, w) is called a *trivial relator*. A letter a of the presentation $(X; R)$ is called *avoidable* if every occurrence of a in a non-trivial relator of R has the form (a, w) or (w, a), where w is a word on $X - \{a\}$. An avoidable letter is never a piece.

3. Let w_m be the word $u_{n-m+1} d_{n-m+1} s_n$ and let $R_m = \{(u_n, d_n), (w_m, w_m)\}$. Prove that $(X; R_m)$ is a presentation of Q_n which satisfies $C(m)$ but not $C(m+1)$ for $1 \leq m < n$.

4. Let $X' = X \cup \{s, t\}$. For $2 \leq k \leq n$ define R_k by

$$R_k = \{(s, u_k), (t, d_k)\}.$$

Show that $(X'; R_k)$ is a presentation for the free semigroup F_X, and that this presentation satisfies $C(k)$, but not $C(k+1)$.

Remark The previous three exercises are due to Jackson (1991). In this paper it is shown that there is an effective way of reducing a given finite presentation $(X; R)$ to an equivalent presentation (that is, one defining an isomorphic semigroup) which is *reduced*, meaning that it is free of trivial relators and avoidable letters. He goes on to prove that for two equivalent presentations $(X_1; R_1)$ and $(X_2; R_2)$ that are finite, reduced, and $C(2)$, one of the presentations satisfies $C(n)$ if and only if the other does also.

5.3 Group diagrams and a theorem of Adjan and Remmers

In this section a short introduction to the so-called group diagrams of combinatorial group theory is given, based on the account of Lyndon and Schupp (1977). The relationship between semigroup and group diagrams will be highlighted by a proof due to Remmers of a theorem of Adjan, that any semigroup S with a cycle-free presentation is group embeddable, in the process augmenting Theorem 1.7.7. The idea of the proof is to show that, given such a presentation, any group diagram relating two positive words is, after all, a semigroup diagram, and so the natural homomorphism v mapping S into the group defined by the same relations is injective. We begin by introducing the notion of diagram over a semigroup S, particular cases of which are the semigroup diagrams used thus far, and group diagrams, to be introduced below.

In Definition 1.7.1 we defined a semigroup diagram Γ as a planar directed graph together with a labelling of the oriented edges of Γ, satisfying certain other conditions. We shall shortly need to consider another type of diagram, known as a group diagram, with different rules for its associated labelling. For this reason we introduce the idea of a directed map, the labelling of which is a separate feature to be appended.

A *(directed) map* M is a finite collection of vertices, oriented edges, and regions. The vertices, edges, and regions of M are to be pairwise disjoint and will satisfy the following:

(i) If e is an edge of M, there are (not necessarily distinct) vertices a and b in M such that the closure of e, $\bar{e} = e \cup \{a, b\}$; these vertices a and b are the *endpoints* of e. The orientation of e specifies a positive direction in which to traverse e, and we may choose notation so that a and b are, respectively, the *initial and terminal endpoints* of e.
(ii) The boundary, ∂D, of each region D of M is connected, and there are edges $e_1, e_2, \ldots, e_n$ in M such that $\partial D = \bar{e}_1 \cup \ldots \cup \bar{e}_n$.

We shall use M to denote the set-theoretic union of the vertices, edges, and regions of M. An edge e is said to be *incident* with its endpoints and a region D is *incident* with the edges and vertices on its boundary. We regard two maps as being the same if one can be mapped onto the other by an orientation-preserving homeomorphism of the plane which induces an injective, incidence-preserving function taking vertices, edges, and regions to vertices, edges, and regions respectively.

Given a map M, the underlying digraph, $\Gamma(M)$, of M is the planar digraph consisting of the vertices and edges of M.

Let Γ be a digraph, let W be the set of walks on Γ, and let W_+ be the set of positive walks on Γ (that is, walks which traverse each arc in the direction of its orientation). In either case, concatenation of walks is a partially defined associative binary operation. We may thus regard W and W_+ as associative partial algebras. For W, we have in addition an obvious unary operation $(\ )^{-1}: \alpha \mapsto \alpha^{-1}$ which has both an involutary property, $(\alpha^{-1})^{-1} = \alpha$, and a partial anti-automorphic property, $(\alpha\beta)^{-1} = \beta^{-1}\alpha^{-1}$. Clearly, the partial algebra W_+ is generated by the positive edges: if we regard W as a partial algebra with involution, then W is also generated by the positive edges.

Let S be an arbitrary semigroup. We say that a function φ is a *labelling function on Γ with values in S* when φ is a partial homomorphism from W_+ to S. (The function φ is a partial homomorphism if $\varphi(\alpha\beta) = \varphi(\alpha)\varphi(\beta)$ whenever the product $\alpha\beta$ is defined in the partial algebra W_+.) Of course, φ is determined once it is defined on the positive edges which generate W_+.

Let S be a semigroup together with some involution $(\ )^{-1}$. Then a partial homomorphism $\varphi: W \to S$ is an *involutary labelling function on Γ* when $(\varphi(\alpha))^{-1} = \varphi(\alpha^{-1})$ for every walk α in W.

A *diagram over the semigroup S* is a pair (M, φ), where M is a map and φ is a labelling of the underlying digraph of M with values in S. When the labelling is clear from context, we simply refer to the diagram M.

Given a walk π and a labelling φ of a digraph Γ, we refer to $\varphi(\pi)$ as the *label of π*. When it is clear from context that we are concerned with a label (and not with topological closure), we write $\bar{\pi}$ for $\varphi(\pi)$.

Let X be a set and let X^{-1} be a disjoint set consisting of elements x^{-1} in one-to-one correspondence with the elements x of X. A *diagram over X* will be taken to mean a diagram over $F(X)$. An *involutary diagram over X* is a labelling over $F(X \cup X^{-1})$, with the additional proviso that the label on each positive edge is in $F(X)$.

Suppose that we have a labelling of a map M. We now define boundary labels for regions of M, and if M is connected and simply connected, a boundary label for M as well. (Recall that simply connected means path connected with trivial fundamental group.)

We can do this in two ways. First, suppose that the label is involutary and define a boundary label for a region, or for a connected and simply connected map, to be a label for some boundary walk of that region or map. This method is used when we have an involutary label with values in a group, and we thus refer to this type of boundary label as a *group label*. For the second method, we must assume that the map or region has a boundary walk of the form $\alpha\beta^{-1}$, where α and β are positive walks. We may then define the *semigroup label* of the map or region to be the ordered pair $(\bar\alpha, \bar\beta)$ of semigroup elements.

Let $G = G_X$ be the free group on a base set X. For each basis element $x \in X$, we call x und x^{-1} *letters*; then every element of G is the value of some word $y_1 y_2 \ldots y_n (n \geq 0;$ each y_i a letter) over X, and the value of a unique *reduced word*; that is, a word which does not contain two adjacent letters of the form xx^{-1} or $x^{-1}x (x \in X)$. A word is *positive* if all its letters are members of X. Two words u and v are *literally equal*, denoted $u \equiv v$, if they are identical sequences of letters, and are *freely equal* if they define the same element of G. A word is *trivial* if it is freely equal to the empty word.

Any relation $r = s$ in a group can be written as $rs^{-1} = 1$, and for this reason the set of relations R in a group presentation can be regarded as forming a set of words, each of which equals the identity element in $[X; R]$. We shall continue to denote the semigroup with presentation $(X; R)$ by $\langle X; R \rangle$, while the group with this same presentation will be denoted by $[X; R]$. The *natural homomorphism* from $\langle X; R \rangle$ into $[X; R]$, where each positive word is mapped to its value in $[X; R]$, will be denoted by v.

We shall assume that any word w representing a member of G_X is reduced, unless otherwise specified. A member $w \in G_X$ is *cyclically reduced* if it cannot be written in the form uru^{-1}, with $u \ne 1$. It follows easily that any $w \in G_X$ can be represented uniquely in the form uru^{-1}, which is a factorization of w that is reduced without cancellation with r cyclically reduced. Note that if $w = u_1 u_2 \ldots u_n$, say, then $u_i u_{i+1} \ldots u_n u_1 u_2 \ldots u_{i-1} (1 \le i \le n)$ is a conjugate, known as a *cyclic conjugate* of w. A subset R of G is *symmetrized* if all the elements of R are cyclically reduced, and $r \in R$ implies that all cyclically reduced conjugates of $r^{\pm 1}$ are also in R. Since any $w \in G_X$ has a cyclically reduced conjugate, it follows that any group presented by generators and relations $[X; R]$ may be taken to have a symmetrized set of words R, and this will be done henceforth.

Let $[X; R]$ be a group presented by a (symmetrized) set of words R so that $[X; R] \simeq G_X / N$ where, since R is symmetrized, $w \in N$ if and only if w is a product of conjugates of members of R and so has the form $w = (u_1 r_1 u_1^{-1})(u_2 r_2 u_2^{-1}) \ldots (u_n r_n u_n^{-1})$ $(r_i \in R)$. We shall construct a diagram M which is a map over the free group G_X satisfying:

(i) each region of M has as boundary label a cyclically reduced conjugate of some $r \in R$;
(ii) M has a boundary label w.

As an alternative to (ii), we can instead construct M satisfying the following:

(ii)' M is labelled by a cyclically reduced conjugate of w.

The proof goes by way of induction on n; if $n = 0$ we can take M to consist of a single vertex 0.

If $n = 1$, $w = uru^{-1}$, take a vertex v and a loop e at v with label r. If $u = 1$, take $O = v$ as the initial vertex of our boundary walk, and M is constructed and satisfies (i) and (ii). If $u \ne 1$, take a vertex O external to the loop e and an arc from O to v with label u, giving a diagram that satisfies (i) and (ii).

---

* The following material, up to and including the proof of Theorem 5.3.3, is taken from *Combinatorial group theory* by R. C. Lyndon and P. Schupp, and appears courtesy of Springer-Verlag publishers. Figures 5.7, 5.8, and 5.11 below are Figs 1.1, 1.2, and 1.4 from pages 236–240 of that book.

For $n > 1$, form M' by taking diagrams for each of $u_1 r_1 u_1^{-1}, \ldots, u_n r_n u_n^{-1}$, and arrange them around a common base point O, as in Fig. 5.7.

If the product $(u_1 r_1 u_1^{-1}) \ldots (u_n r_n u_n^{-1})$ is reduced without cancellation we are finished. If not, (ii) is not presently satisfied. By introducing superfluous vertices if necessary, we may assume that each edge e is labelled by a member of $X \cup X^{-1}$. We shall obtain a required diagram by identifying successive edges whose labels are mutual inverses.

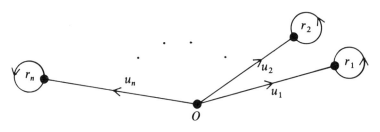

Fig. 5.7

Let α be the boundary cycle of M' which begins at O, so that $\varphi(\alpha) = u_1 r_1 u_1^{-1} \ldots u_n r_n u_n^{-1}$. By assumption α contains a pair of successive edges e and f such that $\varphi(e)$ and $\varphi(f)$ are inverses. Denote the initial and terminal vertices of e and f by v_1, v_2, and v_2, v_3 respectively. Suppose that v_1 is distinct from v_2 and from v_3. The edge e can then be folded onto f without otherwise altering the structure of M', as shown in Fig. 5.8 (whether or not $v_2 = v_3$).

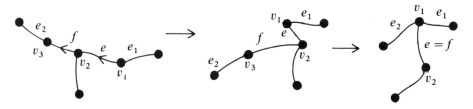

Fig. 5.8

The resulting diagram M'' has a boundary containing fewer edges than α. If v_3 is distinct from v_1 and v_2 we can proceed similarly. Finally, if $v_1 = v_3$ then the closed edges e and f form a loop δ at v_1. Delete $\delta - v_1$ and all of M' interior to δ to form M''. Again, M'' contains fewer edges than α. In both cases the new diagram satisfies (i), and iteration of the above process yields a required diagram also satisfying (ii). If w is not cyclically reduced, further iteration, with suitable choices for the initial vertex of our boundary walk will cyclically reduce the label, giving a diagram satisfying (ii)'.

Semigroup diagrams and word problems | 181

The above procedure shows that, given any finite sequence of words $w_1, w_2, \ldots, w_n$ in the free semigroup on $X \cup X^{-1}$, there is a connected and simply connected diagram M with boundary label the product $w = w_1 \ldots w_n$ reduced without cancellation in G_X and such that any label of any boundary cycle of a region of the diagram is reduced without cancellation and is a cyclically reduced conjugate of one of the w_i. We can, if we wish, insist that M has as boundary label a cyclically reduced conjugate of w.

Example 5.3.1 Let $w = abca^{-1}.ac^{-1}b.b^{-1}a^{-1}$. A sequence of diagrams leading to one satisfying (i) and (ii) is as shown in Fig. 5.9.

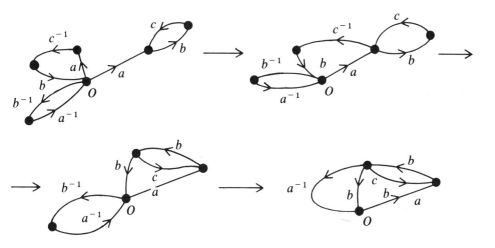

Fig. 5.9

A boundary label of the fourth diagram is aba^{-1}, which is reduced without cancellation so that the diagram satisfies (i) and (ii). However, aba^{-1} is not cyclically reduced, b being the reduced cyclic conjugate of this word. Another application of the procedure yields Fig. 5.10, which satisfies (i) and (ii)'.

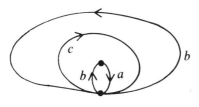

Fig. 5.10

Definition 5.3.2 Let R be a symmetrized subset of the free group G_X. An R-*diagram* is a diagram M such that if δ is any boundary cycle of any region D of M, then $\varphi(\delta) \in R$.

182 | Techniques of semigroup theory

Theorem 5.3.3 *Let R be a symmetrized subset of the free group G_X and let N be the normal subgroup of G_X generated by R. Then for any $w \in G_X$, $w \in N$ if and only if there is a connected, simply connected R-diagram such that the label on ∂M is w [a cyclically reduced conjugate of w].*

Proof Only the converse direction remains to be shown, so, to this end, suppose that M is a connected, simply connected R-diagram with regions $D_1, \ldots, D_m$. Let α be a boundary cycle of M beginning at a vertex v on ∂M, and let $w = \varphi(\alpha)$. The result, including that of the bracketed statement, will follow from showing that there exist elements $u_i \in G_X$ such that

$$w = (u_1 r_1 u_1^{-1}) \ldots (u_m r_m u_m^{-1}), \quad \text{where } r_i \text{ is the label of } D_i.$$

If $m = 0$ there is nothing to prove, for M is a tree and $\varphi(\alpha) = 1$. Assume inductively that our claim is true for maps with k regions, and let M be a map with $k + 1$ regions.

Take a region D of M such that $\partial D \cap \partial M$ contains an edge e. Form the map M' from M by deleting e. Observe that M' is still connected and simply connected (see Fig. 5.11). Write $\alpha = \beta e \gamma$. There is a boundary cycle $e\eta$ of D beginning with e. Let $\varphi(\beta) = b$, $\varphi(e) = z$, $\varphi(\gamma) = c$, and $\phi(\eta) = d$. Then $w = \varphi(\alpha) = bzc$. The boundary cycle μ of M' beginning at v_0 is $\beta \eta^{-1} \gamma$, so that $\varphi(\mu) = bd^{-1}c$. By induction, the regions of M' (that is, the regions of M other than D) can be numbered $D_1, \ldots, D_k$ so that

$$bd^{-1}c = (u_1 r_1 u_1^{-1}) \ldots (u_k r_k u_k^{-1}),$$

where r_i is a label of D_i. Now $w = bzc = (bd^{-1}c)(c^{-1}dzc)$ and dz is a label of D. Take D as D_{k+1}, dz as r_{k+1} and c^{-1} as u_{k+1}. ∎

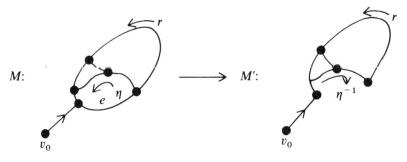

Fig. 5.11

In the remainder of this section we shall be concerned with a given symmetric semigroup presentation $(X; R)$, which will be called a *positive presentation* both for the semigroup $\langle X; R \rangle$ and the group $[X; R]$. For all diagrams (M, φ) considered, φ is assumed to be an involutary labelling function with values in the free semigroup $F(X \cup X^{-1})$. Recall that the label on each positive edge is a positive word. All maps are assumed to be connected and simply connected.

We shall be interested in two types of diagram. The first is the now familiar *semigroup diagram*, or *s-diagram*, over $(X; R)$ for a pair (u, v) of positive words which is a diagram M with the properties:

(S1) Each region of M is labelled by a word rs^{-1} for some $\{r, s\} \in R$.
(S2) The boundary of M is labelled by the word uv^{-1}.
(S3) M has no interior transmitter or receiver.

The second type of diagram that we consider is motivated by Theorem 5.3.3. A *group diagram*, or *g-diagram*, over $(X; R)$ for a word w over X is a diagram M over X with the properties:

(G1) Each region of M is labelled by a cyclically reduced conjugate in G_X of the word rs^{-1} for some $\{r, s\} \in R$.
(G2) The boundary of M is labelled by a cyclically reduced conjugate of w.

The conditions (G1) and (G2) imply that every region of a g-diagram is two-sided or cyclic. In view of Theorem 5.3.3, a g-diagram over $(X; R)$ exists for a word w if and only if w defines the identity element of the group $[X; R]$.

Recall the meaning of left and right cycles for a semigroup presentation $(X; R)$, and of the left and right graphs of a presentation, $LG(X; R)$ and $RG(X; R)$, from Section 1.7. We now develop a series of results culminating in a proof due to Remmers (1980) of a result of Adjan, that a semigroup with a cycle-free presentation is group embeddable.

A pair of regions D, D' of M is *inversely labelled* if their boundaries share an edge e, the removal of which from M consolidates D and D' into a single region, the boundary label of which is freely equal to the empty word: the edge e is the *interface* of D and D' (see Fig. 5.12).

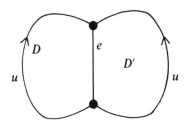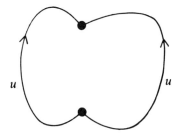

Fig. 5.12

After the removal of the edge called e, a boundary label of the consolidated region is uu^{-1}.

Lemma 5.3.4 *Suppose that $(X; R)$ has no left [right] cycle and that $\theta = (e, f)$ is an interior source-angle [sink-angle] of a group diagram M, such that the* labels of e and f have the same initial [terminal] letter. Then there is a pair of inversely labelled regions included in θ.

Proof We shall prove the case in which $(X; R)$ has no left cycles and in which θ is a source-angle. Let $D_1, D_2, \ldots, D_k$ be the regions included in θ, listed in clockwise cyclic order. Since θ is a source-angle, no D_i is a cycle. By the choice of the ordering of the D_i, the right side of D_{i-1} and the left side of D_i share a common initial edge; let x_i be the initial letter of the label of this edge, and x_0 the common initial letter of the sides of θ. The labels on the left and right sides of D_i are positive words p_i and q_i with distinct terminal letters (since the word $p_i q_i^{-1}$ is cyclically reduced). Since $(X; R)$ has no left cycles, hence for some positive or empty words, t_i, we have $(p_i t_i, q_i t_i) \in R$. The pairs $f_i = \{p_i t_i, q_i t_i\}$ ($1 \leqslant i \leqslant k$) and sequence $x_0, x_1, \ldots, x_{k-1}, x_k$ form a closed left walk of (X, R). Choose indices i and j ($1 \leqslant i < j \leqslant k$) for which $x_i \equiv x_j$, with $j - i$ minimum. Since $(X; R)$ has no left cycles, $j = i + 1$, and the unordered pairs f_i and f_j are identical. Hence

$$p_i t_i \equiv q_j t_j \quad \text{and} \quad q_i t_i \equiv p_j t_j \tag{1}$$

and we may assume, without loss, that t_i is a terminal segment of t_j, so that $t_j = z t_i$ for some positive or empty word z. The relations (1) then become

$$p_i \equiv q_j z \quad \text{and} \quad q_i \equiv p_j z \tag{2}$$

upon cancellation of t_i in each. Let u be the label of the common initial edge e of D_i and D_j, so that $q_i \equiv uv$ and $p_j \equiv uw$, say. From the second relation in (2), it follows that $v \equiv wz$. Deletion of e from M consolidates D_i and D_j into a single region with boundary label $q_j w^{-1} v p_i^{-1} \equiv q_j w^{-1} wzz^{-1} q_j^{-1}$, which is freely equal to the empty word. Hence D_i and D_j are the required inversely labelled regions (see Fig. 5.13). ∎

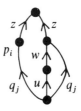

Fig. 5.13

Corollary 5.3.5 *If $(X; R)$ has no left [right] cycle and P is an interior transmitter [receiver] of the group diagram M, then there is a pair of inversely labelled regions incident with P.*

Proof We may list the set of regions of M having P as a boundary point in cyclic order, and apply the same argument as in Theorem 5.3.4. ∎

Given a group diagram M with boundary label w which contains a pair of inversely labelled regions, we may remove their interface and deform M so that the two sides of the consolidated region are identified, thus creating a group diagram for w having at least two fewer regions than M. By iterating this procedure we obtain the following.

Theorem 5.3.6 *If $(X; R)$ has no left [right] cycle and M is a group diagram with boundary label w, then there exists a group diagram M' for w, containing no more regions than M, which has no interior transmitter [receiver] and in which the labels on the sides and included edges of each interior source-angle [sink-angle] have initial [terminal] letters which are all different. If (X, R) is cycle-free then M' can be chosen so as to satisfy both sets of conditions simultaneously.*

A group diagram satisfying the first set, second set, or both sets of conclusions of Theorem 5.3.6 will be called *source-reduced*, *sink-reduced*, or *reduced*, as the case may be. Note that the corresponding definitions and theorem for s-diagrams is our Theorem 1.7.7.

The proof of the next result is due to Remmers (1980), and was first proved by Adjan (1966) with the additional restriction of finiteness of presentation. However, that restriction is unimportant, as the result as stated here can be readily deduced from the special case in which $(X; R)$ is finite.

Theorem 5.3.7 *Suppose that $(X; R)$ is cycle-free, u and v are positive words, and M is a group diagram for uv^{-1} with n regions. Then there is an s-diagram for (u, v) with no more than n regions. In particular, the natural homomorphism of $\langle X; R \rangle$ into $[X; R]$ is an embedding.*

Proof We can, by Lemma 5.3.6, take M to be a reduced group diagram. Since $(X; R)$ is cycle-free, the word rs^{-1} is cyclically reduced, as it stands, for every $\{r, s\} \in R$. Thus each region of M is two-sided. Since M contains no interior transmitters or receivers, M contains no positive cycles, by the proof of Lemma 1.7.3. In particular, ∂M, being labelled by a cyclically reduced conjugate of uv^{-1}, is two-sided. Thus M is an s-diagram with sides labelled by words u' and v', say, where $u \equiv pu'q$ and $v \equiv pv'q$ for some positive or empty words p and q. We construct a required s-diagram for (u, v) by attaching an edge with label p to the initial vertex, and an edge labelled q to the terminal vertex of M. ∎

Remarks An example of a semigroup which is embeddable in a group but does not satisfy Adjan's condition for group embeddability was given by Bush (1963).

186 | Techniques of semigroup theory

As mentioned earlier, Theorem 5.3.7 leads to a solution of the one-relator semigroup problem for the case in which the words u and v of the given relation have different initial letters and different terminal letters to each other. By Theorem 5.3.7 the words $w_1 = w_2$ in $\langle X; u = v \rangle$ if and only if they are also equal in the corresponding group $[X; u = v]$, and the one-relator word problem for groups has long been known to be decidable (Magnus 1932). For a study of conjugacy problems over cycle-free presentations see Goldstein and Teymouri (1992).

Exercises 5.3

1. (Malcev 1937) Let S be the semigroup with generating set $\{a, b, c, d, x, y, u, v\}$ and relations $ax = by$, $cx = dy$, and $au = bv$.
   (a) Show that $cu \neq dv$ in S, but by drawing a suitable group diagram show that $cu = dv$ in the corresponding group.
   (b) Prove that S is cancellative, but not group embeddable.

2. (Ramyantsev 1981) Show that the semigroup
$$S = \langle a, b; a^2 b^2 = b^2 a^2, a^2 bab^2 = b^2 aba^2 \rangle$$
is cancellative but not group embeddable. Furthermore, this example is minimal, in the sense that no semigroup presented by less generators or by less relations is cancellative but not group embeddable.

6 Combinatorial aspects of transformation semigroups

6.1 Probabilistic features of T_n

In this section we shall introduce results of a probabilistic nature concerning finite transformation semigroups. The study of random functions has a wide literature of its own, and many questions along these lines lead to occupancy problems which arise throughout mathematics and physics, particularly in mathematical genetics (Kingman 1978). We concentrate here on results that are of interest to semigroupists and which have elementary proofs. Indeed, one justification for the inclusion of this material is the delightful manner in which classic results from analysis and combinatorics furnish answers to a number of natural questions concerning T_n and O_n (the semigroup of all order-preserving mappings on $\{1, \ldots, n\}$).

Products in finite full transformation semigroups

We begin by investigating random products in T_n, and look at how quickly they tend to 'collapse' into lower $\mathscr{D}$-classes. This will be followed by a study of the mean values of several random variables all of which count some feature of the digraph of a randomly selected $\alpha \in T_n$, including the order of ran α, the order of the stable range of α, and the component number of α. The reader is referred to Section 1.6 for the basic definitions of these concepts.

Consider a randomly chosen $\alpha \in T_n$, by which we mean that every mapping is equally likely to be selected. Perhaps the most obvious feature of α to count is $|\text{ran }\alpha|$, the order of the range of α, also known as its *rank*, which equals the number of non-terminal vertices in its graph. Denote the random variable with value $|\text{ran }\alpha|$ by $R_n(\alpha)$. The random choice of a particular $\alpha \in T_n$ corresponds in an obvious way to a ball-tossing experiment in which n labelled balls are tossed at random into n labelled cells and, thought of in this way, $R_n(\alpha)$ is a random variable of classical probability theory; its distribution appears in Feller (1968) as a sum of terms with alternating sign, using an inclusion–exclusion argument. We shall return to this, but first we show how easily the mean of $R_n(\alpha)$ can be derived. The probability that some $k \notin \text{ran }\alpha (1 \leq k \leq n)$ equals the probability that cell k remains empty in the corresponding experiment, which is obviously $(1 - 1/n)^n$. Since the expectation of R_n, denoted by ER_n, equals $n \Pr(k \in \text{ran }\alpha)$, it follows that $ER_n =$

188 | Techniques of semigroup theory

$n(1 - (1 - 1/n)^n)$, which yields our first asymptotic result:

$$\lim_{n \to \infty} \frac{ER_n}{n} = 1 - e^{-1}. \tag{1}$$

This result was generalized by Kim and Rousch (1980), where it was shown that the limiting mean of the proportion of points in a random product in T_n has value

$$1 - e^{-(1 - e^{-1})}.$$

We shall, amongst other things, generalize this result to k-fold products after analyzing the distribution of R_n.

A little analysis at this stage allows us to prove that not only is it true that $ER_n \sim n(1 - e^{-1})$, but that the difference $ER_n - n(1 - e^{-1})$ approaches a constant:

$$\begin{aligned} ER_n - n(1 - e^{-1}) &= ne^{-1} - n(1 - 1/n)^n = ne^{-1} - ne^{n\log(1 - 1/n)} \\ &= e^{-1}(n - ne^{-(1/2n + O(n^{-2}))}) \\ &= e^{-1}(n - n(1 - (1/2n) + O(n^{-2}))) \\ &= 1/2e + O(n^{-1}), \end{aligned}$$

and thus, in particular,

$$\lim_{n \to \infty} (ER_n - n(1 - e^{-1})) = 1/2e. \tag{2}$$

Later we shall derive analogues of (2) for the mean values of the stable range and component number.

Notation The number of ways of choosing k from n will be denoted in any one of the three ways: $\binom{n}{k}$, C_k^n, or $C(n, k)$.

We now return to the distribution of our random variable R_n; in fact, we shall consider the more general experiment where $r \leq n$ labelled balls are tossed at random into n labelled cells. A random realization of the experiment, or trial, then corresponds to a random choice of mapping α from $X_r = \{1, \ldots, r\}$ to $X_n = \{1, \ldots, n\}$. Let X denote the random variable the value of which is $|\operatorname{ran} \alpha|$, noting that X will depend on n and r. First we shall count the number of mappings α such that $\operatorname{ran} \alpha = X_m$. Let $A_m = \{\alpha : \operatorname{ran} \alpha \subseteq X_m\}$. Using the Principle of Inclusion–Exclusion we claim that

$$|\{\alpha : \operatorname{ran} \alpha = X_m\}| = |A_m| - \binom{m}{1}|A_{m-1}| + \binom{m}{2}|A_{m-2}| + \cdots$$
$$+ (-1)^k \binom{m}{k}|A_{m-k}| + \cdots + (-1)^{m-1}\binom{m}{m-1}|A_1|. \tag{3}$$

Combinatorial aspects of transformation semigroups | 189

To see this, consider the term $C_k^m |A_{m-k}|$. The coefficient $C_k^m = C_{m-k}^m$ counts the number of subsets S of X_m of order $m - k$, and the term $|A_{m-k}|$ counts the number of mappings with ranges contained in S. If $\alpha \in A_m$ with $|\operatorname{ran} \alpha| = m - i \leqslant m - k < m$ (that is, $i \geqslant k > 0$) then the number of times α is counted in this factor is the number of ways $m - k - (m - i) = i - k$ empty cells can be chosen from the $m - (m - i) = i$ available to make up a set S; thus α is counted $C_{i-k}^i = C_k^i$ times. Summing over all terms yields that α contributes a total of

$$1 - \binom{i}{1} + \cdots + (-1)^k \binom{i}{k} + \cdots + (-1)^i \binom{i}{i} = \sum_{k=0}^{i} (-1)^k \binom{i}{k}$$
$$= (1-1)^i = 0$$

times overall; while if $i = 0$, α contributes exactly once in the first term $|A_m|$.

Bearing in mind that there are C_m^n choices for the range of a mapping α of rank m, and that there are n^r possible mappings in all, we deduce that, for $0 \leqslant m \leqslant n$,

$$\Pr(X = m) = \binom{n}{m} \sum_{k=0}^{m} (-1)^k \binom{m}{k} ((m-k)/n)^r. \tag{4}$$

For convenience we write $p_m(r, n)$ for $\Pr(X = n - m)$, and replace n and m by $n - 1$ and $m - 1$, respectively, to obtain from (4)

$$p_{m-1}(r, n-1) = \frac{m}{n} \left(\frac{n}{n-1}\right)^r p_m(r, n). \tag{5}$$

Sum equation (5) over m to obtain

$$1 = \frac{1}{n} \left(\frac{n}{n-1}\right)^r \sum_{m=1}^{n} m p_m(r, n),$$

or, as we have already seen in the $r = n$ case,

$$E(X) = n - n \left(\frac{n-1}{n}\right)^r. \tag{6}$$

All of the factorial moments of $Y = n - X$ can be calculated in this manner. In particular,

$$p_{m-2}(r, n-2) = \frac{m(m-1)}{n(n-1)} \left(\frac{n}{n-2}\right)^r p_m(r, n),$$

whence

$$1 = \frac{1}{n(n-1)} \left(\frac{n}{n-2}\right)^r \sum_{m=1}^{n} m(m-1) p_m(r, n),$$

which implies that

$$E\{Y(Y-1)\} = n(n-1) \left(\frac{n-2}{n}\right)^r.$$

Since Y and X share the same variance we obtain

$$\operatorname{Var} X = n(n-1)\left(\frac{n-2}{n}\right)^r + n\left(\frac{n-1}{n}\right)^r - n^2\left(\frac{n-1}{n}\right)^{2r} \tag{7}$$

or, upon rearrangement,

$$\frac{\operatorname{Var} X}{n} = n\left(\left(1 - \frac{2}{n}\right)^r - \left(1 - \frac{1}{n}\right)^{2r}\right) + \left(1 - \frac{1}{n}\right)^r - \left(1 - \frac{2}{n}\right)^r. \tag{8}$$

Now the first and last terms of the right-hand side of (8) are negative, while the second is bounded above by one. Thus we see that $\operatorname{Var} X < n$, independently of the value of r. Knowing this, we can proceed with the generalization of the result of Kim and Rousch (1980) mentioned above.

Let Z_k be the random variable with value $n^{-1}|\operatorname{ran} \alpha_1 \ldots \alpha_k|$, where α_1, $\alpha_2, \ldots, \alpha_k$ are randomly chosen members of T_n. Let M_k denote the limiting value as n increases without bound of $E(Z_k)$, assuming that it exists, and put $M_0 = 1$. We first show inductively that $M_{k+1} = 1 - e^{-M_k}$, which has been established for the $k = 0$ case in our equation (1).

Suppose that $\alpha_1, \alpha_2, \ldots, \alpha_k$ have been randomly picked from T_n and that $|\operatorname{ran} \alpha_1 \alpha_2 \ldots \alpha_k| = r$. The value of Z_{k+1} then has the same distribution as $n^{-1}X$, where X denotes the number of non-empty cells in a ball-tossing experiment in which r balls are tossed independently and at random into n cells. It can be deduced from (6) and (8) that $\sigma_{Z_{k+1}} \to 0$ as $n \to \infty$.

Now let $\varepsilon, \delta > 0$ be given. Recall Chebyshev's Inequality, which states that if X is any random variable with mean μ and standard deviation σ

$$\Pr(|X - \mu| \geq \alpha\sigma) \leq 1/\alpha^2 \ (\alpha > 0)$$

Applying this and induction on k allows us to state that, for n sufficiently large,

$$\Pr(|Z_k - M_k| < \delta) > 1 - \varepsilon.$$

The claim that $M_{k+1} = 1 - e^{-M_k}$ is now obtained by noting that

$$E(Z_{k+1} | Z_k = \lambda) = 1 - (1 - 1/n)^{n\lambda},$$

where λ lies in the arbitrarily small interval $(M_k - \delta, M_k + \delta)$, with probability approaching 1 as n increases without bound, and λ is bounded (between 0 and 1).

Theorem 6.1.1 $M_{k+1} = 1 - e^{-M_k}$ *for all* $k = 0, 1, 2, \ldots$, $\operatorname{Var} Z_k < 1/n$ *and, in addition,* $M_k \sim 2/k$ *as* $k \to \infty$.

Proof To show the asymptotic property of the sequence (M_k), we give the following elegant argument due to T. Doukakaros.

From $M_{k+1} = 1 - e^{-M_k}$ we obtain

$$\frac{2}{M_{k+1}} - \frac{2}{M_k} = 1 + \coth(M_k/2) - \frac{2}{M_k}.$$

From the series expansion for coth $x - 1/x$ one obtains the inequality

$$0 < \coth x - 1/x < x/3 \qquad \text{for } x > 0.$$

From this and the fact that the M_k's clearly approach zero we deduce that the sequence

$$a_k = \frac{1}{M_{k+1}} - \frac{1}{M_k} \to \frac{1}{2} \qquad \text{as } k \to \infty.$$

Consequently,

$$\frac{a_0 + a_1 + \ldots + a_k}{k} \to \frac{1}{2}$$

also; that is,

$$\frac{1}{k}\left(\frac{1}{M_{k+1}} - 1\right) \to \frac{1}{2},$$

from which the result follows. ∎

In Higgins (1988b) it was proved that the sequence $(kM_k)_{k \geq 1}$ is strictly monotonically increasing to its limit of 2, so that $2/k$ is always an overestimate of M_k. Setting $2/k = 1/n$ yields $k = 2n$, which suggests that, on average, a total of $T = 2n$ randomly chosen maps would be required before a constant product results. This turns out to be true, and is well-known in mathematical genetics, where the problem arises in the so-called Wright–Fisher model for the random 'selection' of the parents of a given gene in a population of order n. The 'time' T then corresponds to the number of generations that one is required to trace back in order to find a common ancestor of all individuals in the population. We quote one result along these lines.

Result 6.1.2 *As $n \to \infty$, $T/n \to U$, where U has the same distribution as*

$$\sum_{k=2}^{\infty} \frac{2V_k}{k(k-1)},$$

where V_k has the standard exponential distribution: $\Pr(V_k \geq x) = e^{-x}$; $x \geq 0$. Also, $\lim_{n \to \infty}(ET/n) = 2$.

For further reference, consult Kingman (1978) or Ewens (1990).

The stable and iterative ranges

We now turn to several features of a randomly selected $\alpha \in T_n$ which can be counted. Let Z_n denote the random variable, with value $|\operatorname{stran}\alpha|$ for a randomly chosen $\alpha \in T_n$ (see Exercise 1.6.1). Intimately connected with the stable range of α is the *iterative range of* α, denoted here by it $\alpha = \{1\alpha^k; k = 0, 1, \ldots\}$, where 1 can be taken to be any fixed member of X_n. Let Y_n denote the random variable with value $|\text{it } \alpha|$. A surprising result shown in Harris (1960) is that Z_n and Y_n have the same distribution. We give an elementary proof of this taken from Brown and Higgins (1988), based on the following identity.

Lemma 6.1.3

$$\sum_{i=1}^{k} \binom{m}{i} \frac{(n-m+i)i!}{n^i} = m - \frac{m(m-1)\ldots(m-k)}{n^k} \quad (1 \leq k \leq m \leq n).$$

Proof We proceed by induction on k. For $k=1$ the statement becomes

$$\binom{m}{1} 1! \frac{(n-m+1)}{n} = m - \frac{m(m-1)}{n},$$

which is evidently true.

In general, the left-hand side of the equality can be written as

$$\sum_{i=1}^{k-1} \binom{m}{i} \frac{(n-m+1)i!}{n^i} + \binom{m}{k} \frac{k!(n-m+k)}{n^k},$$

which, by induction, can be written as

$$m - \frac{m(m-1)\ldots(m-k+1)}{n^{k-1}} + \frac{m(m-1)\ldots(m-k+1)(n-m+k)}{n^k},$$

which simplifies to the required expression. ∎

Lemma 6.1.4 *Let $\alpha \in T_n$ with $|\operatorname{stran}\alpha| = k$. Let t_k be the number of extensions β of α to a member of T_n such that $\operatorname{stran}\beta = \operatorname{stran}\alpha$. Then*

$$t_k = kn^{n-k-1} = \frac{k}{n} \cdot n^{n-k}. \tag{9}$$

Remark This lemma has independent interest, for it says that the proportion of all extensions β of $\alpha|\operatorname{stran}\alpha$ to a member of T_n such that $\operatorname{stran}\beta = \operatorname{stran}\alpha$ is k/n; that is, is directly proportional to the order of the given stable range. Moreover, as will be shown, the use of the notation t_k is justified, as t_k depends only on k and not on the main permutation $\alpha|\operatorname{stran}\alpha$; a plausible fact that can be verified independently with a little effort.

Proof We prove (9) by induction on $m = n - k$. For $m = 0$, the statement (9) says the obvious; $t_n = 1$. Now

$$|\{\beta \in T_n : \beta|\operatorname{stran}\alpha = \alpha|\operatorname{stran}\alpha\}| = n^{n-k}$$

can be expressed as

$$n^{n-k} = t_k + \binom{n-k}{1} 1! t_{k+1} + \ldots + \binom{n-k}{i} i! t_{k+i}$$

$$+ \ldots + \binom{n-k}{n-k}(n-k)! t_n. \tag{10}$$

In detail, the term $C_i^{n-k} i! \, t_{k+i}$ counts the number of mappings β such that $\beta|\operatorname{stran}\alpha = \alpha|\operatorname{stran}\alpha$ and $|\operatorname{stran}\beta| = k + i$. Rearranging (10) gives

$$\frac{t_k}{n^{n-k}} = 1 - \sum_{i=1}^{n-k} \binom{n-k}{i} i! \, \frac{t_{k+i}}{n^{n-k}}. \tag{11}$$

By the inductive hypothesis, we have

$$\frac{t_{k+i}}{n^{n-k-i}} = \frac{k+i}{n} \quad (1 \leq i \leq n-k). \tag{12}$$

By using (12) we rewrite (11) as

$$\frac{t_k}{n^{n-k}} = 1 - \sum_{i=1}^{m} \binom{m}{i} i! \, \frac{(n-m+i)}{n^{i+1}}. \tag{13}$$

To prove the lemma we need to show that

$$t_k / n^{n-k} = k/n, \tag{14}$$

and from (13) we see that (14) is equivalent to

$$\sum_{i=1}^{m} \binom{m}{i} i! \, \frac{(n-m+i)}{n^i} = m. \tag{15}$$

But (15) follows immediately from Lemma 6.1.3 upon putting $k = m$. ∎

Theorem 6.1.5 *The random variables Z_n and Y_n have the same distribution: we write this as*

$$Z_n \stackrel{\mathscr{D}}{=} Y_n \quad \text{for all } n = 1, 2, \ldots .$$

Proof We need to show that

$$|\{\alpha \in T_n : |\operatorname{stran}\alpha| = k\}| = |\{\alpha \in T_n : |\operatorname{it}\alpha| = k\}|, \tag{16}$$

for all $1 \leq k \leq n$. The left-hand side of (16) is given by

$$C_k^n k! \, t_k, \tag{17}$$

while the right-hand side is

$$C_{k-1}^{n-1}(k-1)!kn^{n-k}. \qquad (18)$$

The equality of (17) and (18) now follows easily from Lemma 6.1.4. This completes the proof. ∎

The probability distribution for Y_n, and thus for Z_n, can be written down by inspection:

$$\Pr(Y_n = k) = \frac{k}{n} \prod_{i=0}^{k-1}(1 - i/n) = \frac{k(n-1)!}{(n-k)!n^k} \quad (1 \leq k \leq n). \qquad (19)$$

A pleasing bonus from Theorem 6.1.5 is Cayley's Enumeration Theorem for labelled trees on n vertices (Exercise 6.1.1).

The first two moments of Z_n can be found using the following argument from Higgins and Williams (1988).

Proposition 6.1.6

$$E(Z_n) = \sum_{k=1}^{n} \prod_{i=0}^{k-1}(1 - i/n) = \sum_{k=1}^{n} \frac{n!}{(n-k)!n^k};$$

$$E(Z_n^2) = 2n - EZ_n.$$

Proof $E(Z_n) = \sum_{k=1}^{n} \frac{k^2}{n} \prod_{i=0}^{k-1}(1 - i/n)$. We are required to verify that

$$\sum_{k=1}^{n}(1 - k^2/n) \prod_{i=0}^{k-1}(1 - i/n) = 0. \qquad (20)$$

The following device is due to Bruce Brown. Denote $\prod_{i=0}^{k}(1 - i/n)$ by Q_k, and note that $Q_0 = 1$ and $Q_n = 0$. Also,

$$Q_{k-1} - Q_k = Q_{k-1}(1 - (1 - k/n)) = (k/n)Q_{k-1}.$$

Thus the left-hand side of (20) can be written as

$$\sum_{k=1}^{n}(Q_{k-1} - \frac{k^2}{n}Q_{k-1}) = \sum_{k=1}^{n}(Q_{k-1} - k(Q_{k-1} - Q_k))$$

$$= \sum_{k=1}^{n}(kQ_k - (k-1)Q_{k-1}) = nQ_n = 0.$$

In general, let X be a random variable on the positive integers with probability distribution $\{p_k\}_{k \geq 1}$. Then

$$2\sum_{k=1}^{\infty} k \Pr(X \geq k) = 2\sum_{k=1}^{\infty} \sum_{j=1}^{k} jp_k = 2\sum_{k=1}^{n} \frac{k}{2}(k+1)p_k = E(X^2) + E(X).$$

In our case this yields

$$E(Z_n^2) + E(Z_n) = 2 \sum_{k=1}^{n} k \prod_{i=0}^{k-1} (1 - i/n) = 2n \sum_{k=1}^{n} \frac{k}{n} \prod_{i=0}^{k-1} (1 - i/n) = 2n;$$

and hence $E(Z_n^2) = 2n - E(Z_n)$. ∎

Remark All the moments of Z_n can be recursively calculated in this fashion using the well-known recursion for the calculation of sums of powers of the positive integers (Exercise 6.1.3).

Unlike the expectation of R_n, which increases linearly with n, the expected value of Z_n increases as $\sqrt{n}$. The limiting distribution of $V_n = Z_n/\sqrt{n}$ is surprisingly easy to find. For $v \geq 0$,

$$\Pr(Z_n \geq v\sqrt{n}) = \prod_{j=0}^{i} (1 - j/n), \tag{21}$$

where i is the least positive integer such that $i \geq v\sqrt{n} - 1$. Now, in general,

$$(1 - j/n)(1 - (i-j)/n) = 1 - i/n + j(i-j)/n^2. \tag{22}$$

Also, $j(i-j)/n^2 \leq (i/2n)^2$. Hence, by pairing the terms of (21) in the fashion suggested by (22), and deleting the middle term if i is odd, we obtain the inequality

$$\Pr(Z_n \geq v\sqrt{n}) \leq \left(1 - \frac{i}{n} + \frac{i}{4n^2}\right)^{i/2} = \left(1 - \frac{i}{2n}\right)^i.$$

Taking logarithms, and using the fact that $i < v\sqrt{n} \leq i + 1$, we obtain

$$\log \Pr(V_n \geq v) \leq v\sqrt{n} \log(1 - (v\sqrt{n} - 1)/2n). \tag{23}$$

Since $\log(1 - x) \sim -x$ as $x \to 0$, the right-hand side of (23) is of order

$$-v\sqrt{n}(v\sqrt{n} - 1)/2n \to -v^2/2 \quad \text{as } n \to \infty.$$

Taking exponentials, we obtain that, provided that the limit exists,

$$\lim_{n \to \infty} \Pr(V_n \geq v) \leq e^{-v^2/2}. \tag{24}$$

On the other hand, from (22) it is certainly true that

$$(1 - j/n)(1 - (i-j)/n) \geq 1 - i/n,$$

which gives

$$\Pr(V_n \geq v) \geq \begin{cases} (1 - i/n)^{i/2} & \text{if } i \text{ is even,} \\ \left(1 - \frac{(i+1)/2}{n}\right) \cdot (1 - i/n)^{i/2} & \text{if } i \text{ is odd.} \end{cases}$$

In any case we have $\Pr(V_n \geq v) \geq (1 - (i/n))^{(i+2)/2}$.

Again we take logarithms and use the facts that $i < v\sqrt{n} \leqslant i+1$ and $\log(1-x) \sim -x$ as $x \to 0$ to obtain

$$\lim \Pr(V_n \geqslant v) \leqslant e^{-v^2/2}, \tag{25}$$

provided that the limit exists. Combining the arguments for (24) and (25) furnishes the limiting distribution of V_n. Indeed, the following is true.

Theorem 6.1.7 *The sequence of random variables (V_n) defined above approaches in distribution the random variable V with cumulative distribution function $F(v) = 1 - \exp(-v^2/2)$, $v \geqslant 0$. Moreover, the moments of V_n approach those of V; that is,*

$$\lim_{n \to \infty} E(V_n^k) = E(V^k) = \int_0^\infty v^{k+1} e^{-v^2/2} \, dv$$

and, in particular,

$$\lim_{n \to \infty} E(V_n) = \sqrt{\pi/2}. \tag{26}$$

The statement about moments is not immediate as, in general, convergence in distribution does not imply convergence in moments. The proof can be completed by establishing the uniform integrability of the sequence (V_n^k) for all $k \geqslant 1$ (see, for instance, Brown and Higgins 1988). Statement (26) is the analogue of statement (1) for R_n. The value of $E(Z_n)$ is easily computed recursively using the expression for $E(Z_n)$ in Proposition 6.1.6, and leads to an analogue for (2).

Proposition 6.1.8 (Higgins and Williams 1988)

$$\lim_{n \to \infty} \left(\sqrt{\frac{n\pi}{2}} - E(Z_n) \right) = \tfrac{1}{3}.$$

This fact is a consequence of two classical results, of Stirling and of Ramanujan. The first is the well-known factorial approximation formula, which can be expressed as

$$\frac{n! \, e^n}{\sqrt{2\pi} \, n^n \sqrt{n}} = 1 + O(n^{-1}).$$

The second concerns the 'first half' of the McLaurin series for e^n.

Result 6.1.9 (see Hardy et al. 1962)

$$\sum_{k=0}^{n-1} \frac{n^k}{k!} = \tfrac{1}{2} e^n - \frac{n^n}{n!} \varphi_n,$$

where $\varphi_n = 1/3 + o(n^{-1})$ and φ decreases monotonically from $1/2$ to $1/3$.

Proof of Proposition 6.1.8 By Proposition 6.1.6,

$$E(Z_n) = \sum_{k=1}^{n} \frac{n!}{(n-k)!n^k} = \frac{n!}{n^n} \sum_{k=1}^{n} \frac{n^{n-k}}{(n-k)!} = \frac{n!}{n^n} \sum_{m=0}^{n-1} \frac{n^m}{m!},$$

where $m = n - k$. By Ramanujan's result we obtain

$$E(Z_n) = \frac{n!}{n^n}\left(\tfrac{1}{2}e^n - \frac{n^n}{n!}(\tfrac{1}{3} + o(n^{-1}))\right) = \frac{n!e^n}{2n^n} - \tfrac{1}{3} + o(n^{-1}).$$

By Stirling's formula,

$$\frac{n!e^n}{2n^n} = (1 + O(n^{-1}))\sqrt{\frac{n\pi}{2}},$$

whence

$$\sqrt{\frac{n\pi}{2}} - E(Z_n) = \tfrac{1}{3} + O(n^{-1})\sqrt{\frac{n\pi}{2}} + o(n^{-1}) = \tfrac{1}{3} + O(n^{-1/2}),$$

which proves the result. ∎

The component number

We next investigate the number of components of a randomly chosen $\alpha \in T_n$.

Let G_n denote the symmetric group on $X = \{1, 2, \ldots, n\}$. Let τ_n denote the random variable the value of which is the cycle number of a randomly chosen permutation $\pi \in G_n$. It is shown in Feller (1968) that τ_n is distributed as $Y_1 + Y_2 + \ldots + Y_n$, where the $\{Y_i\}$ are independent indicator random variables (they take on the values 0 and 1 only) with $\Pr(Y_i = 0) = (n - i)/(n - i + 1)$, for all $1 \leq i \leq n$ (Exercise 6.1.7). From this we may infer that

$$E(\tau_n) = \sum_{i=1}^{n} \frac{1}{n} \sim \log n \quad \text{and that} \quad \text{Var}\,\tau_n = \sum_{i=1}^{n} \frac{n-i}{(n-i+1)^2} \sim \log n.$$

Consider the random variable C_n, the value of which is the number of components of a randomly selected $\alpha \in T_n$ or, equivalently, the cycle number of the main permutation of α. Now, writing $\#\alpha$ for the number of cycles of the main permutation of α, we see that

$$\Pr(C_n = k | Z_n = i) = \frac{|\{\alpha \in T_n : |\operatorname{stran}\alpha| = i \text{ and } \#\alpha = k\}|}{|\{\beta \in T_n : |\operatorname{stran}\beta| = i\}|}$$

$$= \frac{C_i^n t_i |\{\pi \in G_i : \#\pi = k\}|}{C_i^n i! t_i}$$

$$= \Pr(\tau_i = k).$$

Another way of stating this is as follows.

Proposition 6.1.10 $C_n \stackrel{\mathscr{D}}{=} \tau_{Z_n}$.

Since $E(\tau_n) \sim \log n$ and $E(Z_n) = O(\sqrt{n})$, one might conjecture that $E(C_n) \sim \log \sqrt{n} = (1/2)\log n$, and indeed it is possible by an elementary argument to show that $(E(C_n))/\log n \to 1/2$, an asymptotic result which should be compared with those given in (1) and (26) above. Indeed, it is known that $(C_n - (1/2)\log n)/((1/2)\log n)^{-1/2}$ approaches in distribution and moments the standard normal distribution (see, for instance, Brown and Higgins, 1988). A more detailed analysis of central limit results for random mappings has recently been carried out by Hansen (1989, 1991); see also Donnelly et al. (1991).

We shall prove an analogue of the results given in equation (2) above and in Theorem 6.1.8.

Proposition 6.1.11 $E(C_n) - (1/2)\log n \to (\gamma + \log 2)/2$, where γ is the Euler–Mascheroni constant.

Remark It follows from the preamble concerning τ_n that

$$E(\tau_n) - \log n \to \gamma, \qquad \text{as } n \to \infty.$$

This result was proved by M. Kruskal (1954) by a method which involves solving a differential equation for a generating function of the probabilities which sum to this mean. We give a derivation taken from Higgins and Williams (1988). First we establish an expression for $E(C_n)$ as a sum which is remarkably similar to that for $E(Z_n)$ in Proposition 6.1.6.

Lemma 6.1.12

$$E(C_n) = \sum_{k=1}^{n} \frac{n!}{k(n-k)!n^k} \qquad \text{for all } k = 1, 2, \ldots.$$

Proof It follows from Proposition 6.1.10 that

$$E(C_n) = \sum_{i=1}^{n} E(\tau_i) \Pr(Z_n = i) = \sum_{i=1}^{n} \left(\sum_{k=1}^{i} (1/k) \right) \Pr(Z_n = i)$$

$$= \sum_{k=1}^{n} (1/k) \Pr(Z_n \geq k) = \sum_{k=1}^{n} (1/k) \prod_{j=0}^{k-1} (1 - j/n)$$

$$= \sum_{k=1}^{n} n!/k(n-k)!n^k. \qquad \blacksquare$$

Remark This summation expression for $E(C_n)$ was also arrived at in Kruskal's 1954 article by quite another route.

Proof of Proposition 6.1.11 From the previous lemma

$$E(C_n) = \sum_{k=1}^{n} C_k^n \frac{(k-1)!}{n^k} = \sum_{k=1}^{n} C_k^n \frac{\Gamma(k)}{n^k} = \sum_{k=1}^{n} C_k^n \int_0^\infty \frac{e^{-x} x^{k-1}}{n^k} dx$$

$$= \int_0^\infty \frac{e^{-x}[(1 + (x/n))^n - 1]}{x} dx$$

$$= \int_0^\infty [e^{-x}(1 + (x/n))^n - e^{-x^2/2n}] \frac{dx}{x} + \int_0^\infty [e^{-x^2/2n} - e^{-x}] \frac{dx}{x}$$

$$= I_1 + I_2,$$

say. Upon integrating by parts and replacing $x^2/2n$ by x in the first term of the integrand the second integral becomes

$$I_2 = \int_0^\infty \left[\frac{x}{n} e^{-x^2/2n} - e^{-x} \right] \log x \, dx$$

$$= \int_0^\infty [(1/2)e^{-x}(\log x + \log 2n) - e^{-x} \log x] dx.$$

Since $\int_0^\infty e^{-x} \log x \, dx = -\gamma$ (see Exercise 6.1.5) we obtain,

$$I_2 = (1/2)\log n + (1/2)(\log 2 + \gamma).$$

As for I_1, the integrand is positive and tending to zero as n increases, and equals

$$(1/x)[\exp(-x + x - x^2/2n + x^3/3n^2 - \ldots) - \exp(-x^2/2n)]$$
$$< \exp(-x^2/2n)(x^2/3n^2 + o(n^{-2}))$$

and is integrable, with the integral tending to zero as $n \to \infty$. The term in $n^{-1/2}$ is in fact

$$\frac{1}{3n^2} \int_0^\infty e^{-x^2/2n} x^2 dx = \frac{2}{3\sqrt{n}} \int_0^\infty e^{-x} x^{1/2} dx = \frac{1}{3}\sqrt{\frac{\pi}{2n}}. \quad \blacksquare$$

A more intractable random variable is $M_n(\alpha)$, the order of a maximal component of a randomly chosen $\alpha \in T_n$. The corresponding group problem, which asks for the distribution of the proportion of points in the largest cycle of a randomly chosen member of the symmetric group on n symbols, was first looked at by Shepp and S. Lloyd (1966). The mean of this random variable was shown to approach a number which to five decimal places is 0.62433. An equivalent problem involves recursive splitting of the unit interval, $[0, 1]$: at the kth stage of the experiment ($k = 0, 1, 2, \ldots$) we have an interval $[x_k, 1]$ remaining ($x_0 = 0$) and we choose a point x_{k+1} in this interval according to the uniform distribution and discard the subinterval $[x_k, x_{k+1})$. The maximum of the lengths of the discards, M, is the subject of the paper by C. Lloyd and Williams (1988). A host of symmetries and identities related to the distribution of M are found: for example, $E(M) = 0.62433\ldots$ equals the probability that the first discard is the longest.

Other related results pertaining to random variables arising from T_n are in Arney and Bender (1982) and Donnelly et al. (1991), while a 'relatively simple formula', in terms of Sterling numbers, for the probability distribution of the number of components of a random function has been given by Kupka (1988). Without describing the deeper results of these papers, we mention a few facts: the probability that $\alpha \in T_n$ has just one component is of order $\sqrt{\pi/2n}$ as $n \to \infty$ (Katz 1955); the asymptotic probability that m given numbers are in the same component is

$$\frac{2.4.6.8.\ldots 2m-2}{3.5.7.9.\ldots 2m-1},$$

and the probability that the largest component has order exceeding $n/2$ has a limiting value of $\log(1 + \sqrt{2}) = 0.88137\ldots$. Relatively little is known about $M_n(\alpha)$, the order of the largest component of α, but Watterson and Guess (1977) calculate a mean value of $0.758\ldots$ for $n^{-1}M_n(\alpha)$.

The semigroup of order-preserving mappings on a finite set

Since the full transformation semigroup is replete with combinatorial gems, one naturally turns to other notable semigroups of mappings for possible analogues and generalizations. In Higgins (1990c) the author presented some similar results concerning O_n, the semigroup of all order-preserving mappings on X_n with the usual ordering. If $R_n(\alpha)$ now denotes the random variable the value of which is the rank of a randomly chosen member of O_n, it is the case that $E(R_n) = n^2/(2n-1)$, and the standard deviation of R_n is given by $\sqrt{(n-1)/2} \cdot n/(2n-1)$, so in particular $E(R_n)/n$ approaches a limiting value of $1/2$ from above. The following analogue of Theorem 6.1.1 is shown there.

Result 6.1.13 *Let Y_n be the random variable with value $n^{-1}|\alpha_1 \alpha_2 \ldots \alpha_k|$, where $\alpha_1, \alpha_2, \ldots, \alpha_k$ are randomly selected members of O_n. Then*

$$M_k = \lim_{k \to \infty} E(Y_n) = 1/(1+k), \quad \text{for all } k = 0,1,\ldots.$$

This can be compared to be asymptotic result $M_k \sim 2/k$ for T_n.

One readily sees that O_n is a combinatorial (i.e. $\mathscr{H}$-trivial) semigroup. In addition, the digraph of $\alpha \in O_n$ is a forest of rooted labelled trees, the root of each tree-component being the unique fixed point of that component. Thus the component number and the order of stran α each equal the number of fixed points of α. Let $Z_n(\alpha)$ now denote the random variable with value |stranα|, where α denotes a random member of O_n. The number of mappings α in O_n with $k\alpha = k (k \in X_n)$ is the product of the number of order-preserving mappings from X_{k-1} to X_k and the number of such maps from X_{n-k} to X_{n-k+1}. Thus

$$\Pr(k\alpha = k) = \frac{C(2k-2, k-1).C(2n-2k, n-k)}{C(2n-1, n-1)}$$

as, in general, the number of order-preserving mappings from the set X_r to the set $X_n (r \leqslant n)$ is $C(r+n-1, n-1)$ (Exercise 6.1.8(a)). Hence

$$E(Z_n) = \sum_{k=1}^{n} \Pr(k\alpha = k)$$

$$= C(2n-1, n-1)^{-1} \sum_{k=1}^{n} C(2k-2, k-1).C(2n-2k, n-k)$$

$$= C(2n-1, n-1)^{-1} \sum_{k=0}^{m} C(2k, k).C(2m-2k, m-k),$$

where $m = n - 1$. Now, in general,

$$\sum_{k=0}^{n} C(2k, k).C(2n-2k, n-k) = 4^n \qquad \text{(Exercise 6.1.8(b))},$$

whereupon we obtain the first statement in the next result.

Proposition 6.1.14 $E(Z_n) = 4^{n-1}/C(2n-1, n-1)$. *Moreover,* $(E(Z_n) - \sqrt{n\pi/2}) \to 0$ *as* $n \to \infty$.

Proof To verify that the mean is of order $\sqrt{n\pi/2}$ one readily shows that

$$\frac{2(E(Z_n)^2)}{n} = \frac{(2.4.6 \ldots (2n-2))^2.2n}{(1.3.5 \ldots (2n-1))^2}.$$

This is the product of Wallis which approaches $\pi/2$, from which we infer that $E(Z_n) \sim \sqrt{n\pi/2}$. In the course of the proof of the Wallis limit it is established that

$$\frac{(2.4.6 \ldots 2n)^2}{(1.3.5 \ldots (2n-1))^2 (2n+1)} \leqslant \frac{\pi}{2} \leqslant \frac{(2.4.6 \ldots 2n)^2}{(1.3.5 \ldots (2n-1))^2.2n},$$

from which we can obtain

$$\frac{4E^2(Z_n)}{2n+1} \leqslant \frac{\pi}{2} \leqslant \frac{4E^2(Z_n)}{2n},$$

which yields,

$$0 \leqslant 4E^2(Z_n) - n\pi \leqslant 4E^2(Z_n)/(2n+1).$$

We know from above that the ratio on the right approaches $\pi/2$; thus we conclude that
$$4E^2(Z_n) - n\pi = O(1),$$
from which it follows that the difference $E(Z_n) - \sqrt{n\pi/2}$ approaches zero. ∎

The survey on combinatorial aspects of semigroups of mappings by the author (1990c) contains about thirty references which lead into various realms of mathematics. For reasons of priority, the papers of Harris (1960), and of Rubin and Sitgreaves (1954) deserve special mention. The note of Riemann (1972) also derives the asymptotic distribution of the order of the stable range. Other papers with a probabilistic flavour and relating to semigroups of mappings are those of Harris (1967, 1973) and Harris and Schoenfeld (1967), together with Kim (1971, 1972), Kim and Rousch (1978), Sioson (1972), Taintier (1968), and Todorov (1979, 1980), this latter paper tackling the problem of determining the possible orders of subsemigroups of T_n. A study of products of idempotents in O_n has been carried out by Howie (1971), in fact, the core of O_n is O_n itself, a fact first noted by Aizenstat (1962), the analogue for partial transformations being obtained by Popova (1962). Corresponding results for infinite totally ordered base sets were obtained by Howie and Schein (1973).

Exercises 6.1

1. Use Theorem 6.1.5 to show that:

    (a) the number of non-isomorphic labelled trees on n vertices is n^{n-2} ($n = 1, 2 \ldots$);
    (b) the number of non-isomorphic forests on n vertices consisting of k rooted labelled trees is $C_{k-1}^{n-1} n^{n-k}$.

2. Consider the model of the iterative action of a random $\alpha \in T_n$ as the following ball-tossing experiment. There are n labelled cells, all of which are empty with the exception of cell 1 (representing $1\alpha^0$). Balls are tossed at random into the cells (representing 1α, $1\alpha^2$, etc.) until some cell contains two balls. The number 1 is then a member of stran α if and only if it is cell number 1 that contains two balls at the conclusion of the experiment. Hence deduce the result in Proposition 6.1.6 that
$$E(Z_n) = \sum_{k=1}^{n} n!/(n-k)! n^k,$$
where $Z_n(\alpha) = |\text{stran}\,\alpha|$.

3. (a) Let $\Sigma_{k,n}$ denote the sum of the nth powers of the first k positive integers. By writing

$$\Sigma_{k,n} = \sum_{m=0}^{k-1} (1+m)^n$$

show that

$$\Sigma_{k,n} = \frac{1}{n+1}\left((1+k)^{n+1} - 1 - \sum_{j=0}^{n-1} C_j^{n+1}\Sigma_{k,j}\right)$$

and deduce that $\Sigma_{k,n}$ is a rational polynomial function of k of degree $n+1$.

(b) Show that $E(Z_n^{m-1}) = n^{-1}\sum_{k=1}^{\infty}(\Sigma_{k,m})p_k$, where Z_n is the random variable of Proposition 6.1.6 and $p_k = \Pr(Z_n = k)$. Hence deduce that all moments of Z_n can be expressed in terms of its mean and, in particular, that $E(Z_n^3) = (3n+1)\mu - 3n$.

4. By use of the McLaurin series for $\cosh x$ and $\sinh x$, verify the inequality invoked in the proof of Theorem 6.1.1:

$$0 < \coth x - 1/x < x/3 \qquad \text{for all } x > 0.$$

5. Use logarithmic differentiation of the Weierstrass product for the gamma function

$$\frac{1}{\Gamma(z)} = z e^{\gamma z} \prod_{n=1}^{\infty} ((1+z/n)e^{-z/n})$$

to evaluate the integral used in Proposition 6.1.8.

$$\int_0^\infty e^{-x}\log x\, dx = -\gamma.$$

6. Let τ_n denote the random variable the value of which is the cycle number of π, a randomly chosen permutation from the full symmetric group G_n. Show that τ_n is distributed as $Y_1 + Y_2 + \ldots + Y_n$, where Y_i takes on the value 0 with probability $(n-i)/(n-i+1)$ $(1 \leq i \leq n)$ and has a value of 1 otherwise. Hence deduce the given expressions for mean and variance:

$$E(\tau_n) = \sum_{i=1}^{n} 1/i \sim \log n,$$

$$\text{Var}(\tau_n) = \sum_{i=1}^{n}(n-i)/(n-i+1)^2 \sim \log n.$$

7. Show that O_n for n finite is a regular combinatorial semigroup.
8. (a) Show that the number of order-preserving mappings from an ordered set of cardinality r to another with n elements is the number of non-negative integer solutions to

$$x_0 + x_1 + \ldots + x_r = n - 1$$

and is therefore equal to $C(r + n - 1, n - 1)$. Deduce that
$$|O_n| = C(2n - 1, n - 1).$$

(b) Show that the coefficients of x^n in the expansions of $(1 - 4x)^{-1/2}$ and $(1 - 4x)^{-1}$ are $C(2n, n)$ and 4^n respectively. Hence verify the identity used in Proposition 6.1.14:
$$\sum_{k=0}^{n} C(2k, k) \cdot C(2n - 2k, n - k) = 4^n.$$

10. (Howie and McFadden 1990b) Let $\alpha \in T_n$ of rank r.

(a) Prove that the left ideal $T_n\alpha$ generated by α is of order r^n, while the corresponding right ideal, αT_n, is of order n^r.

(b) Hence deduce that
$$\frac{|T_n\alpha|}{|\alpha T_n|} = (1 - (d/n))^n \cdot n^d, \qquad \text{where } d = n - r,$$

and so the left ideal generated by α has order greater than the corresponding right ideal except when $r = 2$ and $n \leq 4$, or $r = n$.

(c) For fixed defect d, show that the ratio $|T_n\alpha|/|\alpha T_n|$ is, for large n, asymptotic to $(n/e)^d$.

6.2 The solution of equations in finite full transformation semigroups

Square roots in finite full transformation semigroups

In this section we shall show how the digraph representation of $\alpha \in T_n$ can assist in the calculation of all the square roots of α, a problem first investigated by Snowden and Howie (1982). Indeed, the method introduced here can be extended in a straightforward manner to construct algorithms to solve equations of the form $ax^m b = c$ and $ax = xb$ ($a, b, c \in T_n$) (Higgins 1988a), although their description becomes fairly complicated. The following comes from Higgins (1986c).

Let $\alpha \in T_{15}$ be defined by its digraph, given below:

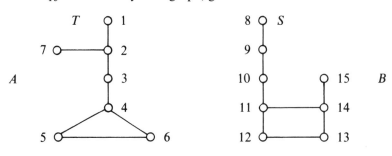

Combinatorial aspects of transformation semigroups | 205

The mapping α partitions X_{15} into two sets corresponding to stran α and its complement, and this partition is invariant under the taking of powers, and therefore also the taking of mth roots. This observation allows the problem of finding the roots of α to be split into two constructions corresponding to the two classes of the partition.

Although our aim is to find a method of constructing square roots in T_n, we first consider α^2 in order to discover how to recognize squares. The graph of α^2 is:

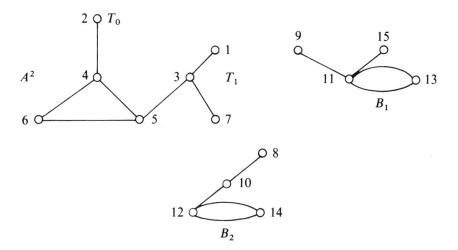

Observe that the component A, the cycle of which is of odd order, has given rise to one component in α^2 the cycle of which is of the same odd order. The tree T has given rise to a pair of trees, T_0 and T_1, each rooted on the cycle of A (in all that follows the word 'tree' shall stand for 'in-tree', as in Section 1.6). The tree T_0 has the same root as T and its vertices are the vertices of T the distance from the root of which is even. Similarly, T_1 has as its points all the points of T an odd distance from the root, plus another point of the cycle as root. We shall not spell out the precise relationship between T_0 and T_1 yet, but note that the height of the 'odd' tree T_1 (that is, the greatest distance of a point of T_1 to its sink) will always be equal to, or one greater than, the height of the corresponding 'even' tree T_0.

In contrast, the component B, upon squaring, gives rise to a pair of components, B_1 and B_2. This occurs because the cycle of B has even order. The tree S 'splits' into an even–odd pair of trees in a similar fashion to T, but the even and odd trees lie on different components.

The foregoing casual analysis does contain all the ideas involved in the solution of the problem. Indeed, we can already say that the map α as given above has no square root, by arguing as follows. Suppose that $\beta \in T_{15}$ and $\beta^2 = \alpha$. The component A of α must have arisen from the squaring of a component of β with a 3-cycle, as the only other way a 3-cycle could be introduced is by squaring a 6-cycle, which would of course create two 3-cycles. The tree T would then be half of an even–odd pair of trees, the partner of which would also lie on the cycle (456). In the absence of such a partner, we infer that no such β exists. The argument is even quicker if we focus our attention on the component B, for in a square the components with cycles of even order must occur in pairs. Hence α is not a square as it has but one component with a 4-cycle.

As another example consider the member α of T_{13} given by

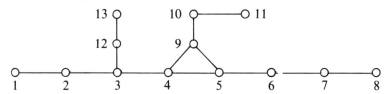

We can state immediately that α is not a square as there are an odd number of trees rooted on its cycle, and so they cannot be associated in even–odd pairs (we must be a little more careful—a tree with one arrow upon squaring gives rise to a pair in which the 'even tree' has no arrows; however, this is not a possibility here as α has no single-arrow trees).

The construction of parent trees

For a component A [tree T] of $\alpha(\alpha \in T_n)$ we shall write A^2 [T^2] for the corresponding subgraph of α^2.

To examine the relationship between a tree T and its square we introduce the idea of even–odd offspring. Let T be a tree with sink 0 and other points $1, 2, \ldots, m$, say. We define the *even–odd offspring of T* as an ordered pair of trees (T_0, T_1). The points of T_0, the *even tree*, are the points of T an even distance from the sink (including the sink) and jk is an arc of T_0 if $d(j, k) = 2$ in T ($d(j, k)$ denotes the length of the dipath from j to k). The points of the *odd-tree* T_1 are the points of T an odd distance from the sink, together with a new point $0'$, and jk is an arc of T_1 if $d(j, k) = 2$ in T, or $k = 0'$ and $d(j, 0) = 1$ in T. We call T a *parent tree* of the ordered pair (T_0, T_1).

One of the constructions that will need to be performed in order to find all square roots of a given $\alpha \in T_n$ will be the construction of all parent trees (if any) of a given pair of trees (T_0, T_1). To this end we shall investigate the relationship between a tree T and its offspring.

Take a maximal directed path P of T from an endpoint u of T to the sink, and label the points of P by $k, k-1, \ldots, 0$, where $d(u, 0) = k \geq 1$. The path P corresponds to maximal directed paths (P_0, P_1) in (T_0, T_1) respectively, in which either $|P_0| = |P_1|$ or $|P_1| = |P_0| + 1$ according as k is even or odd. ($|P|$ denotes the length of the path P.) Now consider a sub-tree T' of T rooted at the point $2r$ on P ($0 \leq 2r \leq k - 1$). The tree T' corresponds to a pair of trees

T'_0, T'_1 rooted on P_0 and P_1 respectively. The pair (T'_0, T'_1) is the even–odd offspring of T'; each member of the pair is rooted at a distance r from the sink of P_0 and of P_1 respectively. On the other hand, a tree T' rooted at a point $2r + 1$ of P $(1 \leqslant 2r + 1 \leqslant k - 1)$ gives rise to a pair of trees T'_0 and T'_1 rooted on P_1 and P_0 respectively. Furthermore, T'_0 is rooted at a distance $r + 1$ from the sink of P_1, while T'_1 is rooted as a distance r from the sink of P_0; the pair (T'_0, T'_1) is again an even–odd offspring pair of T'.

These observations allows us to construct all parent trees of a given pair (T_0, T_1) of trees with no common non-root points. We assume inductively that we may construct all parent trees of any such pair (T'_0, T'_1) for which the total number of points is less than that of (T_0, T_1). There is no difficulty in beginning this induction, for the pair

$$(\circ\ 0,\ 1\ \circ\!\!\longrightarrow\!\!\circ\ 0') \quad \text{has a unique parent in} \quad 1\ \circ\!\!\longrightarrow\!\!\circ\ 0.$$

If (T_0, T_1) is an even–odd offspring pair of some tree T, it must be possible to choose maximal paths P_0, P_1 to the sinks 0 and $0'$ of T_0 and T_1 respectively, such that $|P_1| = |P_0|$ or $|P_1| = |P_0| + 1$. Furthermore, it must be possible to make this choice in such a way that the rooted trees of (P_0, P_1) can be listed in even–odd offspring pairs so that for any such pair (T'_0, T'_1) either T'_0 is rooted on P_0 and T'_1 is rooted on P_1 at some distance r from their respective roots, or T'_0 is rooted on P_1 at a distance $r + 1$ from $0'$ and T'_1 is rooted on P_0 at a distance r from 0 $(r \geqslant 0)$.

We then construct a path P from (P_0, P_1) as follows. Label the points of P_0 by $0, 1, 2, \ldots, k$ (where $k = |P_0|$) and those of P_1 by $0', 1', \ldots$, up to k' or $(k + 1)'$, as the case may be. The points of P from the sink outwards are then $0, 1', 1, 2', 2, \ldots$ ending either k', k or $k', k, (k + 1)'$, as the case may be. The trees of P_0 and P_1 have been associated in pairs according to the criterion of the previous paragraph. For each such pair (T'_0, T'_1) construct a parent tree T' which will then have its sink on P at either the point r or r', as the case may be. The tree T so constructed is then a parent of (T_0, T_1), and all such parent trees can be so constructed.

The theory described above will be illustrated by means of the following example. Let $\alpha \in T_{20}$ be defined by the digraph.

Example 6.2.1

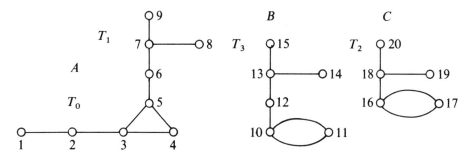

We shall eventually find all square roots of α, but for moment let us calculate the parent trees of (T_0, T_1) and of (T_2, T_3) beginning with the former pair.

There are two choices for a maximal path P_1 from T_1:

$$\underset{9}{\circ}\!\!\rightarrow\!\!\underset{7}{\circ}\!\!\rightarrow\!\!\underset{6}{\circ}\!\!\rightarrow\!\!\underset{5}{\circ} \quad \text{or} \quad \underset{8}{\circ}\!\!\rightarrow\!\!\underset{7}{\circ}\!\!\rightarrow\!\!\underset{6}{\circ}\!\!\rightarrow\!\!\underset{5}{\circ}$$

If we choose the latter, then the only possible choice for P_0 is

$$\underset{1}{\circ}\!\!\rightarrow\!\!\underset{2}{\circ}\!\!\rightarrow\!\!\underset{3}{\circ}$$

There is only one non-trivial tree on P_1 or P_0, the tree

$$\underset{9}{\circ}\!\!\rightarrow\!\!\underset{7}{\circ}$$

occurring at the point of P_1 labelled by 7, which can be paired with the trivial tree at the point of P_0 labelled by 1, to give an even–odd pair

$$(T'_0, T'_1) = (\circ\, 1, 9\, \circ\!\!\rightarrow\!\!\circ\, 7)$$

in accord with the criteria laid down above (since T'_1 lies on P_1 at a distance 2 from the root, its partner, T'_0, must also lie at a distance of 2 from the root of P_0). Our parent tree is then:

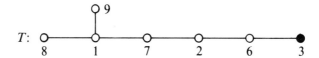

where the root, labelled by 3, is shaded. This parent tree is unique, as the alternative choice for P_1 leads also to T.

For the pair (T_2, T_3) drawn from the components B and C respectively, we select the path

$$\underset{14}{\circ}\!\!\rightarrow\!\!\underset{13}{\circ}\!\!\rightarrow\!\!\underset{12}{\circ}\!\!\rightarrow\!\!\underset{10}{\circ}$$

for our P_1 (the other choice leads to the same set of parents). There are two possible choices for P_0; let us first choose

$$\underset{19}{\circ}\!\!\rightarrow\!\!\underset{18}{\circ}\!\!\rightarrow\!\!\underset{16}{\circ}$$

We may regard

$$(15\, \circ\!\!\rightarrow\!\!\circ\, 13, \quad 20\, \circ\!\!\rightarrow\!\!\circ\, 18)$$

as an even–odd pair positioned at the second point of P_1 and the first point of P_0, whence our parent tree is T:

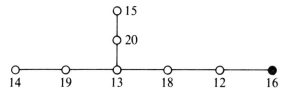

Alternatively, we may regard

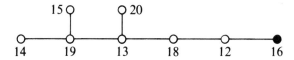

as even–odd pairs yielding the parent tree T':

15 ○ ○ 20
○——○——○——○——○——●
14   19   13   18   12   16

The alternative choice for P_0 gives another two parent trees, making four in all.

The construction of square roots

Let $\alpha \in T_n$. We shall say that a component A of α is *odd* (*even*) if its cycle, denoted by $Z(A)$, is of odd (even) order. We examine the relationship between the components of α and those of its square.

Let A be an odd component of α. As noted before, A^2 is also an odd component of α^2. Each tree T rooted on $Z(A)$ gives rise to an even–odd offspring pair (T_0, T_1) on A^2. The remaining question is to determine the point $0'$, the sink of T_1. Clearly, if we label the sink of T by 0, then $0'$ is the point of $Z(A)$ adjacent to 0 as we travel anticlockwise. If $Z(A)$ has $2t - 1$ points ($t \geq 1$) then the dipath from 0 to $0'$ in $Z(A^2)$ has length t. We call such a positioning of the roots of T_0 and T_1 around $Z(A^2)$ *consistent*.

Finally, let A be an even component of α with $Z(A)$ of order $2t$ ($t \geq 1$). Then A consists of two components, A_0 and A_1, each of the cycles of which has order t. A tree T of A gives rise to even–odd offspring T_0, T_1 rooted on different components, and given the roots of T_0 and T_1 the cycle $Z(A)$ can be uniquely reconstructed: if T_0 and T_1 are rooted at 0 and $0'$ on $Z(A_0)$ and $Z(A_1)$ respectively with $Z(A_0) = (0, 1, \ldots, t)$ and $Z(A_1) = (0', 1', \ldots, t')$, then $Z(A) = (0, 0', 1, 1', \ldots, t, t')$. Hence all offspring pairs of the trees of A must be rooted on A_0 and A_1 so as to determine the same cycle $Z(A)$. We call a list of pairs of the trees of A_0, A_1 *consistent* if each pair determines the same cycle $Z(A)$.

Theorem 6.2.2 *Let $\alpha \in T_n$. Then α is a square if and only if the components of α can be grouped in pairs, (A_0, A_1) such that either:*

(i) $A_0 = A_1 = A$, *say, A is odd and the trees of A can be consistently listed in offspring pairs; or*
(ii) $A_0 \neq A_1$, $|Z(A_0)| = |Z(A_1)|$, *and the trees of A_0, A_1 can be grouped consistently in offspring pairs.*

Furthermore, each such grouping allows construction of a square root, and all square roots can be so constructed.

Proof It remains to check that given α and such a positioning of its components we may construct a square root. Suppose an odd component is paired with itself, as in condition (i). The cycle $Z(A)$ of A, which we take as $1\,2\,3 \ldots 2t-1)$, has a unique square root in $Z(\bar{A}) = (1\,t+1\,2t+(2\ldots t\,2t-1)$. For each pair of trees (T_0, T_1) construct a parent tree T. We then construct the component $\bar{A}$ with cycle $Z(\bar{A})$ and one parent tree for each offspring pair. The consistency of the pairing guarantees that the reconstructed component $\bar{A}$ is such that $\bar{A}^2 = A$.

Finally, suppose that A_0, A_1 are paired in accordance with (ii). Take an offspring pair (T_0, T_1) and construct the unique cycle $Z(\bar{A})$ the square of which is $Z(A)$, and such that the roots of T_0 and T_1 are anticlockwise adjacent on $Z(\bar{A})$. Consistency of the pairing allows construction of a component $\bar{A}$ with cycle $Z(\bar{A})$ the square of which is the pair (A_0, A_1).

Therefore a square root of α may be constructed, and we obtain distinct roots for each choice of pairings of components and of trees. ∎

We finish this section by calculating the square roots of the mapping α of T_n, as given in Example 6.2.1.

The only possible pairing of the components is (A, A) and (B, C). For the (A, A) case the only possible pairing of trees of A is (T_0, T_1). Note that this pairing is consistent (if T_0 was rooted at the point labelled 4, the pairing would be inconsistent and we would conclude that α was not a square). The unique parent tree of this pair has been calculated in the working of Example 6.2.1. For the (B, C) pairing the only possible pairing of the trees is (T_2, T_3). Since there is just one pair to consider, consistency is automatic. The four parent trees of (T_2, T_3) were calculated in Example 6.2.1, giving $1 \times 4 = 4$ square roots in all, one of which is

Combinatorial aspects of transformation semigroups | 211

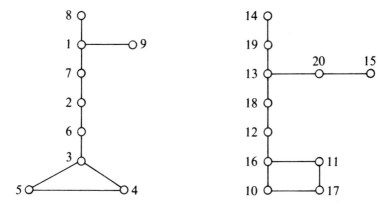

where we have chosen the tree labelled T in Example 6.2.1 as the parent tree for (T_2, T_3).

One facet of the algorithm which does not arise in the above calculation is present in our next example.

Example 6.2.3 Consider $\alpha \in T_{11}$ defined by the digraph

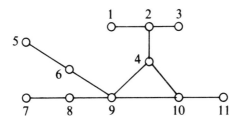

The vertex 9 can be regarded as the root of one or of two trees. If thought of as the root of one tree only, this tree can be the even half of a pair, the partner of which is the tree rooted at 4. The tree at 10 can then be paired with a trivial even tree at 4, and one of the square roots which arises from such pairings is given below:

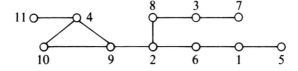

On the other hand, if we regard 9 as the root of two trees we can take one to be even and pair it with the tree rooted at 4, and the other to be odd and pair it with the tree at 10. A square root that results from this pairing is:

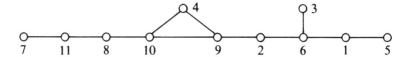

Theorem 6.2.2 can be used to construct all square roots of $\alpha \in PT_n$. We identify PT_n with the subsemigroup of all members of $\mathcal{T}_X$ which fix 0, where $X = \{0, 1, 2, \ldots, n\}$. We now calculate the square roots of α, regarding it as a member of $\mathcal{T}_X$, but only roots that fix 0 need be considered.

The method given here can be extended to the problem of all pth roots for any prime P which then, as mentioned in the introduction, allows the problem of finding all mth roots of a given $\alpha \in T_n$ to be solved.

A theorem of Schein

We complete this section with a proof of a result due to Schein (1971), which is akin to the above technique of cycle and tree manipulation of the digraph.

Theorem 6.2.4 *A finite full transformation semigroup, T_n, is covered by its inverse subsemigroups.*

We prove the theorem by showing that each $a \in T_n$ has at least one *strong inverse* b, meaning an inverse b of a such that $\langle a, b \rangle$ is an inverse subsemigroup of T_n. In fact, we shall determine all members of $S(a)$, the set of all strong inverses of a.

Remark The result is false for infinite full transformation semigroups: see Exercise 6.2.8 below. However, the proof is valid under the weaker assumption that the mapping $a \in \mathcal{T}_X$ is of finite index.

The next result, attributed by Schein to Gluskin, identifies the characteristic property of a pair of strong inverses in any semigroup.

Lemma 6.2.5 (Gluskin's Lemma) *Let S be an arbitrary semigroup with $(a, b) \in V(S)$. Then $\langle a, b \rangle$ is an inverse subsemigroup of S if and only if:*

6.2.6 $a^m b^m$ *commutes with* $b^n a^n$ *for all positive integers $m, n \geq 1$.*

Proof In the forward direction the statement is clearly true. The stages of argument in the reverse direction are conveniently presented as a series of short lemmas. We proceed under the hypothesis given in 6.2.6.

Lemma 6.2.7 $b^n \in V(a^n)$ *for all $n \geq 1$.*

Proof The result is given for $n = 1$. For $n > 1$ we use induction to deduce that

$$a^n b^n a^n = a \cdot \underline{a^{n-1} b^{n-1}} \cdot \underline{ba} \cdot a^{n-1} = aba \cdot a^{n-1} b^{n-1} a^{n-1} = a \cdot a^{n-1} = a^n,$$

where the underlined terms have been commuted. (Such commutation will be indicated in this way throughout the proof.) Similarly, $b^n a^n b^n = b^n$. □

Lemma 6.2.8 *Any $w \in \langle a, b \rangle^1$ has the form $a^x b^y a^z$ for some $x, y, z \geq 0$.*

Proof We shall show that the statement holds for any w of the form $a^m b^n a^p b^q$ ($m, n, p, q \geq 0$), from which the general statement easily follows. First we establish a result symmetric in a and b, namely that w has a factorization in one of the forms $a^x b^y a^z$ or $b^x a^y b^z$.

If $n \geq p \geq q$ we obtain

$$a^m b^n a^p b^q = a^m b^{n-p+q} \underline{b^{p-q} a^{p-q}} a^q b^q = a^m b^{n-p+q} a^q b^p a^{p-q}$$
$$= a^m b^{n-p} b^p a^p a^{p-q} = a^m b^n a^{p-q}.$$

By symmetry, if $m \leq n \leq p$, then $w = b^{n-m} a^p b^q$.

If $m \geq n$ then we may assume that $n > p > q$ for, otherwise, if $n \leq p$:

$$w = a^{m-n} a^n b^n a^n a^{p-n} b^q = a^{m+p-n} b^q,$$

while if $n > p \leq q$ then

$$w = a^m b^{n-p} b^p a^p b^p b^{q-p} = a^m b^{n+q-p}.$$

It follows that we can produce a required factorization if $m \geq n$ or if $q \geq p$. On the other hand, if $m < n$ and $q < p$ at least one of the cases $m \leq n \leq p$ or $q \leq p \leq n$ applies, which have been dealt with above.

To complete the proof consider w' of the form $b^x a^y b^z$. If $y \leq x, z$ the product collapses to b^{x+z-y}, so without loss take $x < y$. Then

$$b^x a^y b^z = b^x a^x a^{y-x} b^z = \underline{b^x a^x} \cdot \underline{a^{y-x} b^{y-x}} \cdot a^{y-x} b^z = a^{y-x} b^y a^y b^z.$$

If $y \leq z$ this yields $a^{y-x} b^z$; otherwise, $y > z$ and we obtain

$$a^{y-x} b^z \underline{b^{y-z} a^{y-z}} a^z b^z = a^{y-x} b^z a^z b^z b^{y-z} a^{y-z} = a^{y-x} b^y a^{y-z},$$

as required. □

Lemma 6.2.9 *Any $w \in \langle a, b \rangle$ can be written in one of the forms:*

(i) $a^x b^y$ $(x \geq 1, y \geq 0)$;
(ii) $b^y a^z$ $(y, z \geq 1)$;
(iii) $a^x b^y a^z$ $(0 \leq x, z; y > x, z)$.

Proof By Lemma 6.2.8 we have $w = a^x b^y a^z$, say. If $y \leq x, z$ then $w = a^{x+z-y}$. Next, if $x \leq y \leq z$ then

$$a^x b^y a^z = \underline{a^x b^x} \underline{b^{y-x} a^{y-x}} \cdot a^{z-y+x} = b^{y-x} a^{y-x} a^x b^x a^{z-y+x} = b^{y-x} a^z.$$

214 | Techniques of semigroup theory

By symmetry, if $z \leq y \leq x$ then $w = a^x b^{y-z}$. The remaining case is where w has the form given in (iii).

Lemma 6.2.10 *The subsemigroup $\langle a, b \rangle$ of S is regular.*

Proof Suppose that $w \in \langle a, b \rangle$ has the form as in (iii) of Lemma 6.2.9. Then

$$a^x b^y a^z . b^z a^y b^x . a^x b^y a^z = a^x b^y a^y b^y a^z = a^x b^y a^z,$$

and another similar calculation shows that $(a^x b^y a^z, b^z a^y b^x) \in V(\langle a, b \rangle)$.

If w has the form $a^x b^y$ $(x \geq 1, y \geq 0)$ then $a^x b^y . a^y b^x . a^x b^y = a^x b^m a^m b^y$ (where $m = \max(x, y)) = a^x b^y$, and then one easily deduces that $(a^x b^y, a^y b^x) \in V(\langle a, b \rangle)$. Finally, if $w = b^y a^z$ $(y, z \geq 1)$ then one similarly verifies that $b^z a^y \in V(b^y a^z)$. □

We now complete the proof of Gluskin's Lemma. Observe that the proof of Lemma 6.2.8 showed that any word w in the letters $\{a, b\}$ represents a member of $\langle a, b \rangle$ of the form $a^x b^y a^z$ such that $|w|_a - |w|_b = x - y + z$, (where $|w|_a$ denotes the number of occurrences of a in the given factorization of w). It is now routine to verify that the set

$$E = \{a^m b^{m+n} a^n; m + n \geq 1\}$$

forms a subsemilattice of $\langle a, b \rangle$, and furthermore, for any $w \in \langle a, b \rangle$ taken to be in one of the forms given in Lemma 6.2.9, $w\bar{w}, \bar{w}w \in E$, where $\bar{w}$ is the inverse of w given in the proof of Lemma 6.2.9. The result now follows from Exercise 1.2.10(iii). ■

We now return to the problem of constructing all strong inverses b of a given member a of T_n. The basic idea is to construct the digraph of b by 'reversing all arrows' of the digraph of a. The difficulty lies in discovering the correct treatment of the points of a with in-degree unequal to one.

We shall introduce several concepts pertaining to a point x of X_n which will require the extended integers, by which we mean the integers together with a new symbol ω endowed with the properties that $n < \infty$, and $\infty \pm n = \infty$ for all integers n.

For each $x \in X_n$ the *depth of x*, denoted by $d(x)$, is the length of a longest dipath ending at x; if $x \in \text{stran } a$ we conventionally define $d(x) = \infty$. Note that $d(x) = k < \infty$ if and only if $x \in \text{ran } a^k \setminus \text{ran } a^{k+1}$. The *height of x*, denoted by $h(x)$, is the least non-negative integer k such that $d(xa^k) > k + d(x)$; again we take $h(x) = \infty$ if $x \in \text{stran } a$. The height of x is the length of the dipath which begins at x and ends at the first point u the depth of which exceeds that of its predecessor by more than 1: in other words, u is the first vertex we meet on the dipath from x which is also the endpoint of some longer dipath.

Lemma 6.2.11 *Let $a \in T_n$ and suppose that $b \in S(a)$. Suppose that x_1, x_2 are members of X_n with $d(x_1), d(x_2) \geq k$ and $x_1 b^k = x_2 b^k$. Then $x_1 = x_2$.*

Remark If $k = \infty$ this lemma has the interpretation that all non-negative powers of b can be cancelled in the given circumstances. In the sequel we shall not state the interpretation of similar statements in cases involving infinite height or depth, as they are generally clear and of little importance to our purposes.

Proof By hypothesis there exists $y_1, y_2 \in X_n$ such that $y_1 a^k = x_1$ and $y_2 a^k = x_2$. Since $b^k \in V(a^k)$ we obtain

$$x_1 = y_1 a^k = y_1 a^k b^k a^k = x_1 b^k a^k = x_2 b^k a^k = y_2 a^k b^k a^k = y_2 a^k = x_2. \blacksquare$$

Lemma 6.2.12 *Let $a \in T_n$ and $b \in S(a)$. Then:*

(i) *for any x with $d(x) > 0$, xb is a member of xa^{-1} of maximal depth;*
(ii) *if $d(x) = 0$ then $(xb)a^{h+1} = xa^h$, where $h = h(x)$, the height of x.*

Proof (i) Since $b \in V(a)$ we know that $xb \in xa^{-1}$. Suppose that $y \in xa^{-1}$ but $d(y) > d(xb) = k$, say $(k \geq 0)$. Then $d(yab \cdot b^{k+1} a^{k+1}) \geq k + 1$, and since $b \in S(a)$, $ab \cdot b^{k+1} a^{k+1} = b^{k+1} a^{k+1} \cdot ab$. Now there exists some $z \in X_n$ such that $y = za^{k+1}$, and thus

$$yab \cdot b^{k+1} a^{k+1} = yb^{k+1} a^{k+1} \cdot ab = za^{k+1} b^{k+1} a^{k+1} \cdot ab$$
$$= za^{k+1} \cdot ab = y \cdot ab = xb;$$

but $d(xb) = k$, a contradiction.

(ii) Let $y = xa^h$ be the first point in the sequence $x, xa, xa^2, \ldots, y = xa^h, \ldots$ such that $d(xa^k) > k$. Now $xa^h b^h \cdot ba = xa^h b^h$ as $d(xa^h b^h) = d(yb^h) > 0$ by part (i). Since idempotents commute with each other, we obtain $xa^h b^h = xa^h b^h \cdot ba = (xb)a^{h+1} b^h \Rightarrow xa^h = y = (xb)a^{h+1}$ (by Lemma 6.2.11), whence the condition of (ii) is satisfied. $\blacksquare$

It follows from (i) and (ii) that if a ånd b are a pair of mutual strong inverses, then a and b share common orbits and stable range. In fact, it C is a component of the digraph of a with $Z = Z(C)$ its cycle then $b|V(Z) = Z^{-1}$, the reverse cycle of Z.

Lemma 6.2.13 *Let b satisfy the condition of Lemma 6.2.12(i). If $d(x) \geq k$ then $xb^k a^k = x$. Furthermore, if $xa^s b^s = x$ $[xb^s a^s = x]$, then $xa^r b^r = x$ $[xb^r a^r = x]$ for all $0 \leq r \leq s$.*

Proof Since $d(x) \geq k$ we can express x as za^k for some $z \in X_n$. The statement holds trivially if $k = 0$, so assume that $k > 0$ and that the statement holds for all $0 \leq l < k$. Then $xb^k a^k = xb^{k-1} baa^{k-1}$, and by the given condition $d(xb^{k-1}) = d(za^k b^{k-1}) \geq 1$, so that $xb^{k-1} baa^{k-1} = xb^{k-1} a^{k-1} = x$ by the inductive hypothesis.

To prove the second statement we first observe that

$$xa^r b^r = xa^s b^s . a^r b^r = xa^s b^{s-r} b^r a^r b^r.$$

Now $d(xa^s b^{s-r}) \geq s - (s-r) = r$, so by the first part above we obtain

$$xa^r b^r = xa^s b^{s-r} b^r = xa^s b^s = x.$$

As for the bracketed statement, we have

$$xb^r a^r = xb^s a^s b^r a^r = xb^s a^s \quad \text{(by the first statement)} = x,$$

as required. ∎

Remark Note that it follows from Lemma 6.2.13 that the statement of Lemma 6.2.11 holds under the condition (i) of Lemma 6.2.12.

The greatest extended integer k such that $xb^k a^k = x$ is clearly $d(x)$: we define the *grasp of* x to be the greatest extended integer g such that $xa^g b^g = x$. Although $g(x)$ will depend on b, it is always the case that $g(x) \leq h(x)$, with the inequality strict unless $g(x) = \infty$. The extended integer $g + 1$ will occur frequently, so we name it the *reach of* x, and write $r(x)$ to denote it (as an aid to memory, the reach exceeds the grasp).

We next introduce a pre-order on X_n that will be useful in some inductive arguments that follow.

Once again, let $a \in T_n$, and let b be a strong inverse of a. Let S denote the (inverse) subsemigroup $\langle a, b \rangle$ of T_n. Define a pre-order $\prec$ on X_n by saying that $x \prec y$ if there exists $w \in S^1$, such that $xw = y$. Let $\sim$ be the equivalence relation on X_n induced by $\prec$; that is, $x \sim y$ if $xw_1 = y$ and $yw_2 = x$ for some $w_1, w_2 \in S^1$. We shall write $[x]$ for the $\sim$-class associated with x, and denote by $\leq$ the partial order on the family of $\sim$-classes of X_n whereby $[x] \leq [y]$ if and only if $x \prec y$.

Notation We shall have occasion to write x^t for xa^t $(t > 0)$ and for xb^{-t} $(t < 0)$; we take $x^0 = x$.

Define the *level* of $x \in X_n$ as $l(x) = d(x) + g(x)$.

Lemma 6.2.14 $d(xa) = d(x) + 1$, $d(xb) \geq d(x) - 1$, $g(xa) \geq g(x) - 1$, and $g(xb) \geq g(x) + 1$ for all $x \in X_n$. Hence, for any $w \in S$, $l(xw) \geq l(x)$; indeed, if $d(x) = 0$ then $l(xb) > l(x)$. Also $g(xb^k) \geq g(x) + k$.

Proof The first claim is obvious, and the second follows from Lemma 6.2.12(i). To see the third let $g = g(x) \geq 1$ (the inequality being trivially satisfied if $g = 0$); then $xa \cdot a^{g-1} b^{g-1} \cdot b = xab$ which, by Lemma 6.2.11, implies $xa \cdot a^{g-1} b^{g-1} = xa$, and so $g(xa) \geq g - 1$. For the next inequality we

Combinatorial aspects of transformation semigroups | 217

use the fact that idempotents commute in $\langle a, b \rangle$ to deduce that $xb \cdot a^{g+1} b^{g+1} = xba \cdot a^g b^g \cdot b = xa^g b^g \cdot bab = xbab = xb$, whence $g(xb) \geq g + 1$.

The penultimate statement now follows easily. The final statement follows by induction on k and the fact that $g(xb) \geq g(x) + 1$. ∎

Lemma 6.2.15 $[x] = \{x^t: -d \leq t \leq g\}$.

Proof It follows from Lemma 6.2.13 that the set $U \subseteq [x]$, where U is the set on the right of the equation, so we prove the reverse containment, and only the case where $x \notin \operatorname{stran} a$ presents any difficulties.

Lemma 6.2.14 yields that if $x \sim y$ then $l(x) = l(y)$. Now, if $xw \notin U$ then w can be factored in S as $w_1 w_2$, where either $xw_1 = xa^{g+1}$, or $xw_1 = xb^{d+1}$. However, $l(xb^{d+1}) > l(x)$, which implies that $l(xw) > l(x)$ by Lemma 6.2.14, whence it follows that $xw \notin [x]$. Hence we are left with the case where $xw_1 = xa^{g+1}$. We shall derive a contradiction from the supposition that $xw \sim x$.

Write y for $xa^{g+1} = xw_1$ and replace w_2 by $v \in S$ such that $yv = yw_2$ and v has a factorization $v = c_1 c_2 \ldots c_p (c_i \in \{a, b\})$, which is minimal in length amongst all members u of S such that $yu = yw_2$. The sequence of points $y, yc_1, yc_1 c_2, \ldots, yv$ then represents a path in the underlying graph of the mapping a because, with the possible exception of yv, none of these vertices are endpoints (by Lemma 6.2.14 and the minimality of the length of v). Since the edge (xa^g, xa^{g+1}) is a bridge of the underlying graph, it follows from the minimality condition on v that $yc_1 = xa^g$, but this is not so for either possible value a or b of c_1, thus completing the proof. ∎

Remark It follows from the previous two lemmas that $x \sim y$ implies that $l(x) = l(y)$, and thus we can define the *level of* $[x]$ by $l([x]) = l(x)$. Indeed, $x \prec y$ implies $l([x]) \leq l([y])$, with equality if and only if $x \sim y$. Moreover, $|[x]| = l(x) + 1$, for all $x \notin \operatorname{stran} a$, so that if $x \prec y$ and $y \notin \operatorname{stran} a$ then $|[x]| \leq |[y]|$.

We are now in a position to state a construction which we shall show yields all the strong inverses of a given mapping $a \in T_n$.

Theorem 6.2.16 *Let $a \in T_n$. Then $S(a)$ is not empty. In fact, any mapping $b \in T_n$ constructed as follows is a strong inverse of a and, conversely, all members of $S(a)$ arise in this fashion:*

(i) *if $d(x) > 0$, let xb be any member of xa^{-1} of maximal depth;*
(ii) *if $d(x) = 0$, $r(x) = r$, and $h(x) = h$, put $xb = y$, where $g(y) \geq r$, $ya^{h+1} = xa^h$, and $xba \cdot a^r b^r = xa^r b^r \cdot ba$.*

Proof Let $b \in S(a)$. Then b satisfies (i), by Lemma 6.2.12(i). Next, if $d(x) = 0$

218 | Techniques of semigroup theory

then $ya^{h+1} = xa^h$, by Lemma 6.2.12(ii), and $g(y) \geq r$, by Lemma 6.2.14, while the last condition is satisfied because idempotents commute in $\langle a, b \rangle$.

Next we show that we can always construct at least one b satisfying the above conditions. Obviously, we can construct b to satisfy (i). For all x such that $d(x) = 0$ put $xb = y = xa^h b^{h+1}$, which is well-defined by condition (i). Now $xa^h = za^{h+1}$ for some $z \in X_n$, whence $ya^{h+1} = za^{h+1}b^{h+1}a^{h+1} = za^{h+1} = xa^h$, where the second equality follows from Lemma 6.2.13.

The grasp of y is at least r, because $ya^r b^r = xa^h b^{h+1}.a^r b^r = xa^h b^{h+1-r}.b^r a^r.b^r$, and $d(xa^h b^{h+1-r}) \geq h+1-(h+1-r) = r$, so by Lemma 6.2.13 we obtain $ya^r b^r = xa^h b^{h+1-r}.b^r = xa^h b^{h+1} = y$, as required.

Finally, consider $xba.a^r b^r = xa^h b^{h+1}a.a^r b^r = xa^h b^h.a^r b^r$ (as $d(xa^h b^h) > 0$) $= xa^h b^h$, again by Lemma 6.2.13. On the other hand, if $r = h$, $xa^r b^r.ba = xa^h b^{h+1}a = xa^h b^h$; if $r < h$ then $h(xa^r b^r) = h$ and

$$xa^r b^r.ba = xa^r b^r.a^h b^{h+1}.a = xa^h b^{h+1}.a = xa^h b^h.$$

Hence in either case, the final condition of (ii) is satisfied by y.

Remark It remains to show that any b satisfying conditions (i) and (ii) above is a strong inverse of a. Since we now know that at least one such b exists, our original Theorem 6.2.4 will follow from this. We note that Lemmas 6.2.13 and 6.2.14 are valid under these hypotheses on b: the statement that $g(xb) \geq g(x) + 1$ being a hypothesis of Theorem 6.2.16 above, the other arguments go through unaltered. Hence the pre-order $\prec$ can be defined on X_n using b satisfying the hypotheses of Theorem 6.2.16 and Lemma 6.2.15 also remains valid. However, before completing the proof we pause for an example.

Example 6.2.17 Consider the following mapping $a \in T_{25}$ with digraph:

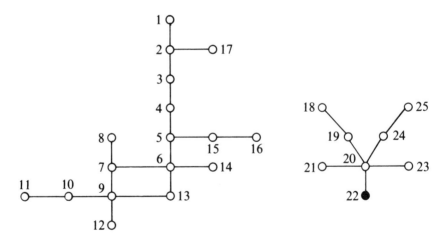

One inverse, b, constructed in accord with the construction of Theorem 6.2.16, is the following mapping with digraph:

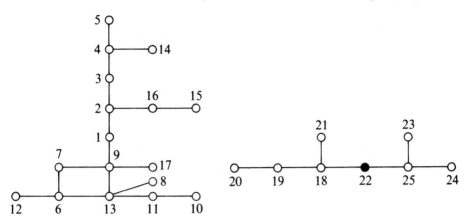

This map is not the unique strong inverse of a. The points which have a plurality of possible images under a member of S(a) are: 2 (1 and 17), 12 (6 and 11), 14 (4 and 9), 20 (19 and 24), 21 (18 and 25), and 23 (18 and 25). In general, such choices cannot be made independently.

The length of the word w in the letters a and b will be denoted by $|w|$, while the number of occurrences of each of the letters a and b will be denoted by $|w|_a$ and $|w|_b$ respectively.

We now return to the proof of Theorem 6.2.16. Henceforth let $a \in T_n$ and let b be a mapping satisfying conditions (i) and (ii) of the statement of the theorem. The elementary properties of a and b given by Lemmas 6.2.13–6.2.15 will be used, sometimes without further comment.

Lemma 6.2.18 $b^m \in V(a^m)$ *for all* $m = 1, 2, \ldots$.

Proof For any $x \in X_n$ we have, by Lemma 6.2.13, that $xa^m b^m a^m = xa^m$. By Lemma 6.2.14, $g(xb^m) \geq m$, whence $xb^m a^m b^m = xb^m$. □

Lemma 6.2.19 *Let* $x \in X_n$, $w \in S^1$, *and suppose that* $x \sim xw$. *Then* $xw = x^t$, *where* $t = t_w = |w|_a - |w|_b$ *for some factorization of* w.

Proof Write w as $c_1 c_2 \ldots c_p$ $(c_i \in \{a, b\})$. By definition of $[x]$ it follows that $xc_1 \ldots c_i \sim x$ for all $i = 1, 2, \ldots, p$. The result now readily follows by induction on p. □

Lemma 6.2.20 *Let* $x \in X_n$ *with* $d(x) = d$, $g(x) = g$, $w \in S^1$ *such that* $xw \notin [x]$. *Let* $w = c_1 c_2 \ldots c_p$ $(c_i \in \{a, b\})$ *and let* q *be the least index such that* $xc_1 \ldots c_q \notin [x]$. *Then there exists* $y \in X_n$ *such that* $[x] < [y]$ *and* $xw = yw$. *Indeed, either:*

(i) $xc_1 \ldots c_q = xa^r$, *whence we may take* $y = y_1 = xa^r b^r$, *with* $d(y) \geq d$ *and* $g(y) \geq r$; *or*
(ii) $xc_1 \ldots c_q = xb^{d+1}$, *whence we may take* $y = y_2 = xb^{d+1} a^{d+1}$, *with* $d(y) \geq d + 1$, $g(y) \geq g$.

Proof (i) $d(y_1) \geq d + r - r = d$, $g(y_1) \geq r$. As in the proof of Lemma 6.2.19, $y_1 c_1 \ldots c_i = y^t$ (where $t = |c_1 \ldots c_i|_a - |c_1 \ldots c_i|_b$ for all $1 \leq i \leq q$). In particular,

$$y_1 c_1 \ldots c_q = xa^r b^r a^r = xa^r = y_1 a^r,$$

whence,

$$y_1 w = y_1 a^r c_{q+1} \ldots c_p = xa^r c_{q+1} \ldots c_p = xw.$$

(ii) In this case $g(xb^d) = g + d$, so that $g(xb^{d+1}) \geq g + d + 1$. Hence $d(y_2) \geq d + 1$ and $g(y_2) \geq g + d + 1 - (d+1) = g$ and, as before,

$$y_2 w = y_2 c_1 \ldots c_q \ldots c_p = y_2 b^{d+1} c_{q+1} \ldots c_p = xb^{d+1} c_{q+1} \ldots c_p = xw. \quad \square$$

By Gluskin's Lemma, we can now complete the proof of Theorem 6.2.16 by proving that $w_1 = a^s b^s$ and $w_2 = b^l a^l$ commute in $\langle a, b \rangle$, for $b \in V(a)$ by Lemma 6.2.18.

It is clear that $xw_1 w_2 = xw_2 w_1$ for all $x \in Z(C)$, the cycle of some component C of the digraph of a, so let us assume that this equation holds for all y such that $[x] < [y]$. We proceed by induction on the length of a $\sim$-chain, anchoring the induction on the maximal $\sim$-class $V(Z)$ and ending in the arbitrary fixed $\sim$-class, $[x]$.

If $xw_i \sim x$ for either value of i the result follows readily: Suppose $xw_1 \sim x$, whence $xw_1 = x$ as $t_w = 0$; either $xw_2 = x$ also, in which case $xw_1 w_2 = x = xw_2 w_1$ or, by Lemma 6.2.20, there exists $y \in V(C)$ such that $[x] < [y]$, and $xw_2 = yw_2$, $d(x) \leq d(y)$, and $g(x) \leq g(y)$. Lemma 6.2.19 then yields $yw_1 = y$ also, whence

$$xw_1 w_2 = xw_2 = yw_2 = yw_1 w_2 = yw_2 w_1 = xw_2 w_1,$$

and the same argument applies if $xw_2 \sim x$.

Henceforth we may assume that $g < s$ and $d < l$. If $s \geq h$ then it follows by Condition (ii) on b in the statement of Theorem 6.2.16 that

$$y_2 a^s = xb^{d+1} a^{d+1} a^s = (xb^d) b \cdot a^{h+d+1+s-h} = xb^d a^{h+d+s-h} = xb^d a^{d+s} = xa^s,$$

whence it follows that $y_2 w_1 = xw_1$; since $y_2 w_2 = xw_2$, as above, this yields $xw_1 w_2 = xw_2 w_1$. Hence we shall assume that $s < h$. We make the following claim.

6.2.21 $\quad xb^{d+1} a^{d+1} \cdot a^s b^s = xa^s b^s \cdot b^{d+1} a^{d+1}, \qquad s < h.$

Assuming the truth of the claim we obtain:

$$xw_2 w_1 = y_2 w_2 w_1 = y_2 w_1 w_2 = xb^{d+1} a^{d+1} \cdot a^s b^s \cdot w_2 = xa^s b^s \cdot b^{d+1} a^{d+1} w_2$$

$$= xw_1 \cdot b^{d+1} a^{d+1} w_2 = xw_1 w_2,$$

since $d + 1 \leq l$ and $b^{d+1} \in V(a^{d+1})$, by Lemma 6.2.18, and the required result follows. We therefore set about proving 6.2.21.

The claim is true for all $s < r$, as in this case both sides equal $xb^{d+1} a^{d+1}$, by the given conditions on b and Lemma 6.2.20. If $s = r$, then applying the conditions given in the statement of Theorem 6.2.16 to xb^d yields:

$$(xb^d)ba \cdot a^{d+r}b^{d+r} = (xb^d)a^{d+r}b^{d+r} \cdot ba$$

$$\Rightarrow xb^{d+1}a^{d+1}a^rb^r \cdot b^d = xa^rb^r \cdot b^{d+1}a$$

$$\Rightarrow xb^{d+1}a^{d+1}a^rb^r \cdot b^d a^d = xa^rb^r \cdot b^{d+1}a^{d+1}$$

$$\Rightarrow xb^{d+1}a^{d+1} \cdot a^rb^r = xa^rb^r \cdot b^{d+1}a^{d+1},$$

as $d(xb^{d+1}a^{d+1}a^rb^r) \geq d+1$. Finally, let $r < s (< h)$. Define a finite sequence $x = x_0, x_1, \ldots x_m$, of members of $V(C)$ recursively, as follows.

Suppose that the sequence has been defined up to x_{i-1} ($i \geq 1$). If $r_{i-1} = r(x_{i-1}) = h$, we stop. Otherwise, $r_{i-1} < h$ and we define

$$x_i = x_{i-1}a^{r_{i-1}}b^{r_{i-1}}.$$

Then all the x_i have depth d, height h, and $r_{i-1} < r_i \leq h$. We next show by induction on i that $x_i = xa^{r_{i-1}}b^{r_{i-1}}$. This is true by definition for $i = 1$. For $i > 1$ we obtain

$$x_i = x_{i-1}a^{r_{i-1}}b^{r_{i-1}} = xa^{r_{i-2}}b^{r_{i-2}} \cdot a^{r_{i-1}}b^{r_{i-1}} = xa^{r_{i-1}}b^{r_{i-1}}.$$

Now let p be a non-negative integer such that $r_{i-1} + p < r_i$. Then

$$x_i = x_i a^{r_{i-1}+p}b^{r_{i-1}+p} = xa^{r_{i-1}}b^{r_{i-1}} \cdot a^{r_{i-1}+p}b^{r_{i-1}+p} = xa^{r_{i-1}+p}b^{r_{i-1}+p}.$$

Therefore:

6.2.22 $x_i = xa^{r_{i-1}+p}b^{r_{i-1}+p}$, for $r_{i-1} + p < r_i$, and all $i = 1, \ldots, m$.

Let y_i denote $x_i b^{d+1}a^{d+1}$. We show inductively that $y_i = y_0 a^{r_{i-1}}b^{r_{i-1}}$.

$$y_1 = x_1 b^{d+1}a^{d+1} = x_0 a^{r_0}b^{r_0} \cdot b^{d+1}a^{d+1} = x_0 b^{d+1}a^{d+1} \cdot a^{r_0}b^{r_0} = y_1 a^{r_0}b^{r_0},$$

where the third equality follows from the '$s = r$' case of 6.2.21, which has been verified. This establishes our statement for the case $i = 1$. If $i > 1$ then

$$y_i = x_i b^{d+1}a^{d+1} = x_{i-1}a^{r_{i-1}}b^{r_{i-1}} \cdot b^{d+1}a^{d+1} = x_{i-1}b^{d+1}a^{d+1} \cdot a^{r_{i-1}}b^{r_{i-1}}$$

$$= y_{i-1}a^{r_{i-1}}b^{r_{i-1}} = y_0 a^{r_{i-2}}b^{r_{i-2}} \cdot a^{r_{i-1}}b^{r_{i-1}}$$

$$= y_0 a^{r_{i-1}}b^{r_{i-1}},$$

as required.

Since $g(y_i) \geq g(x_i)$ we can show, in the same way as we proved 6.2.22, that for any non-negative integer p such that $r_{i-1} + p < r_i$:

6.2.23 $y_i = y_0 a^{r_{i-1}+p}b^{r_{i+1}+p}$ for $r_{i-1} + p < r_i$, and all $i = 1, \ldots, m$.

Now to complete the proof of the claim 6.2.21 (and thus of Theorem 6.2.16) we write s in the form $r_{i-1} + p < r_i$, say. Then, using 6.2.22 and 6.2.23, we obtain

$$xa^sb^s \cdot b^{d+1}a^{d+1} = x_i b^{d+1}a^{d+1} = y_i = y_0 a^sb^s = xb^{d+1}a^{d+1} \cdot a^sb^s,$$

as required. ∎

Exercises 6.2

1. Does every $\mathcal{H}$-class of T_n contain a square?
2. Find both square roots of the map in T_8 given by (2 5 2 3 6 2 6 7).
3. Find all square roots of the partial mapping in PT_{10} given by
   $(-2\ 4\ 3\ 3\ 5\ 5\ 9\ 8\ 8)$.
4. (a) Describe algrithms to solve the equations $ax = bx$, and $xa = xb$ for unknown x, where $a, b, x \in T_n$.
   (b) Given the existence of an algorithm to find all pth roots of a given mapping $\alpha \in T_n$ for any prime p, describe an algorithm to solve the equation $ax^m b = c$ for any positive integer m and any given mappings a, b, c of T_n.
5. For $a \in T_n$ define $C_a = \{x \in T_n : ax = xa\}$. Show that
   $$C_a = \{a^m : m = 0, 1, 2, \ldots\}$$
   if a has a unique component and a unique endpoint, or every component of a is cyclic, and the orders of these components are pairwise relatively prime (the converse is also true; Higgins 1988a).
6. Find generators of an inverse semigroup that contains the mapping α as given in Example 6.2.1.
7. (Schein written communication) Let X be a non-empty set, let $x \in X$, and put $Y = X \setminus \{x\}$. Consider the symmetric inverse semigroup $\mathscr{I}_Y$, and extend each $\alpha \in \mathscr{I}_Y$ to a member α^* of $\mathscr{I}_X$ by defining
   $$a\alpha^* = \begin{cases} a\alpha, & \text{if defined,} \\ x, & \text{otherwise.} \end{cases}$$
   Then $\{\alpha^* : \alpha \in \mathscr{I}_Y\}$ forms a maximal inverse subsemigroup of $\mathscr{I}_X$.

Remark Maximal inverse subsemigroups of the semigroups of linear transformations have been described by Shneperman (1974), while a large class of maximal inverse subsemigroups of $\mathscr{I}_X$ was identified by Reilly (1977).

8. (Schein written communication)
   (a) Let X be an infinite set and let α be a member of $\mathscr{I}_X$ which has a component of the form

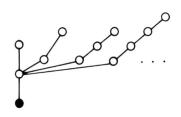

Show that α is not contained in any inverse subsemigroup of $\mathcal{T}_X$, and conclude that no infinite full transformation semigroup is covered by its inverse subsemigroups.

(b) For which sets X is $\mathcal{PT}_X$ covered by its inverse subsemigroups?

6.3 Products of idempotents in T_n

It was shown by Howie (1966) that S, the semigroup of all singular mappings on X_n (maps of positive defect), is generated by E, the set of idempotents of T_n of defect 1, a set of order $n(n-1)$. The minimum number of such idempotents required to express a particular $\alpha \in T_n$ as a product can be expressed succinctly in terms of the digraph of α: exactly $n + c(\alpha) - f(\alpha)$ are needed, where $f(\alpha)$ is the number of fixed points of α and $c(\alpha)$ is the number of cyclic orbits of α, meaning components which are cycles of order greater than 1. By maximizing this quantity, $g(\alpha) = n + c(\alpha) - f(\alpha)$, called the *gravity of* α, we shall discover that the least positive integer k such that $E^k = S$ is the greatest integer not exceeding $(3/2)(n-1)$. In addition, we shall solve the relatively simple problem of identifying all the minimal sets of idempotent generators for S. This section is based on Howie (1978, 1980), but the gravity formula, repeated as our Theorem 6.1.1, was also found independently by Iwahori (1977).

From the above it follows that the core of T_n is S^1. Corresponding results for PT_n are due to Evssev and Podran (1970, 1972). Other combinatorial and graph-theoretic aspects of transformation semigroups have been explored by Denes: see, for instance, Denes (1970, 1981), the latter paper employing T_n as a framework to study random number generators. Also see Biggs *et al.* (1976) and Howie *et al.* (1988).

Theorem 6.3.1 *Let S denote the semigroup of singular mappings on X_n and let E denote the set of idempotents of defect 1 in S. For each $\alpha \in S$, the least positive integer k for which $\alpha \in E^k$ is*

$$g(\alpha) = n + c(\alpha) - f(\alpha).$$

Remark and notation The first stage of the proof consists of identifying $g(\alpha)$ members of E, the product of which is the given map α. The more difficult second stage proves that a lesser number of idempotents will not suffice.

We denote a typical orbit (component of the digraph) by Ω, and the cycle points of Ω by $K(\Omega)$. An orbit is *cyclic* if $|K(\Omega)| > 1$ and $K(\Omega) = \Omega$; if $|\Omega| > 1$ and $|K(\Omega)| = 1$ we call the orbit *acyclic*, while if $|\Omega| = 1$ then Ω is a *singleton orbit*. If an orbit is not of one of these special kinds we shall refer to it as being of *general type*.

224 | Techniques of semigroup theory

For each $p \in \Omega$ define the subset
$$p\alpha^{-N} = \{x \in X_n : x\alpha^i = p \text{ for some } i > 0\}.$$
For an idempotent ε of rank $n - 1$, there is a unique i ($1 \leq i \leq n$) such that $i\varepsilon = j \neq i$, which allows us to write ε as
$$\varepsilon = \binom{i}{j},$$
referring to i and j as the *upper* and *lower* entries of ε respectively.

For any $\alpha \in T_n$ we denote the complement of its range, $X_n \setminus \operatorname{ran} \alpha$ by $Z(\alpha)$.

For $\varepsilon \in E$ we denote the equivalence $\varepsilon \circ \varepsilon^{-1}$ by $[ij]$, and this notation specifies the unique non-trivial equivalence class in this relation.

Proof of Theorem 6.3.1 Let us suppose that α has orbits as follows:

orbits of general type $\Omega_1, \ldots, \Omega_m$;

acyclic orbits $\Omega_{m+1}, \ldots, \Omega_{m+a}$ $(m + a \geq 1)$;

cyclic orbits $\Omega_{m+a+1}, \ldots, \Omega_{m+a+c}$;

singleton orbits $\Omega_{m+a+c+1}, \ldots, \Omega_{m+a+c+s}$.

Then α has $a + s$ fixed points. Also
$$\sum_{i=1}^{m+a+c+s} |\Omega_i| = n, \quad \text{and so} \quad \sum_{i=1}^{m+a+c} |\Omega_i| = n - s. \tag{1}$$

We shall show that
$$\alpha = \beta_1 \ldots \beta_m \gamma_{m+1} \ldots \gamma_{m+a} \delta_{m+a+1} \ldots \delta_{m+a+c},$$
where each β_i is a product of $|\Omega_i|$ idempotents, each γ_i is a product of $|\Omega_i| - 1$ idempotents, and each δ_i is a product of $|\Omega_i| + 1$ idempotents. The total number of idempotents is thus
$$\sum_{i=1}^{m} |\Omega_i| + \sum_{i=m+1}^{m+a} (|\Omega_i| - 1) + \sum_{i=m+a+1}^{m+a+c} (|\Omega_i| + 1)$$
$$= n + c - (a + s) \quad \text{(by (1))}$$
$$= g(\alpha).$$

To avoid excessive use of subscripts, let us now consider a typical orbit Ω from $\{\Omega_1, \ldots, \Omega_n\}$. Let $x_1 \in \Omega \setminus \operatorname{ran} \alpha$ and let
$$K(\Omega) = \{x_1 \alpha^t, \ldots, x_1 \alpha^{t+r-1}\},$$
where $x_1 \alpha^{t+r} = x_1 \alpha^t$. Let
$$\zeta_1 = \binom{x_1 \alpha^{t+r-1}}{x_1 \alpha^{t-1}} \binom{x_1 \alpha^{t+r-2}}{x_1 \alpha^{t+r-1}} \cdots \binom{x_1 \alpha^{t-1}}{x_1 \alpha^t} \cdots \binom{x_1}{x_1 \alpha}, \tag{2}$$

a product of $t + r$ idempotents, and let Z_1 be the set of elements of Ω appearing along the top of this product:

$$Z_1 = \{x_1, x_1\alpha, \ldots, x_1\alpha^{t+r-1}\}.$$

If $Z_1 = \Omega$ then ζ_1 is the product β of idempotents we require (as we shall verify below); otherwise choose $x_2 (\neq x_1)$ in $\Omega \setminus \operatorname{ran} \alpha$. Since $K(\Omega) \subseteq Z_1$ there is a least p such that $x_2 \alpha^p \in Z_1$. Let

$$\zeta_2 = \begin{pmatrix} x_1 \alpha^{p-1} \\ x_2 \alpha^p \end{pmatrix} \cdots \begin{pmatrix} x_2 \\ x_2 \alpha \end{pmatrix}$$

and let

$$Z_2 = \{x_2, \ldots, x_2 \alpha^{p-1}\}.$$

If $Z_1 \cup Z_2 = \Omega$ then $\beta = \zeta_1 \zeta_2$ is the required product of idempotents; otherwise we continue, finding

$$\zeta_3 = \begin{pmatrix} x_3 \alpha^{q-1} \\ x_3 \alpha^q \end{pmatrix} \cdots \begin{pmatrix} x_3 \\ x_3 \alpha \end{pmatrix}$$

and $Z_3 = \{x_3, \ldots, x_3 \alpha^{q-1}\}$; and so on. Eventually we have Ω as the disjoint union of $Z_1, \ldots, Z_k$ and a product $\beta = \zeta_1 \ldots \zeta_k$ of $|Z_1| + \ldots + |Z_k| = |\Omega|$ idempotents.

Now notice that each element z of Ω appears exactly once as an upper entry in the product $\zeta_1 \ldots \zeta_k$. Moreover, with the sole exception of $z = x_1 \alpha^{t-1}$, an element z appearing as a lower entry never subsequently reappears as an upper entry. Hence each $z \neq x_1 \alpha^{t+r-1}$ is moved by exactly one of the idempotents appearing in $\beta = \zeta_1 \ldots \zeta_k$ and, moreover, is moved to $z\alpha$. The exceptional element $x_1 \alpha^{t+r-1}$ is moved to $x_1 \alpha^{t-1}$ by the first idempotent in the product ζ_1, and then is moved by

$$\begin{pmatrix} x_1 \alpha^{t-1} \\ x_1 \alpha^t \end{pmatrix}$$

to $x_1 \alpha^t (= x_1 \alpha^{t+r})$. Thus $z\beta = z\alpha$ for every $z \in \Omega$, while $x\beta = x$ for every $x \notin \Omega$.

This argument applies to each of the orbits $\Omega_1, \ldots, \Omega_m$ of general form. If we now suppose that Ω is an acyclic orbit then exactly the same argument applies, except that the formula (2) simplifies to

$$\zeta_1 = \begin{pmatrix} x_1 \alpha^{t-1} \\ x_1 \alpha^t \end{pmatrix} \begin{pmatrix} x_1 \alpha^{t-2} \\ x_1 \alpha^{t-1} \end{pmatrix} \cdots \begin{pmatrix} x_1 \\ x_1 \alpha \end{pmatrix},$$

a product of t idempotents. (In effect, it is not necessary to provide an idempotent to specify the image of $x_1 \alpha^t$, since this is a fixed point.) We thus obtain a product γ of $|\Omega| - 1$ idempotents, with the property that $z\gamma = z\alpha$ for all $z \in \Omega$, while $x\gamma = x$ for all $x \notin \Omega$.

226 | Techniques of semigroup theory

Now suppose that we have found $\beta_1, \ldots, \beta_m, \gamma_{m+1}, \ldots, \gamma_{m+a}$. Then the last idempotent in the product $\beta_1 \ldots \beta_m \gamma_{m+1} \ldots \gamma_{m+a}$ is

$$\begin{pmatrix} y \\ y\alpha \end{pmatrix},$$

where $y \notin \operatorname{ran} \alpha$. Let $\Omega = \{x, x, \ldots, x\alpha^{r-1}\}$ be a cyclic orbit, where $x\alpha^r = x$. Define

$$\delta = \begin{pmatrix} x\alpha^{r-1} \\ y \end{pmatrix} \begin{pmatrix} x\alpha^{r-2} \\ x\alpha^{r-1} \end{pmatrix} \cdots \begin{pmatrix} x \\ x\alpha \end{pmatrix} \begin{pmatrix} y \\ x \end{pmatrix},$$

a product of $|\Omega| + 1$ idempotents. Now

$$y \notin \operatorname{ran}(\beta_1 \ldots \beta_m \gamma_{m+1} \ldots \gamma_{m+a})$$

and so $z\delta = z$ for all z in $\operatorname{ran}(\beta_1 \ldots \beta_m \gamma_{m+1} \ldots \gamma_{m+a}) \setminus \Omega$. Also, $z\delta = z\alpha$ for all z in Ω.

Doing this (using the same y) for each of the cyclic orbits gives us

$$z\alpha = z\beta_1 \ldots \beta_m \gamma_{m+1} \ldots \gamma_{m+a} \delta_{m+a+1} \ldots \delta_{m+a+c}$$

for all z in X_n, since clearly the fixed points of α are left fixed by both sides. Thus α is expressible as a product of $g(\alpha)$ idempotents from E, as required.

For the opportunity to work through an example of this construction the reader should see Exercise 6.3.1.

It remains to prove that if $\alpha \in E^k$ then $k \geq g(\alpha)$. We shall approach this by showing that, for every α in S and every ε in E,

$$g(\alpha\varepsilon) \leq g(\alpha) + 1. \tag{3}$$

This, together with the observation that $g(\varepsilon) = 1$ for all ε in E, shows that if $\varepsilon_1, \ldots, \varepsilon_k \in E$ then

$$g(\varepsilon_1 \ldots \varepsilon_k) \leq k;$$

thus E^k contains no element of gravity greater than k. This will be sufficient to prove the required result.

To prove (3) we must consider the ways in which the numbers of fixed points and cyclic orbits of α can differ from the corresponding numbers for $\alpha\varepsilon$. Now let us suppose that the orbits of α are $\Omega, \Theta, \ldots$, and that

$$\varepsilon = \begin{pmatrix} p \\ q \end{pmatrix}.$$

We shall proceed by means of a series of lemmas, gathering together the information that we need. We begin with three simple lemmas.

Lemma 6.3.2 *For all α in S and all ε in E,*

$$f(\alpha\varepsilon) \geq f(\alpha) - 1.$$

In fact, $f(\alpha\varepsilon) \geq f(\alpha)$ unless p is a fixed point of α.

Proof If x is a fixed point of α then $x\alpha\varepsilon = x\varepsilon$, which equals x except when $x = p$. Both assertions of the lemma are now immediate. □

The next lemma is obvious.

Lemma 6.3.3 *If $p \notin \operatorname{ran} \alpha$ then $\alpha\varepsilon = \alpha$.*

Lemma 6.3.4 *If Ω is an orbit of α and $p, q \in X_n \setminus \Omega$, then Ω is an orbit of $\alpha\varepsilon$. The $\alpha\varepsilon$-orbit Ω is not cyclic unless the Ω-orbit is cyclic.*

Proof We have $x(\alpha\varepsilon) = x\alpha$ except when $x \in p\alpha^{-1}$ (in which case $x(\alpha\varepsilon) = q$); also $x(\alpha\varepsilon)^{-1} = x\alpha^{-1}$ except when $x = q$. Now $p\alpha^{-1}$ is contained in the orbit of p and so $p\alpha^{-1} \cap \Omega = \emptyset$; also $q \notin \Omega$. Hence, for all $x \in \Omega$, we have

$$x(\alpha\varepsilon) = x\alpha, \qquad x(\alpha\varepsilon)^{-1} = x\alpha^{-1}.$$

It follows that Ω is an orbit of $\alpha\varepsilon$, and of exactly the same type (that is, general, acyclic, cyclic, or singleton) as before. □

Next we have two less simple lemmas.

Lemma 6.3.5 *Let Ω, Θ be distinct orbits of α and let $p \in \Omega$, $q \in \Theta$. If $p \in K(\Omega)$ then $\Omega \cup \Theta$ is an orbit of $\alpha\varepsilon$. If $p \in \Omega \setminus K(\Omega)$ then $\alpha\varepsilon$ has orbits $p\alpha^{-N} \cup \Theta$ and $\Omega \setminus p\alpha^{-N}$. The other orbits of $\alpha\varepsilon$ are identical to those of α. No new cyclic orbits are introduced.*

Proof Let $x \in p\alpha^{-N}$. Then $x\alpha^w = p$ for some $w > 0$; let us suppose that w is the least positive integer for which this holds. Since $x\alpha\varepsilon = x$ for all $x \notin p\alpha^{-1}$ we thus have

$$x(\alpha\varepsilon)^i = x\alpha^i \ (i = 1, \ldots, w-1).$$

Hence

$$x(\alpha\varepsilon)^w = x(\alpha\varepsilon)^{w-1}\alpha\varepsilon = x\alpha^{w-1}\alpha\varepsilon = x\alpha^w\varepsilon = p\varepsilon = q. \tag{4}$$

Thus, if we denote that $\alpha\varepsilon$-orbit containing q by Φ, we have $p\alpha^{-N} \subseteq \Phi$.

Next, $x\alpha\varepsilon = x\alpha$ for all $x \in \Theta$. Hence, since for every x in Θ there exist $u, v \geq 0$ such that $x\alpha^u = q\alpha^v$, we may conclude that

$$x(\alpha\varepsilon)^u = q(\alpha\varepsilon)^v.$$

Hence $\Theta \subseteq \Phi$ and so $p\alpha^{-N} \cup \Theta \subseteq \Phi$.

To show the opposite inclusion, notice first that if $z \in K(\Theta)$ then $z(\alpha\varepsilon)^r = z$, where $r = |K(\Theta)|$. Hence, by Lemma 1.6.6, the kernel of the $\alpha\varepsilon$-orbit Φ is $\{z, z\alpha, \ldots, z\alpha^{r-1}\} = K(\Theta)$. If $x \in \Phi$ it then follows by Lemma 1.6.7 that $x(\alpha\varepsilon)^w = z$ for some $w > 0$. This implies either that $x\alpha^w = z$, in which case $x \in \Theta$; or for some u such that $0 < u \leq w$,

$$x\alpha^u = p, \qquad q\alpha^{w-u} = z,$$

in which case $x \in p\alpha^{-N}$. Thus $\Phi \subseteq p\alpha^{-N} \cup \Theta$.

If $p \in K(\Omega)$ then $p\alpha^{-N} = \Omega$ by Lemma 1.6.7, and so $\Phi = \Omega \cup \Theta$.

If $p \in \Omega \setminus K(\Omega)$ then $\Phi = p\alpha^{-N} \cup \Theta$. Let Ψ denote the $\alpha\varepsilon$-orbit containing p. If $x \in \Omega \setminus p\alpha^{-N}$ then $x\alpha^i = x(\alpha\varepsilon)^i$, for all $i \geq 0$. Hence, since for all x in $\Omega \setminus p\alpha^{-N}$ there exist $u, v \geq 0$ such that

$$x\alpha^u = p\alpha^v,$$

it is equally the case that

$$x(\alpha\varepsilon)^u = p(\alpha\varepsilon)^v.$$

Hence $\Omega \setminus p\alpha^{-N} \subseteq \Psi$.

Conversely, if $x \in \Psi$ then there exist $u, v \geq 0$ such that

$$x(\alpha\varepsilon)^u = p(\alpha\varepsilon)^v = p\alpha^v. \tag{5}$$

Now $x(\alpha\varepsilon)^u = x\alpha^u$ unless for some $t \leq u$ it is the case that $x\alpha^t = p$. In that case $x(\alpha\varepsilon)^t = q$ and $x(\alpha\varepsilon)^u = q\alpha^{u-t}$. But since p and q are in different α-orbits, it is not possible for $p\alpha^u$ and $q\alpha^{u-t}$ to be equal. Hence from (5) the only possible conclusion is that $x\alpha^u = p\alpha^v$, which implies that $x \in \Omega$. Since $\Phi = p\alpha^{-N} \cup \Theta$ has been shown to be an orbit, we have that

$$\Psi \cap (p\alpha^{-N} \cup \Theta) = \emptyset;$$

hence $\Psi \subseteq \Omega \setminus p\alpha^{-N}$. This, together with the conclusion of the last paragraph, gives us that $\Psi = \Omega \setminus p\alpha^{-N}$.

In the case in which $p \in K(\Omega)$ we have that $p \in (\Omega \cup \Theta) \setminus \operatorname{ran} \alpha\varepsilon$. It follows that the $\alpha\varepsilon$-orbit $\Omega \cup \Theta$ is not cyclic. In the same way, when $p \in \Omega \setminus K(\Omega)$ then $p \in (\Omega \setminus p\alpha^{-N}) \setminus \operatorname{ran} \alpha\varepsilon$, and so the orbit $\Omega \setminus p\alpha^{-N}$ is not cyclic. If we exclude the trivial case in which $p \notin \operatorname{ran} \alpha$ (when $p\alpha^{-N} = \emptyset$, in which case $\alpha\varepsilon = \alpha$) then there exists z in $p\alpha^{-N}$ such that $z \notin \operatorname{ran} \alpha$. Then $z \in \Phi$ and $z \notin \operatorname{ran} \alpha\varepsilon$. Hence the orbit $\Phi = p\alpha^{-N} \cup \Theta$ is not cyclic, even if Θ is cyclic. In any event, $\alpha\varepsilon$ cannot have more cyclic orbits than α. $\square$

Lemma 6.3.6 *Let Ω be an α-orbit and let $p, q \in \Omega$:*

(i) *If either p or q is in $K(\Omega)$, or if $p, q \in \Omega \setminus K(\Omega)$ and $q \notin p\alpha^{-N}$, then Ω is an $\alpha\varepsilon$-orbit and is not cyclic.*

(ii) *If $p, q \in \Omega \setminus K(\Omega)$ and $q \in p\alpha^{-N}$ then $p\alpha^{-N}$ and $\Omega \setminus p\alpha^{-N}$ form separate $\alpha\varepsilon$-orbits. The $\alpha\varepsilon$-orbit $\Omega \setminus p\alpha^{-N}$ is never cyclic.*

Remark The $\alpha\varepsilon$-orbit $p\alpha^{-N}$ of part (ii) may be cyclic: see Exercise 6.3.2.

Proof (i) We begin by remarking that the $\alpha\varepsilon$-orbit Φ containing q cannot properly contain Ω, since by Lemma 6.3.4 all other orbits are unaffected by ε.

Suppose first that $p \in K(\Omega)$. Then, by Lemma 1.6.7, there exists for each x in Ω a least positive integer w such that $x\alpha^w = p$. It then follows as in (4) that $x(\alpha\varepsilon)^w = q$. Hence $x \in \Phi$. We have shown that $\Omega \subseteq \Phi$; hence $\Omega = \Phi$, by the remark in the last paragraph.

Next, suppose that $q \in K(\Omega)$. Then, again by Lemma 1.6.7, for each $x \in \Omega$ there exists $w > 0$ such that $x\alpha^w = q$. If $p \notin \{x\alpha, \ldots, x\alpha^{w-1}\}$ then $x(\alpha\varepsilon)^w = q$.

If $p = x\alpha^u$, where $1 \leq u \leq w - 1$, then $x(\alpha\varepsilon)^u = q$. In either case $x \in \Phi$. Hence $\Phi = \Omega$ in this case also.

Next suppose that $p, q \in \Omega \backslash K(\Omega)$, with $q \notin p\alpha^{-N}$. Then, since $p \notin \{q, q\alpha, q\alpha^2, \ldots\}$, we have $q\alpha^w = q(\alpha\varepsilon)^w$ for all $w \geq 0$. If $x \in \Omega$ then there exist $v, w \geq 0$ such that

$$x\alpha^v = q\alpha^w.$$

If $p \notin \{x\alpha, \ldots, x\alpha^{v-1}\}$ then $x(\alpha\varepsilon)^v = x\alpha^v$, and thus $x(\alpha\varepsilon)^v = q(\alpha\varepsilon)^w$. On the other hand, if $p = x\alpha^u$, where $1 \leq u \leq v - 1$, then $x(\alpha\varepsilon)^u = q$. In either case x is in the same $\alpha\varepsilon$-orbit as q and so $x \in \Phi$. Thus $\Phi = \Omega$ in this case as well.

In the three cases considered so far, $p \in \Omega \backslash \operatorname{ran} \alpha\varepsilon$, and so the $\alpha\varepsilon$-orbit Ω is not cyclic.

(ii) Now suppose that $p, q \in \Omega \backslash K(\Omega)$, with $q \in p\alpha^{-N}$, and let us again denote the $\alpha\varepsilon$-orbit containing q by Φ. If $x \in p\alpha^{-N}$ and if u is the (unique) positive integer such that $x\alpha^u = p$ then, again, $x(\alpha\varepsilon)^u = q$. Hence $p\alpha^{-N} \subseteq \Phi$.

Conversely, let us suppose that $x \in \Phi$. Let w be the (unique) positive integer such that $q\alpha^w = p$. Then $q(\alpha\varepsilon)^w = q$. Since $x \in \Phi$ there exist $u, v \geq 0$ such that $x(\alpha\varepsilon)^u = q(\alpha\varepsilon)^v$. If $(i-1)w < v \leq iw$ we may infer that

$$x(\alpha\varepsilon)^{u+iw-v} = q(\alpha\varepsilon)^{iw} = q.$$

Let us now suppose that t is the least non-negative integer such that $x(\alpha\varepsilon)^t = q$. If $x = q$ there is nothing to prove; so we may assume that $t > 0$. Then $p \notin \{x\alpha, \ldots, x\alpha^{t-1}\}$ (since if we could choose a least j, with $1 \leq j \leq t-1$, such that $x\alpha^j = p$ we would have $x(\alpha\varepsilon)^j = q$, with $j < t$). Hence

$$x(\alpha\varepsilon)^{t-1} = x\alpha^{t-1}, \quad (x\alpha^t)\varepsilon = q.$$

Therefore, either $x\alpha^t = p$, giving $x \in p\alpha^{-N}$; or $x\alpha^t = q$, in which case $x\alpha^{t+w} = q\alpha^w = p$, again giving $x \in p\alpha^{-N}$. Thus $\Phi = p\alpha^{-N}$.

Observe now that for every x in $\Omega \backslash p\alpha^{-N}$ we have $x\alpha\varepsilon = x\alpha$. It follows that all the elements of $\Omega \backslash p\alpha^{-N}$, being in the same α-orbit, are in the same $\alpha\varepsilon$-orbit. Thus, denoting by Ψ the $\alpha\varepsilon$-orbit containing p, we have that $\Omega \backslash p\alpha^{-N} \subseteq \Psi$.

Conversely, we certainly have $\Psi \subseteq \Omega$, by the remark in the first paragraph of the proof. We also have that $\Psi \cap p\alpha^{-N} = \emptyset$, since $p\alpha^{-N}$ has been shown to be an orbit. Hence $\Psi \subseteq \Omega \backslash p\alpha^{-N}$ and so $\Psi = \Omega \backslash p\alpha^{-N}$, are required.

Finally, observe that $p \in (\Omega \backslash p\alpha^{-N}) \backslash \operatorname{ran} \alpha\varepsilon$, and so $\Omega \backslash p\alpha^{-N}$ cannot be a cyclic orbit. $\square$

Our formula (3) can now be easily deduced (thus completing the proof of Theorem 6.3.1). Lemmas 6.3.3–6.3.6 list the various ways in which the orbital structure of α is affected by postmultiplication by ε. The conclusion is that, in all circumstances

$$c(\alpha\varepsilon) \leq c(\alpha) + 1.$$

Moreover, we have $c(\alpha\varepsilon) = c(\alpha) + 1$ only in the situation described in Lemma 6.3.6(ii). In this situation p is certainly not a fixed point of α and so $f(\alpha\varepsilon) \geqslant f(\alpha)$ by Lemma 6.3.2. Hence

$$c(\alpha\varepsilon) - f(\alpha\varepsilon) \leqslant c(\alpha) - f(\alpha) + 1$$

in all cases, and so $g(\alpha\varepsilon) \leqslant g(\alpha) + 1$, as required. ∎

In a series of recent papers, Saito has generalized results of Howie, Iwahori, Schein, and others concerning products of idempotents in finite full transformation semigroups (Saito 1988, 1989a,b). In particular, he turned his attention to finding the least integer $k(\alpha)$ such that α is a product of k idempotents of T_n (and not just idempotents of defect 1, as has been considered here). In Saito (1989b) the following inequality is established. (For extensions of these results to PT_n see Garba 1990.)

Result 6.3.7 *Let $\alpha \in T_n$, let k be the least integer such that $\alpha \in E^k$ ($E = E(T_n)$), and let g and d denote the gravity and defect of α. Then*

$$g/d \leqslant k \leqslant g/d + 2.$$

The global counterpart to the problem solved by our Theorem 6.3.1 is to find the least k such that $E^k = S$. This is seen to be equivalent to maximizing the gravity $g(\alpha)$ as α ranges over S. This result is contained in the next theorem.

For a real number x, denote the integral part of x, that is the unique integer m such that $x - 1 < m \leqslant x$, by $[x]$.

Theorem 6.3.8 *Let S be the semigroup of singular mappings on X_n. Then*

$$\max\{g(\alpha): \alpha \in S, \alpha \text{ of rank } r\} = n + [(1/2)(r - 2)],$$

$$\max\{g(\alpha): \alpha \in S\} = [(3/2)(n - 1)].$$

Proof From the definition of gravity it is clear that we maximize $g(\alpha)$ by making $c(\alpha)$ as large as possible while keeping $f(\alpha)$ as small as possible. Evidently, we maximize $c(\alpha)$ by using 2-cycles exclusively. If r ($r \geqslant 1$) is odd we may use $(1/2)(r - 1)$ 2-cycles, accounting for $r - 1$ of the elements of ran α. The remaining $n - r + 1$ elements of X_n must then all map to a single element chosen from among their number. This gives a gravity of

$$n + (1/2)(r - 1) - 1 = n + [(1/2)(r - 2)].$$

Alternatively, still for odd r, provided that $r \geqslant 3$, we might form an element α of rank r by using $(1/2)(r - 3)$ 2-cycles, leaving $n - r + 3$ elements $x_1, \ldots,$

x_{n-r+3} of X_n to be mapped to three elements. Doing this so as to avoid fixed points, for example by defining

$$x_1\alpha = \ldots = x_{n-r}\alpha = x_{n-r+1},$$

$$x_{n-r+1}\alpha = x_{n-r+3}, \quad x_{n-r+2}\alpha = x_{n-r+3},$$

$$x_{n-r+3}\alpha = x_{n-r+1},$$

we obtain a gravity of

$$n + (1/2)(r - 3) = n + [(1/2)(r - 2)].$$

It is now clear that no other element of odd rank r has gravity larger than this.

If r is even then the unique configuration giving maximum gravity is where we have $(1/2)(r - 2)$ 2-cycles, leaving the remaining $n - r + 2$ elements to map to two of their number. Doing this so as to avoid fixed points we obtain a gravity of

$$n + (1/2)(r - 2) = n + [(1/2)(r - 2)].$$

The second formula is obtained from the first upon putting $r = n - 1$. ∎

This result and Theorem 6.3.1 combine to yield the following.

Corollary 6.3.9 *Let S be the semigroup of all singular mappings on X_n and let E be the set of idempotents of S of defect 1. Then $E^k = S$ whenever $k \geq [(3/2)(n - 1)]$. This bound is best possible; that is, E^k is properly contained in S for all $k < [(3/2)(n - 1)]$.*

From Theorems 6.3.1 and 6.3.8 it also follows that a mapping $\alpha \in S$ of rank r is expressible as a product $\varepsilon_1 \ldots \varepsilon_k$ of idempotents of rank $n - 1$, where

$$k \leq n + [(1/2)(r - 2)].$$

Indeed, it follows from Exercise 1.4.13 that α is expressible as a product $\eta_1 \ldots \eta_k$ of idempotents of rank r, where $\eta_i \leq \varepsilon_i$ for $i = 1, \ldots, k$. Thus if we denote the set of rank r by S_r, and the set of idempotents of rank r by E_r, then $E_r^k \supseteq S_r$ whenever $k \geq n + [(1/2)(r - 2)]$, but this bound is no longer best possible. For example, for $r = 1$ we have $E_1 = S_1$, but the formula gives only that $E_1^k \supseteq S_1$ for all $k \geq n - 1$.

Having found the least k such that $E^k = S$, another line of inquiry results from asking for minimal sets of idempotents M which generate S. First, let M be any set of idempotents of S and let $\varepsilon \in E$ so that $|\text{ran } \varepsilon| = n - 1$. If $\varepsilon \in \langle M \rangle$ then ε is a product of idempotents from M of defect 1. It follows that if M is a minimal idempotent generating set for S then $M \subseteq E$. It is easily seen that $|E| = n(n - 1)$; we shall show that minimal idempotent generating sets have exactly half this number of elements. In answering this question we shall follow Howie (1978). The key observations are as follows.

232 | Techniques of semigroup theory

Lemma 6.3.10 *Let α be an element of S of defect 1 and suppose that α is expressed as a product of idempotents from E*

$$\alpha = \begin{pmatrix} i_1 \\ j_1 \end{pmatrix}\begin{pmatrix} i_2 \\ j_2 \end{pmatrix}\cdots\begin{pmatrix} i_m \\ j_m \end{pmatrix} \quad (m \geq 2)$$

of minimum length m. Then:
(i) $\alpha \circ \alpha^{-1} = [i_1 j_1]$; $Z(\alpha) = \{i_m\}$;
(ii) $i_{r-1} = j_r, j_{r-1} \neq i_r$ $(r = 2, \ldots, m)$.

Proof (i) Certainly $(i_1, j_1) \in \alpha \circ \alpha^{-1}$, and since $|\operatorname{ran} \alpha| = n - 1$ the first part of (i) follows. Similarly,

$$\operatorname{ran} \alpha \subseteq \operatorname{ran}\begin{pmatrix} i_m \\ j_m \end{pmatrix} \quad \text{so that} \quad \{i_n\} \subseteq Z\begin{pmatrix} i_m \\ j_m \end{pmatrix} \subseteq Z(\alpha),$$

and equality follows as the defect of α is just 1.
 (ii) Consider a typical subproduct $\beta = ef$, where

$$e = \begin{pmatrix} i_{r-1} \\ j_{r-1} \end{pmatrix} \quad \text{and} \quad f = \begin{pmatrix} i_r \\ j_r \end{pmatrix}.$$

Since $|\operatorname{ran} \beta| = n - 1$, it follows that f is one-to-one on $\operatorname{ran} e = X_n \setminus \{i_{r-1}\}$, but since m is minimum, we infer that f is not the identity when restricted to $\operatorname{ran} e$. To reconcile these facts we must have $j_r = i_{r-1}$. Furthermore, we must also have $j_{r-1} \neq i_r$, for otherwise the product ef would equal

$$\begin{pmatrix} j_{r-1} \\ i_{r-1} \end{pmatrix},$$

which would contradict the minimality of m, for ef could then be replaced by this single member of E. ∎

The effect of the previous lemma is to show that a product of idempotents of defect 1 of minimum length can itself be of defect 1 only if it has the form

$$\begin{pmatrix} i_1 \\ i_2 \end{pmatrix}\begin{pmatrix} i_3 \\ i_1 \end{pmatrix}\begin{pmatrix} i_4 \\ i_3 \end{pmatrix}\begin{pmatrix} i_5 \\ i_4 \end{pmatrix}\cdots\begin{pmatrix} i_{m-1} \\ i_{m-2} \end{pmatrix}\begin{pmatrix} i_m \\ i_{m-1} \end{pmatrix},$$

with $i_{r+1} \neq i_r$ $(r = 1, 2, \ldots, m-1)$, $i_{r+2} \neq i_r$ $(r = 1, 3, 4, \ldots, m-2)$, $i_4 \neq i_1$.

Theorem 6.3.1 assures us that S is generated by the idempotents of defect 1. To discover whether any of these idempotents are superfluous, we must study the ways in which idempotents of defect 1 can multiply to produce new idempotents also of defect 1. The next lemma shows how this can occur.

Combinatorial aspects of transformation semigroups | 233

Lemma 6.3.11 *If $m \geqslant 3$ and if $i_1, i_2, \ldots, i_m \in X_n$ are all distinct, then*

$$\left[\binom{i_1}{i_2}\binom{i_3}{i_1}\binom{i_4}{i_3}\cdots\binom{i_m}{i_{m-1}}\binom{i_2}{i_m}\right]^{m-1} = \binom{i_2}{i_1}.$$

Proof The product

$$\beta = \binom{i_1}{i_2}\binom{i_3}{i_1}\binom{i_4}{i_3}\binom{i_5}{i_4}\cdots\binom{i_m}{i_{m-1}}\binom{i_2}{i_m}$$

is equal to

$$= \begin{pmatrix} i_1 & i_2 & i_3 & \ldots & i_{m-1} & i_m \\ i_m & i_m & i_1 & \ldots & i_{m-2} & i_{m-1} \end{pmatrix}$$

(all other elements of X_n remaining fixed), and it is now fairly easy to see that

$$\beta^{m-1} = \begin{pmatrix} i_1 & i_2 & i_3 & \ldots & i_m \\ i_1 & i_1 & i_3 & & i_m \end{pmatrix} = \binom{i_2}{i_1}. \blacksquare$$

The proof of our next theorem characterizes generating sets of S in terms of a certain directed graph associated with the given set of idempotents (see Section 1.6 for definitions).

Given a set I of idempotents of defect 1 in S, we form a digraph $\Gamma(I)$ with n vertices labelled $1, 2, \ldots, n$, in which the arcs are those (j, i) for which the idempotent $\binom{i}{j}$ is in I. Thus, for example, if $n = 4$ and if

$$I = \left\{\binom{1}{3}, \binom{1}{4}, \binom{3}{1}, \binom{2}{4}\right\},$$

then $\Gamma(I)$ is

Theorem 6.3.12 *Let S be the semigroup of all singular mappings on X_n with $n \geqslant 3$. A set I of idempotents of defect 1 in S is a generating set if and only if the associated digraph $\Gamma(I)$ is strong and complete.*

Proof Suppose first that I is a generating set. If I is the set of all idempotents of defect 1, then strength and completeness are immediate. Otherwise, consider an arbitrary $\binom{i}{j}$ not in I. By hypothesis there is a product of idempotents of minimum length whereby

$$\binom{i}{j} = \binom{x_1}{x_2}\binom{x_3}{x_1}\binom{x_4}{x_3}\cdots\binom{x_m}{x_{m-1}} \quad (m \geqslant 3)$$

for $\binom{i}{j}$ as a product of idempotents in I. Since all members appearing in this expression are idempotents of defect 1, it follows from Lemma 6.3.10(i) that

$$Z\binom{i}{j} = Z\binom{x_m}{x_{m-1}} = i = x_m \quad \text{and that} \quad [ij] = [x_1 x_2];$$

thus

$$\binom{x_1}{x_2} = \binom{j}{i} \quad \text{and so} \quad \binom{j}{i} \in I.$$

we have thus shown that

$$(j, i) \notin \Gamma(I) \Rightarrow (i, j) \in \Gamma(I),$$

and so $\Gamma(I)$ is complete. Moreover, under the same assumption that $\binom{i}{j} \notin I$ we have established the existence of a path

$$(j, x_3), (x_3, x_4), \ldots, (x_{m-1}, i)$$

in $\Gamma(I)$ from j to i. This effectively establishes that $\Gamma(I)$ is strong, since under the alternative assumption that $\binom{i}{j} \in I$ we have $(j, i) \in \Gamma(I)$, whereupon it trivially follows that there is a path from j to i in $\Gamma(I)$.

Conversely, suppose that $\Gamma(I)$ is strong and complete. To show that I is a generating set for S it is sufficient by Theorem 6.3.1 to show that every idempotent of defect one not in I is a product of elements in I. Accordingly, suppose that $\binom{i}{j} \notin I$. It follows that $(j, i) \notin \Gamma(I)$ and hence, $\Gamma(I)$ being complete, that $(i, j) \in \Gamma(I)$. Since $\Gamma(I)$ is strong there exists a path

$$(j, x_1), (x_1, x_2), \ldots, (x_{m-1}, i)$$

in $\Gamma(I)$ from j to i, and since any path with repetitions can be shortened, we may assume that $j, x_1, x_2, \ldots, x_{m-1}, i$ are all distinct. We now have idempotents

$$\binom{j}{i}, \binom{x_1}{j}, \binom{x_2}{x_1}, \ldots, \binom{i}{x_{m-1}}$$

in I and, by Lemma 6.3.11,

$$\binom{i}{j} = \left[\binom{j}{i}\binom{x_1}{j}\binom{x_2}{x_1} \cdots \binom{i}{x_{m-1}}\right]^{m-1},$$

a product of idempotents in I. This completes the proof. ∎

We can now settle the question as to the size of a minimal set M of idempotent generators of S. The completeness of $\Gamma(M)$ implies that $|M| \geq (1/2)n(n-1)$, and we do in fact have equality.

Theorem 6.3.13 *If M is a minimal set of idempotent generators of the semigroup S of singular mappings of T_n, where $n \geq 3$, then $|M| = (1/2)n(n-1)$. The number of distinct sets M is*

$$2^{n-2} \prod_{r=2}^{n-2} (2^r - 1).$$

Proof If $|M| = (1/2)n(n-1)$ and if $\Gamma(M)$ is complete, then for every pair (i,j) with $i \neq j$ in X exactly one of $(i,j) \in \Gamma(M)$, $(j,i) \in \Gamma(M)$ holds. Thus $\Gamma(M)$ is a tournament, and so the number of ways of choosing a minimal set M is the number of strong labelled tournaments with n vertices. This equals the number, $\lambda(n)$, of ways of assigning arrows to the edges of the complete (undirected) graph K_n so as to make it a strong digraph.

Consider K_n with vertices $1, 2, \ldots, n$. There are $\lambda(n-1)$ ways of assigning arrows to the edges of the subgraph K_n generated by the vertices $1, 2, \ldots, n-1$. There are $n-1$ edges emanating from the vertex n, and of the 2^{n-1} ways of assigning arrows to these edges only two (namely, the one that makes all the arrows point towards n, and that which makes them all point away from n) fail to make K_n into a strongly connected digraph. Hence

$$\lambda(n) = (2^{n-1} - 2)\lambda(n-1).$$

From this and the observation that $\lambda(3) = 2$, the result follows readily by induction. ∎

Remark The case $n = 2$ is exceptional. Here

$$S = \left\{ \begin{pmatrix} 1 \\ 2 \end{pmatrix}, \begin{pmatrix} 2 \\ 1 \end{pmatrix} \right\}$$

and so consists entirely of idempotents of defect 1. Obviously, both are necessary to generate S.

In a recent paper (Howie and McFadden 1990a) the above result was extended to the subsemigroup $K(n, r)$ consisting of elements of T_n of rank r or less. There it is proved that the idempotent rank (defined as the cardinality of a minimal generating set of idempotents) of $K(n, r)$ is $S(n, r)$, the corresponding Stirling number of the second kind; also see Howie et al. (1990). For extensions to PT_n see Garba (1990).

Exercises 6.3

1. Express the map α given by

$$\begin{pmatrix} 1 & 2 & 3 & 4 & 5 & 6 & 7 & 8 & 9 & 10 & 11 & 12 & 13 & 14 & 15 & 16 & 17 & 18 & 19 & 20 \\ 2 & 4 & 2 & 6 & 6 & 7 & 8 & 4 & 12 & 11 & 12 & 12 & 14 & 13 & 16 & 17 & 15 & 19 & 18 & 20 \end{pmatrix}$$

as a product of $g(\alpha) = 21$ idempotents of defect 1.

2. Let α be the member of T_8 given by

$$\begin{pmatrix} 1 & 2 & 3 & 4 & 5 & 6 & 7 & 8 \\ 2 & 3 & 4 & 5 & 6 & 7 & 8 & 7 \end{pmatrix} \quad \text{and} \quad \varepsilon = \begin{pmatrix} 5 \\ 1 \end{pmatrix}.$$

Show that α has a single non-cyclic orbit Ω, while $\alpha\varepsilon$ has one cyclic and one non-cyclic orbit (see Remark following Lemma 6.3.6).

3. Use Exercise 1.4.13 to show that any $\alpha \in T_n$ of rank $r < n$ is expressible as a product of idempotents of rank r.

4. Show that any finite semigroup S can be embedded in a finite regular semiband.

Remark It was pointed out by Howie (1981c) that it follows from an embedding result in Pastijn (1977) that if $|S| = n$, then S may be embedded in a semiband with no more than $2n^2 + 4n + 1$ members. Three other relevant papers in this area are Giraldes and Howie (1984), Howie (1990), and Laffey (1983).

5. (Howie 1989) A semigroup S is *nilpotent-generated* (NG) if every member of S is a product of nilpotents. (Recall that $a \in S$ is nilpotent if some power of a is zero.) Suppose that S is finite of order n, and let T be the semigroup

$$T = (\{1, 2\} \times S^1 \times \{1, 2\}) \cup \{0\},$$

where 0 is the zero of T, and products are otherwise defined by

$$(i, a, j)(k, b, l) = \begin{cases} (i, ab, l) & \text{if } j = k, \\ 0 & \text{if } j \neq k. \end{cases}$$

(a) Use T to prove that if S is a monoid then S can be embedded in an NG-semigroup of order $4n + 1$.

(b) Prove that in any case S can be embedded in an NG-semigroup of order no more than $4n + 2$.

(c) Show that if S is a group then S cannot be embedded in an NG-semigroup of order less than $4n + 1$.

6. Show that

$$M = \left\{ \begin{pmatrix} i \\ j \end{pmatrix} : i < j, i+j \text{ odd} \right\} \cup \left\{ \begin{pmatrix} i \\ j \end{pmatrix} : i > j, i+j \text{ even} \right\}$$

is a minimal generating set of idempotents for the subsemigroup of all singular mappings of T_n ($n > 2$).

References

Adjan, S. I. (1966). Defining relations and algorithmic problems for groups and semigroups. *Trudy Mat. Inst. Steklov* **85** (in Russian); *Translations of the Am. Math. Soc.* **152** (1967).

Alimpic, B. P. and Krgovic, D. N. (1988). Some congruences on regular semigroups. In *Semigroup theory and applications, Proceedings of a Conference, Oberwolfach, 1986*. Lecture Notes in Mathematics 1320. Springer-Verlag, Berlin, pp. 1–10.

Aizenstat, A. Ja. (1962). Defining relations of the semigroup of endomorphisms of a finite linearly ordered set. *Sibirsk. Mat. Z.* **3**, 161–9 (in Russian).

Allen, D. Jr. (1971). A generalization of the Rees Theorem to a class of regular semigroups. *Semigroup Forum* **2**, 321–31.

Arney, J. and Bender, E. A. (1982). Random mappings with constraints on coalescence and number of origins. *Pac. J. Math.* **103**(2), 269–94.

Ash, C. J. (1980). Embedding theorems using amalgamation bases, In *Semigroups, Proceedings of the Conference on Semigroups, Monash University, 1979*. Academic Press, New York, pp. 167–76.

Biggs, R. G., Rankin, S. A., and Reis, C. M. (1976). A study of graph closed subsemigroups of a full transformation semigroup. *Trans. Am. Math. Soc.* **219**, 211–23.

Birget, J. C. (1988a). The synthesis theorem for finite regular semigroups, and its generalization. *J. Pure Appl. Algebra* **55**, 1–79.

Birget, J. C. (1988b). Stability and J-depth of expansions. *Bull. Austral. Math. Soc.* **38**, 41–54.

Blyth, T. S. and Gomes, G. M. S. (1983). On the compatibility of the natural order on a regular semigroup. *Proc. R. Soc. Edinb.* **94**, 79–84.

Blyth, T. S. and Hickey, J. B. (1984). RP-dominated regular semigroups. *Proc. R. Soc. Edinb.* Ser. A **99**, 185–91.

Brown, B. M. and Higgins, P. M. (1988). Finite full transformation semigroups as collections of random functions. *Glasgow Math. J.* **30**, 203–11.

Bulman-Fleming, S. D. and McDowell, K. (1983). Absolutely flat semigroups. *Pac. J. Math.* **107**, 319–33.

Bulman-Fleming, S. D. and McDowell, K. (1984). Flatness and amalgamation in semigroups. *Semigroup Forum* **29**, 337–42.

Burgess, W. and Raphael, R. (1978). On Conrad's partial order relation on semiprime rings and on semigroups. *Semigroup Forum* **16**, 133–40.

Bush, G. C. (1963). Note on an embedding theorem of Adjan. *Proc. Am. Math. Soc.* **14**, 597–9.

Byleen, K. (1984). Embedding a countable semigroup in a 2-generated bisimple monoid. *Glasgow Math. J.* **25**, 153–61.

Byleen, K. (1988). Embedding any countable semigroup without idempotents in an 2-generated simple semigroup without idempotents. *Glasgow Math. J.* **30**, 121–8.

Byleen, K., Meakin, J., and Pastijn, F. (1978). The fundamental four-spiral semigroup. *J. Algebra* **54**, 6–26.

Byleen, K., Meakin, J., and Pastijn, F. (1980). Building bisimple idempotent-generated semigroups. *J. Algebra* **65**, no. 1, 60–83.

Clifford, A. H. (1941). Semigroups admitting relative inverses. *Ann. Math.* **42**, 1037–49.

Clifford, A. H. (1949). Semigroups without nilpotent ideas. *Am. J. Math.* **71**, 834–44.

Clifford, A. H. and Preston, G. B. (1961). *The algebraic theory of semigroups, vol. I.* Mathematics Surveys of the American Mathematics Society 7, Providence, R.I.

Clifford, A. H. and Preston, G. B. (1967). *The algebraic theory of semigroups, vol. II.* Mathematics Surveys of the American Mathematics Society 7, Providence, R.I.

Cohn, P. M. (1959). On the free product of associative rings. *Math. Z.* **71**, 380–98.

Cowan, D. and Rankin, S. (1987). A generalization of the Bruck–Reilly construction. *Semigroup Forum* **35**, 135–49.

Denes, J. (1970). Some combinatorial properties of transformations and their connections with the theory of graphs. *J. Combin. Theory* **9**, 108–16.

Denes, J. (1981). *Unary algebras and random number generators.* Notes on Algebraic Systems III, Department of Mathematics, Karl Marx University of Economics, Budapest, 1981-3, 1–22.

Donnelly, P. J., Ewens, W. J., and Padmadisastra, S. (1991). Functionals of random mappings: exact and asymptotic results (to appear).

Drazin, M. P. (1986). A partial order in completely regular semigroups, *J. Algebra* **98**(2), 362–74.

Easdown, D. (1984a). Biordered sets of bands. *Semigroup Forum* **29**, 241–6.

Easdown, D. (1984b). Biordered sets of eventually regular semigroups. *Proc. Lond. Math. Soc.* (3) **49**, 483–503.

Easdown, D. (1984c). A new proof that regular biordered sets come from regular semigroups. *Proc. R. Soc. Edinb.*, Sec. A, **96**, 109–16.

Easdown, D. (1984d). Biordered sets are biordered subsets of semigroups. *J. Austral. Math. Soc.* **37**, 258–68.

Easdown, D. (1985). Biordered sets come from semigroups. *J. Algebra* **96**, 581–91.

Easdown, D. (1986). Biordered sets of some interesting classes of semigroups. In *Proceedings of the International Symposium on Theory of Regular Semigroups and Applications*, University of Kerala.

Easdown, D. (1988). Biorder preserving coextensions of fundamental semigroups. *Proc. Edinb. Math. Soc.* (2) **31**(3), 463–7.

Easdown, D. (1991). Biordered sets, a tool for constructing semigroups. In *Monoids and semigroups with applications.* World Scientific Publishers, New Jersey.

Easdown, D. and Hall, T. E. (1984). Reconstructing some idempotent-generated semigroups from their biordered sets. *Semigroup Forum* **29**, 207–16.

Eberhart, C., Williams, W., and Kinch, L. (1973). Idempotent-generated regular semigroups. *J. Austral. Math. Soc.* **15**, 27–34.

Edwards, P. M. (1983). Eventually regular semigroups. *Bull. Austral. Math. Soc.* **28**, 23–38.

Edwards, P. M. (1985a). Fundamental semigroups. *Proc. R. Soc. Edinb.* Sec. A **99**, 313–17.

Edwards, P. M. (1985b). On the lattice of congruences of an eventually regular semigroup. *J. Austral. Math. Soc.* Ser. A **38**, 281–6.

Edwards, P. M. (1986). Eventually regular semigroups that are group-bound. *Bull. Austral. Math. Soc.* **34**, 127–32.

Edwards, P. M. (1987). Congruences and Green's relations on eventually regular semigroups. *J. Austral. Math. Soc. Ser. A* **43**, 64–9.

Ehresmann, C. (1984). *Oeuvres completes et commentees* (ed. A. C. Ehresmann). Amiens, 1980–1984.

Evans, T. (1971). The lattice of semigroup varieties. *Semigroup Forum* **2**, 1–43.

Evssev, A. E. and Podran, N. E. (1970). Semigroups of transformations generated by idempotents with given projection characteristics. *Izv. Vyss. Ucebn. Zaved. Mat.* **12**, 103, 30–36 (in Russian).

Evssev, A. E. and Podran, N. E. (1972). Semigroups of transformation generated by idempotents of given defect. *Izv. Vyss. Ucebn. Zaved. Mat.*, **2** (117), 44–50 (in Russian).

Ewens, W. J. (1990). Population genetics theory—the past and the future. In *Mathematical and statistical developments of evolutionary theory* (ed. S. Lessard) Kluwer Academic Publishers, pp. 177–227.

Feigenbaum, R. (1979). Regular semigroup congruences. *Semigroup Forum* **17**, 373–7.

Feller, W. (1968). *An introduction to probability theory and its applications vol.* 1 (3rd edn). John Wiley, New York.

Fitzgerald, D. G. (1972). On inverses of products of idempotents in regular semigroups. *J. Austral. Math. Soc.* **13**, 335–7.

Fleischer, I. (1990). Absolutely flat monoids are amalgamation bases. *C. R. Math. Rep. Acad. Sci. Can.* **12** (1), 17–20.

Fountain, J. (1977). A class of right PP monoids, *Q. J. Math. Oxford Ser.* (2), **28**, no. 111, 285–300.

Fountain, J. (1982). Abundant semigroups. *Proc. Lond. Math. Soc.* (3), **44**, 103–29.

Gagnon, F.-P. (1981). A representation theorem for I-medial and for generalized inverse semigroups. *Semigroup Forum* **23**, 49–72.

Garba, G. U. (1990). Idempotents in partial transformation semigroups. *Proc. R. Soc. Edinb. Sec. A* **116A**, 359–66.

Gerhard, J. A. (1983). Free completely regular semigroups I and II. *J. Algebra* **82**, 135–42, 143–56.

Giraldes, E. and Howie, J. M. (1984). Embedding semigroups in semibands of minimal depth. *Semigroup Forum* **28**, 135–42.

Gluskin, L. M. (1957). Elementary generalized groups. *Mat. Sbornik* **41**(83), 23–36 (in Russian).

Gratzer, G. and Lasker, H. (1971). The structure of pseudo-complemented distributive lattices II: congruence extension and amalgamation. *Trans. Am. Math. Soc.* **156**, 30–3.

Green, J. A. (1951). On the structure of semigroups. *Ann. Math.* **54**, 163–72.

Green, J. A. and Rees, D. (1952) On semigroups in which $x^r = x$. *Proc. Cambridge Phil. Soc.* **48**, 35–40.

Goberstein, S. M. (1980). Fundamental order relations on inverse semigroups and their generalizations. *Semigroup Forum* **21**, 285–328.

Goberstein, S. M. (1985). Partial automorphisms of inverse semigroups. In *Proceedings of Marquette Conference on Semigroups*, Milwaukee, pp. 29–43.

Golab, S. (1939). Ueber den Begriff der 'Pseudogruppe von Transformationen'. *Math. Ann.* **116**, 768–80.

Goldstein R. and Teymouri, J. (1992). Adjan's theorem and conjugacy in semigroups. Semigroup Forum (to appear).
Hall, T. E. (1969a). On regular semigroups whose idempotents form a subsemigroup. *Bull. Austral. Math. Soc.* **1**, 195–208.
Hall, T. E. (1969b). On the lattice of congruences on a regular semigroup. *Bull. Austral. Math. Soc.* **1**, 231–5.
Hall, T. E. (1972). Congruences and Green's relations on regular semigroups. *Glasgow Math. J.* **20**, 167–75.
Hall, T. E. (1973). On regular semigroups. *J. Algebra* **24**, 1–24.
Hall, T. E. (1978). Representation extension and amalgamation for semigroups. *Q. J. Math.* Oxford Ser. **29**, 309–34.
Hall, T. E. (1980). Inverse and regular semigroups and amalgamation: a brief survey. In *Proceedings of* 1979 *Symposium of Regular Semigroups at Dekalb*, pp. 7–17.
Hall, T. E. (1982a). Epimorphisms and dominions. *Semigroup Forum* **24**, 271–83.
Hall, T. E. (1982b). Some properties of local subsemigroups inherited by larger subsemigroups. *Semigroup Forum* **25**, 35–49.
Hall, T. E. (1986). On regular semigroups II: an embedding. *J. Pure Appl. Algebra* **40**, 215–28.
Hall, T. E. (1988). Amalgamation for inverse and generalized inverse semigroups. *Trans. Am. Math. Soc.* **310**(1), 93–123.
Hall, T. E. and Jones, P. R. (1983). Epis are onto for finite regular semigroups. *Proc. Edinb. Math. Soc.* **26**, 151–62.
Hall, T. E. and Munn, W. D. (1979). Semigroups satisfying minimal conditions II. *Glasgow Math. J.* **20**, 133–40.
Hansen, J. C. (1989). A functional central limit theorem for random mappings. *Ann. Probability* **17**(1), 317–32.
Hansen, J. C. (1991). Functional limit theorems for random labelled combinatorial structures (to appear).
Hanumantha Rao, S. and Lakshmi, P. (1988). Group congruences on eventually regular semigroups. *J. Austral. Math. Soc.* Ser. A **45**, 320–5.
Harary, F. (1959). The number of functional digraphs. *Math. Ann.* **138**, 203–10.
Harary, F. (1969). *Graph theory*. Addison-Wesley, Reading, Mass.
Hardy, G. H., Seshu Aiyar, P. V., and Wilson, B. M. (ed.) (1962). *Collected papers of Srinivasa Ramanujan*, Chelsea, New York.
Harris, B. (1960). Probability distributions related to random mappings. *Ann. Math. Stat.* **31**, 1045–62.
Harris, B. (1967). A note on the number of idempotent elements in symmetric semigroups. *J. Combin. Theory* **3**, 1234–5.
Harris, B. (1973). The asymptotic distribution of the order of elements in symmetric semigroups. *J. Combin. Theory* **15**, 66–74.
Harris, B. and Schoenfeld, L. (1967). The number of idempotent elements in symmetric semigroups. *J. Combin. Theory* **3**, 122–35.
Hartwig, R. (1980). How to partially order regular elements. *Math. Japon.* **35**, 1–13.
Hickey, J. B. (1983). Semigroups under a sandwich operation. *Proc. Edinb. Math. Soc.* **26**, 371–82.
Hickey, J. B. (1986). On variants of a semigroup. *Bull. Austral. Math. Soc.* **34**, 447–59.
Higgins, P. M. (1981). Epis are onto for generalised inverse semigroups. *Semigroup Forum* **23**, 151–62.

Higgins, P. M. (1983a). The determination of all absolutely closed varieties of semigroups. *Proc. Am. Math. Soc.* **87**, 419–21.

Higgins, P. M. (1983b). The commutative varieties of semigroups for which epis are onto. *Proc. Edinb. Math. Soc.* Ser. A **94**, 1–7.

Higgins, P. M. (1983c). A semigroup with an epimorphically embedded subband. *Bull. Austral. Math. Soc.* **27**, 231–42.

Higgins, P. M. (1984a). Saturated and epimorphically closed varieties of semigroups. *J. Austral. Math. Soc.* Ser. A **36**, 153–75.

Higgins, P. M. (1984b). Epimorphisms, permutation identities and finite semigroups. *Semigroup Forum* **29**, 87–97.

Higgins, P. M. (1985a). Semigroup epimorphisms. *Proceedings of the 1984 Marquette Conference on Semigroups* Milwaukee, pp. 51–64.

Higgins, P. M. (1985b). Epimorphisms, dominions and semigroups. *Algebra Universalis* **21**, 225–33.

Higgins, P. M. (1986a). Completely semisimple semigroups and epimorphisms. *Proc. Am. Math. Soc.* **96**(3), 387–90.

Higgins, P. M. (1986b). Dense subsets of some common classes of semigroup. *Semigroup Forum* **34**, 5–19.

Higgins, P. M. (1986c). A method for constructing square roots in finite full transformation semigroups. *Can. Math. Bull.* **29**(3), 344–51.

Higgins, P. M. (1988a). Digraphs and the semigroup of all functions on a finite set. *Glasgow Math. J.* **30**, 41–57.

Higgins, P. M. (1988b). The range order of a product of i transformations from a finite full transformation semigroup. *Semigroup Forum* **37**, 31–6.

Higgins, P. M. (1988c). Epimorphisms and amalgams. *Colloquium Mathematicum* **LVI**, 1–17.

Higgins, P. M. (1990a). Embedding in bisimple semigroups. *Semigroup Forum* **40**, 105–7.

Higgins, P. M. (1990b). A short proof of Isbell's Zigzag Theorem. *Pac. J. Math.* **144**(1), 47–50.

Higgins, P. M. (1990c). Random products in semigroups of mappings. In *Lattices, semigroups and universal algebra* (ed. J. Almeida *et al.*). Plenum Press, New York, pp. 89–100.

Higgins, P. M. and Williams, E. J. (1988). Random functions on a finite set. *Ars Combinatoria*, 26A (1988), 93–102.

Howie, J. and Pride, S. J. (1986). The word problem for one-relator semigroups. *Math. Proc. Camb. Phil. Soc.* **99**, 33–44.

Howie, J. M. (1964). The maximum idempotent-separating congruence on an inverse semigroup. *Proc. Edinb. Math. Soc.* **14**, 71–79.

Howie, J. M. (1966). The subsemigroup generated by the idempotents of a full transformation semigroup. *J. Lond. Math. Soc.* **41**, 707–16.

Howie, J. M. (1971). Products of idempotents in certain semigroups of transformations. *Proc. R. Soc. Edinb.* **17**, 223–36.

Howie, J. M. (1975). Semigroup amalgams whose cores are inverse semigroups. *Q. J. Math.* Oxford (2), **26**, 23–45.

Howie, J. M. (1976). *An introduction to semigroup theory*. Academic Press, London.

Howie, J. M. (1978). Idempotent generators in finite full transformation semigroups, *Proc. R. Soc. Edinb.* Ser. A **81**, 317–23.

Howie, J. M. (1980). Product of idempotents in finite full transformation semigroups. *Proc. R. Soc. Edinb.* **86**, 243–54.

Howie, J. M. (1981a). A class of bisimple, idempotent-generated congruence -free semigroups. *Proc. R. Soc. Edinb* Ser A, **88**, 169–84.

Howie, J. M. (1981b). Epimorphisms and amalgamations: a survey of recent progress. *Colloquia Mathematica Societatis Janos Bolyai*, 39: *Semigroups.* Szeged, Hungary.

Howie, J. M. (1981c). Embedding semigroups in semibands: some arithmetical results. *Q. J. Math.* Oxford Ser. 2, **32**, no. 127, 323–37.

Howie, J. M. (1989). Embedding semigroups in nilpotent-generated semigroups. *Math. Slovaca* **39**, 47–54.

Howie, J. M. (1990). Arithmetical aspects of semigroup embeddings. In *Lattices, semigroups and universal algebra* (ed. J. Almeida *et al.*) Plenum Press, New York, pp. 101–4.

Howie, J. M. (1991). *Automata and languages.* Oxford University Press.

Howie. J. M. and Isbell, J. R. (1967). Epimorphisms and dominions II. *J. Algebra* **6**, 7–21.

Howie, J. M. and Lallement, G. (1966). Certain fundamental congruences on a regular semigroup, *Proc. Glasgow Math. Assoc.* **7**, 145–56.

Howie, J. M., Lusk, E. L., and McFadden, R. B., (1990). Combinatorial results relating to products of idempotents in finite full transformation semigroups, *Proc. Roy. Soc. Edinburgh* Sec. A, **115**, 289–99.

Howie, J. M. and McFadden, R. B. (1990a) Idempotent rank in finite full transformation semigroups. *Proc. R. Soc. Edinb*, Sec. A, **114**(3–4), 161–7.

Howie, J. M. and McFadden, R. B. (1990b). Ideals are greater on the left. *Semigroups Forum* **40**, 247–8.

Howie, J. M., Robertson, E. F., and Schein, B. M. (1988). A combinatorial property of finite full transformation semigroups. *Proc. R. Soc. Edinb.*, Sec. A **109**(3–4), 319–28.

Howie. J. M. and Schein, B. M. (1973). Products of idempotent order-preserving transformations. *J. Lond. Math. Soc.* **7**, 357–66.

Imaoka, T. (1976). Free products with amalgamation of bands, *Mem. Fac. Lit. Sci. Shimane Univ. Natur. Sci.* **10**, 7–17.

Imaoka, T. (1987). Free products and amalgamation of generalized inverse ∗-semigroups, *Mem. Fac. Sci. Shimane Univ.* **21**, 55–64.

Isbell, J. R. (1966). Epimorphisms and dominions, In *Proceedings of the Conference on Categorical Algebra, La Jolla*, 1965. Lange and Springer-Verlag, Berlin, pp. 232–46.

Isbell, J. R. (1968). Epimorphisms and dominions III. *Am. J. Math.* **90**, 1025–30.

Isbell, J. R. (1969). Epimorphisms and dominions IV. *J. Lond. Math. Soc.* (2) **1**, 265–73.

Isbell, J. R. (1974). Notes on semigroup dominions. *Semigroup Forum* **7**, 364–8.

Iwahori, N. (1977). A length formula in a semigroup of mappings. *J. Fac. Sci. Univ. Tokyo* Sec. 1A, Math. **24**(2), 255–60.

Jackson, D. A. (1986). Some one relator semigroup presentations with solvable word problems. *Math. Proc. Camb. Phil. Soc.* **99**, 433–4.

Jackson, D. A. (1991). Invariance of small overlap hypotheses. *Semigroup Forum* **43**, 299–304.

Jackson, D. A. (1992). Achievable relators (to appear).

Johnston, K. G. and Jones, P. R. (1984). The lattice of full regular subsemigroups of a regular semigroup. *Proc. R. Soc. Edinb.* Ser. A **98**, 203–14.

Jones, P. R. (1982). A graphical representation of the free product of E-unitary inverse semigroups. *Semigroup Forum* **24**, 195–222.

Jones, P. R. (1983). On congruence lattices of regular semigroups. *J. Algebra* **82**(1), 18–39.

Jones, P. R. (1984). Joins and meets of congruences on a regular semigroup. *Semigroup Forum* **30**, 1–16.

Jonsson, B. (1965). Extensions of relational structures in the theory of models. In *Proceedings of the* 1963 *Symposium at Berkeley* (ed. J. W. Addison, L. and A. Tarski). North Holland, Amsterdam, pp. 146–57.

Joubert, G. (1966). Contribution à l'étude des catégories ordonnées, applications aux structures feuilletées. *Cahier Top. Diff.,* **VIII.**, +117 pp.

Kasincev, E. V. (1970). Graphs and the word problem for finitely presented semigroups. *Tul. Gos. Ped. Inst. Ucen. Zap. Mat. Kaf. Vyp.* **2** (*Geom.* 1 *Algebra*), 290–302 (in Russian).

Katz, L. (1955). Probability of indecomposability of a random mapping function. *Ann. Math. Stat.* **26**, 512–17.

Khan, N. M. (1982). Epimorphisms, dominions and varieties of semigroups. *Semigroup Forum* **25**, 331–7.

Khan, N. M. (1983). Some saturated varieties of semigroups. *Bull. Austral. Math. Soc.* **27**, 419–25.

Khan, N. M. (1985a). On saturated permutative varieties and consequences of permutation identities. *J. Austral. Math. Soc.* Ser. A. 38, no 2, 186–97.

Khan, N. M. (1985b). Epimorphically closed permutative varieties. *Trans. Am. Math. Soc.* **287**(2), 507–28.

Kim, J. B. (1971). Mutants in symmetric semigroups. *Czech. Math. J.* **21** (96), 355–63.

Kim, J. B. (1972). Idempotents in symmetric semigroups. *J. Combin. Theory,* Ser. A **13**, 155–61.

Kim, K. H. and Rousch, F. W. (1978). Order of subsemigroups of $\mathscr{T}_X$, *Semigroup Forum* **16**, 203.

Kim, K. H. and Rousch, F. W. (1980). The average rank of a product of transformations. *Semigroup Forum* **19**, 79–85.

Kingman, J. F. C. (1978). The representation of partition structures. *J. Lond. Math. Soc.* **18**, 374–80.

Koch, R. J. (1984). Sandwich sets and partial order. *Semigroup Forum* **30**(1), 53–66.

Kolchin, V. F. (1976). A problem of the allocation of particles in cells and random mappings. *Theory Prob. Appl.* **21**, 48–63.

Kowol, G. and Mitsch, H. (1986). Naturally ordered transformation semigroups. *Mh. Math.* **102**, 115–38.

Kruskal, M. D. (1954). The expected number of components under a random mapping function. *Am. Math. Monthly* **61**, 392–7.

Kimura, N. (1957). Unpublished doctoral dissertation. The Tulane University of Louisiana.

Kupka, J. (1988). The distribution and moments of the number of components. Statistics Research Report no. 179, Monash University.

Lallement, G. (1966). Congruences et equivalences de Green sur un demi-groupe regulier. *C. R. Acad. Sci. Paris,* Ser. A **262**, 613–16.

Lallement, G. (1979). *Semigroups and combinatorial applications.* John Wiley, New York.

Lallement, G. (1986). Some algorithms for semigroups and monoids presented by a single relation. *Semigroup Theory and Applications, Oberwolfach, Lecture Notes in Math.,* 1320, Springer-Verlag, pp. 176–82.

Laffey, T. J. (1983). A note on embedding finite semigroups in finite semibands. *Q. J. Math. Oxford Ser* (2), **34**, no. 136, 453–4.

La Torre, D. R. (1982). Group congruences on regular semigroups. *Semigroup Forum* **24**, 327–40.

La Torre, D. R. (1983). The least semilattice of groups congruence on a regular semigroup. *Semigroup Forum* **27**, 319–29.

Lawson, M. V. (1986). The structure of type A semigroups. *Q. J. Math.* Oxford Ser. (2), **37**, no. 147, 279–98.

Lawson, M. V. (1987). The natural partial order on an abundant semigroup. *Proc. Edinb. Math. Soc.* (2) **30**, no. 2, 169–86.

Lawson, M. V. (1989). An order theoretic characterization of locally orthodox semigroups. *Semigroup Forum* **39**, 113–16.

Lawson, M. V. (1991). Some observations on the McAlister Covering Theorem (to appear).

Levi, I. (1986) Green's relations on Croisot–Teissier semigroups. *Semigroup Forum*, **33**, 299–307.

Lloyd, C. J. and Williams, E. J. (1988). Recursive splitting of an interval when the proportions are identical and independent random variables. *Stoch. Processes Applic.* **27**, 111–22.

Lyapin, E. S. (1974). *Semigroups.* Translations of Mathematical Monographs, Vol. 3. American Mathematical Society, Providence, R. I.

Lyndon, R. C. and Schupp, P. (1977). *Combinatorial group theory.* Springer-Verlag, Heidelberg.

McAlister, D. B. (1974a). Groups, semilattices and inverse semigroups. *Trans. Am. Math. Soc.* **192**, 227–44.

McAlister, D. B. (1974b). Groups, semilattices and inverse semigroups II. *Trans. Am. Math. Soc.* **196**, 351–70.

McAlister, D. B. (1978). E-unitary inverse semigroups over semilattices. *Glasgow Math. J.* **19**, 1–12.

McAlister, D. B. (1980). A random ramble through inverse semigroups. In *Semigroups,* Proceedings of the Monash Conference on Semigroups. Academic Press New York, pp. 1–20.

McAlister, D. B. (1984). Rees matrix covers for regular semigroups. *J. Algebra* **89**, 264–79.

McAlister, D. B. and Reilly, N. R. (1977). E-unitary covers for inverse semigroups. *Pac. J. Math.* **68**, 161–74.

Madhaven, S. (1978). Some results on generalized inverse semigroups. *Semigroup Forum* **16**, 355–67.

Magnus, W. (1932). Das Identitats-Problem fur Gruppen mit einer definierended Relation. *Math. Ann.* **106**, 295–307.

Malcev, A. (1937). On the immersion of an algebraic ring into a field. *Math. Ann.* **113**, 686–91.

Margolis, S. W. and Pin, J. E. (1987a). Inverse semigroups and extensions of groups by semilattices. *J. Algebra* **110**, 277–97.

Margolis, S. W. and Pin, J. E. (1987*b*). Expansions, free inverse semigroups and Schützenberger products. *J. Algebra* **110**, 298–305.
Margolis, S. W., Meakin, J. C., and Stephen, J. B. (1987). Some decision problems for inverse monoid presentations. In *Semigroups and their Applications* (ed. S. Goberstein and P. Higgins). D. Reidel, Dordrecht, pp. 99–110.
Masat, F. E. (1973). Right group and group congruences on a regular semigroup. *Duke Math. J.* **40**, 393–402.
Masat, F. E. (1978). Proper regular semigroups. *Proc. Am. Math. Soc.* **71**, 189–92.
Masat, F. E. (1982). The structure of the minimum group kernel of a regular semigroup. *Czech. Math. J.* **32** (107), no. 3, 377–83.
Meakin, J. C. (1972*a*). Congruences on orthodox semigroups II. *J. Austral. Math. Soc.* **11**, 259–66.
Meakin, J. C. (1972*b*). The maximum idempotent-separating congruence on a regular semigroup. *Proc. Edinb. Math. Soc.* (2), **18**, 159–63.
Meakin, J. C. (1980). Constructing biordered sets. In *Semigroups*, Proceedings of the Monash Conference on Semigroups. Academic Press, New York, pp. 67–84.
Miller, D. D. and Clifford, A. H. (1956). Regular $\mathscr{D}$-classes in semigroups. *Trans. Am. Math. Soc.* **82**, 270–80.
Mitsch, H. (1986). A natural partial order for semigroups. *Proc. Am. Math. Soc.* **97**, 384–8.
Munn, W. D. (1961). A class of irreducible matrix representations of an arbitrary inverse semigroup. *Proc. Glasgow Math. Assoc.* **5**, 41–8.
Munn, W. D. (1974). Free inverse semigroups. *Proc. Lond. Math. Soc.* (3) **29**, 385–404; announced in Semigroup Forum 5 (1973), 262–269.
Munn, W. D. (1976). A note of *E*-unitary inverse semigroups. *Bull. Lond. Math. Soc.* **8**, 71–6.
Munn, W. D. (1981). An embedding theorem for free inverse semigroups. *Glasgow Math. J.* **22**(2), 217–22.
Nambooripad, K. S. S. (1974). Structure of regular semigroups I. *Semigroup Forum* **9**, 354–63.
Nambooripad, K. S. S. (1975). Structure of regular semigroups II. The general case. *Semigroup Forum* **9**, 364–71.
Nambooripad, K. S. S. (1979). The structure of regular semigroups I. *Mem. Am. Math. Soc.* **22**, No. 224, vii + 119 pp.
Nambooripad, K. S. S. (1980). The natural partial order on a regular semigroup. *Proc. Edinb. Math. Soc.* **23**, 249–60.
Nambooripad, K. S. S. and Sitarman, Y. (1979). Some congruences on regular semigroups. *J. Algebra* **57**(1), 10–25.
Nico, W. R. (1983). On the regularity of semidirect products. *J. Algebra* **80**, 29–36.
Nivat, M. and Perrot, J. F. (1970). Une generalisation du monoide bicyclique. *C.R. Acad. Sci. Paris* **217A**, 824–7.
O'Carroll, L. (1976). Embedding theorem for proper inverse semigroups. *J. Algebra* **42**, 26–40.
Oganesyan, G. U. (1984). Isomorphism problems for semigroups with one defining relation. *Math. Notes* **35**, 360–3.
Pastijn, F. (1977). Embedding semigroups in semibands. *Semigroup Forum* **14**, 247–63.
Pastijn, F. (1980). The biorder on the partial groupoid of idempotents of a semigroup. *J. Algebra* **65**, 147–87.

Pastijn, F. (1982). The structure of pseudo-inverse semigroups. *Trans. Amer. Math. Soc.*, **273**, 631–55.
Pastijn, F. (1985). Congruences on regular semigroups—a survey. In *Proceedings of the 1984 Marquette Conference on Semigroups*. Milwaukee, pp. 159–75.
Pastijn, F. (1990). The kernel of an idempotent-separating congruence on a regular semigroup. In *Lattices, semigroups, and universal algebra* (ed. J. Almeida *et al.*). Plenum Press, New York, 203–10.
Pastijn, F. and Petrich, M. (1986). Congruences on regular semigroups. *Trans. Am. Math. Soc.* **295**(2), 607–33.
Pastijn, F. and Petrich, M. (1987). Congruences on regular semigroups associated with Green's relations, *Boll. U.M.I.*, **7** (1-B), 591–603.
Pastijn, F. and Petrich, M. (1988). The congruence lattice of a regular semigroup. *J. Pure Appl. Algebra* **53**(1–2), 93–123.
Penrose, R. (1989). *The emperor's new mind*. Oxford University Press.
Petrich, M. (1971). A construction and a classification of bands. *Math. Nachr.* **48**, 263–71.
Petrich, M. (1973). *Introduction to semigroups*. Merrill, Columbus, Ohio.
Petrich, M. (1974). The structure of completely regular semigroups. *Trans. Am. Math. Soc.* **189**, 211–36.
Petrich, M. (1977). *Lectures in semigroups*. John Wiley, New York.
Petrich, M. (1984). *Inverse semigroups*. John Wiley, New York.
Petrich, M. (1985). Locally inverse semigroups. In *Proceedings of the 1984 Marquette Conference on Semigroups*. Milwaukee, pp. 177–82.
Petrich, M. (1987). Cayley theorems for semigroups. In *Semigroups and their applications* (ed S. M. Goberstein and P. M. Higgins), D. Reidel, pp. 133–8.
Philip, J. M. (1974). A proof of Isbell's Zigzag Theorem. *J. Algebra* **32**, 328–31.
Pin, J. E. (1986). *Varieties of formal languages*. Plenum Press, London.
Polak, L. (1986). On the word problem for free completely regular semigroups. *Semigroup Forum* **34**(2), 127–38.
Popova, L. M. (1962). Defining relations of the semigroup of partial endomorphisms of a finite linearly ordered set. *Leningrad Gos. Ped. Inst. Ucen. Zap.* **238**, 78–88.
Post, E. L. (1947). Recursive unsolvability of a problem of Thue. *J. Symbolic Logic* **12**, 1–11.
Power, A. J. (1990). A 2-categorical pasting theorem. *J. Algebra* **129**, 439–45.
Premchand, S. (1984). Independence of axioms for biordered sets. *Semigroup Forum* **28**, 249–63.
Preston, G. B. (1954). Inverse semi-groups. *J. Lond. Math. Soc.* **29**, 396–403.
Preston, G. B. (1959). Embedding any semigroup in a $\mathscr{D}$-simple semigroup. *Trans. Am. Math. Soc.* **93**, 351–5.
Preston, G. B. (1986*a*). Monogenic inverse semigroups. *J. Austral. Math. Soc.*, Ser. A **40**, 321–42.
Preston, G. B. (1986*b*). Semidirect products of semigroups. *Proc. R. Soc. Edinb. Sec. A* **102** (1–2), 91–102.
Preston, G. B. (1986*c*). The semidirect product of an inverse semigroup and a group. *Bull. Austral. Math. Soc.* **33**(2), 261–72.
Protic, P. (1987). The lattice of r-semiprime idempotent-separating congruences

on *r*-semigroups. *Proceedings of a Conference on Algebra and Logic, Cetinje*, pp. 157–65.
Ramyantsev, Y. (1981). A minimal example of a semigroup with cancellation which is not imbeddable in a group. *Algebraic systems, Ivanov*, Gos. Univ. Ivanovna, 198–202.
Reilly, N. R. (1965). Embedding inverse semigroups in bisimple inverse semigroups. *Q. J. Math.* Oxford (2) **16**, 183–7.
Reilly, N. R. (1972). Free generators of free inverse semigroups. *Bull. Austral. Math. Soc.* **7** (1972), 407–24; Corrigenda. ibid. **9** (1973), 479.
Reilly, N. R. (1977). Maximal inverse subsemigroups of $\mathcal{T}_X$. *Semigroup Forum* **15**, 319–26.
Reilly, N. R. (1989). Free combinatorial strict inverse semigroups. *J. Lond. Math. Soc.* (2), **39**(1), 102–20.
Reilly, N. R. and Munn, W. D. (1976). *E*-unitary congruences on inverse semigroups. *Glasgow Math. J.* **17**, 57–75.
Reilly, N. R. and Scheiblich, H. E. (1967). Congruence on regular semigroups. *Pac. J. Math.* **23**, 349–60.
Remmers, J. H. (1980). On the geometry of semigroup presentations. *Adv. Math.* **36**, 283–96.
Renshaw, J. (1986*a*). Extensions and amalgamations in monoids and semigroups. *Proc. London Math. Soc.* **52**, no. 1, 119–41.
Renshaw, J. (1986*b*). Flatness and amalgamation in monoids, *J. Lond. Math. Soc.* (2) **33**(1), 73–88.
Renshaw, J. (1991*a*). Perfect amalgamation bases. *J. Algebra* (to appear).
Renshaw, J. (1991*b*). Subsemigroups of free products of semigroups (to appear).
Reynolds, M. A. (1984). A new construction for free inverse semigroups. *Semigroup Forum* **30**, 291–6.
Rhodes, J. and Allen, D. Jr. (1976). Synthesis of classical and modern semigroup theory. *Adv. Math.* **11**, 238–66.
Riemann, J. (1972). Statistical investigations of symmetrical semigroups. *Proceedings of a Conference on Semigroup Theory, Szeged* pp. 61–3.
Robinson, D. W. (1962). On the generalized inverse of an arbitrary linear transformation's Amer. Math. Monthly **69**, 412–416
Rubin, H. and Sitgreaves, R. (1954). Probability distributions related to random transformations on a finite set. Tech. Report No. 19a, Applied Mathematics and Statistics Laboratory, Stanford, January
Saito, T. (1965). Proper ordered inverse semigroups. *Pac. J. Math.* **15**, 649–66.
Saito, T. (1988). Some remarks on finite full transformation semigroups. *Semigroup Forum*, **37**(1), 37–43.
Saito, T. (1989*a*). Products of four idempotents in finite full transformation semigroups. *Semigroup Forum* **39**(2), 179–93.
Saito, T. (1989*b*). Products of idempotents in finite full transformation semigroups. *Semigroup Forum* **39**(3), 295–309.
Scheiblich, H. E. (1971). A characterization of a free elementary inverse semigroups. *Semigroup Forum* **2**, 76–9.
Scheiblich, H. E. (1972). Free inverse semigroups. *Semigroup Forum* **4**, 351–9.

Scheiblich, H. E. (1973). Free inverse semigroups. *Proc. Am. Math. Soc.* **38**, 1–7.
Scheiblich, H. E. (1974). Kernels of inverse semigroup homomorphisms. *J. Austral. Math. Soc.* **18**, 289–92.
Scheiblich, H. E. (1976). On epics and dominions of bands. *Semigroup Forum* **13**, 103–14.
Scheiblich, H. E. (1982). Generalized inverse semigroups with involution. *Rocky Mountain J. Math.* **12**(2), 205–11.
Scheiblich, H. E. and Hsieh, S. C. (1982). Some closed subbands. *Tamkang J. Math.* **15** (2), 213–17.
Scheiblich, H. E. and Moore, K. C. (1973). $\mathcal{T}_X$ is absolutely closed. *Semigroup Forum* **6**, 216–26.
Schein, B. M. (1963). On the theory of generalized groups. *Dokl. Akad. Nauk SSSR* **153**, 296–9 (in Russian).
Schein, B. M. (1971). A symmetric semigroup of transformations is covered by its inverse subsemigroups. *Acta Mat. Acad. Sci. Hung.* **22**, 163–71.
Schein, B. M. (1972). Pseudosemilattices and pseudolattices (in Russian). *Izv. Vyss. Ucebn. Zaved. Math.* No. 2, **117**, 81–94; *Trans. Am. Mat. Soc.* (2), **119**, 1–16.
Schein, B. M. (1975a). A new proof of the McAlister P-Theorem. *Semigroup Forum* **10**, 185–8.
Schein, B. M. (1975b). Free inverse semigroups are not finitely presentable. *Acta Math. Acad. Sci. Hung.* **26**, 41–52.
Schein, B. M. (1986). Prehistory of the theory of inverse semigroups, *Proceedings of the 1986 Semigroup Conference, Baton Rouge*. Louisiana State University, pp. 72–6.
Schützenberger, M. P. (1957). $\mathcal{D}$-representation des demi-groups. *C.R. Acad. Sci. Paris* **244**, 1994–6.
Shepp, L. A. and Lloyd, S. P. (1966). Ordered cycle lengths in a random permutation. *Trans. Am. Math. Soc.* **121**, 340–57.
Shoji, K. (1980). Right self-injective semigroups are absolutely closed. *Mem. Fac. Sci. Shimane Univ.* **14**, 35–9.
Shoji, K. (1988). Absolute flatness of the full transformation semigroup. *J. Algebra* **118** (2), 477–86.
Shoji, K. (1990). Amalgamation bases for semigroups. *Math. Japon.* **35** (3),473–83.
Shneperman, L. B. (1974). Maximal inverse subsemigroups of the semigroup of linear transformations (in Russian) *Izv. Vyssh. Ucebn. Zaved. Mat.* no. 11. 93–100.
Simon, I. (1980) Conditions de finitude pour des semigroupes, *C.R. Acad. Sci.*, Sér. A, 290 1081–1082.
Sioson, F. M. (1972). Counting elements in semigroups, *J. Combin. Theory.* Ser. A, **12**, 339–45.
Snowden, M. and Howie, J. M. (1982). Square roots in finite full transformation semigroups. *Glasgow Math. J.* **23**, 137–49.
Stamenkovic, B. and Protic, P. (1987). The natural partial order on an r-cancellative semigroup. *Mat. Vesnik*, **39**(4), 455–62.
Stenstrom, B. (1971). Flatness and localization over monoids. *Math. Nachr.* **48**, 315–34.
Stephen, J. B. (1990). Presentation of inverse monoids. *J. Pure Appl. Algebra* **63**(1), 81–112.
Storrer, H. H. (1976). An algebraic proof of Isbell's Zigzag Theorem. *Semigroup Forum* **12**, 83–8.

References

Suschkewitsch, A. (1928). Untersuchungen uber verallgemeinerte Substitutionen. *Atti del Congresso Internazionale dei Matematici Bologna*, pp. 147–57.

Szendrei, M. B. (1987). A generalization of McAlister's *P*-theorem for *E*-unitary regular semigroups. *Acta. Sci. Math.* **51**, 229–49.

Taintier, M. (1968). A characterization of idempotents in semigroups. *J. Combin. Theory* **5**, 370–3.

Tret'yakova, E. G. (1986). Biordered sets of idempotents of semigroups of certain types, *Izv. Vyssh. Ucehebn. Zaved. Mat.* no . 3, 77–80.

Todorov, K. (1979). Inverse elements and their average number in a finite symmetric semigroup. *Semigroup Forum* **18**, 279–81.

Todorov, K. (1980). On the orders of the subsemigroups of the symmetrical semigroup. *Semigroup Forum* **21**, 329–35.

Trotter, P. G. (1978). Normal partitions of idempotents of regular semigroups. *J. Austral. Math. Soc.*, Ser. A **26**, 110–114.

Trotter, P. G. (1982). Congruences on regular and completely regular semigroups. *J. Austral. Math. Soc.*, Ser. A **32**, 388–98.

Trotter, P. G. (1984). Free completely regular semigroups. *Glasgow Math. J.* **25**(2), 241–54.

Trotter, P. G. (1986). A non-surjective epimorphism of bands, *Algebra Universalis* **22**, 109–16.

Tully, E. J. (1961). Representation of a semigroup by transformations acting transitively on a set. *Am. J. Math.* **83** (1961), 533–41.

Veeramony, R. (1984). Proper pseudo-inverse semigroups. *Simon Stevin* **58**(1–2), 65–86.

Venkatesan, P. S. (1976). On right unipotent semigroups. *Pac. J. Math.* **63**, 555–61.

Wagner, V. V. (1961). Generalized heaps and generalized groups with a transitive compatibility relation. *Ucen. Zap. Sarat. Gos. Univ., Meh-Mat.* **70**, 25–39 (in Russian).

Warne, R. J. (1972). $\mathscr{L}$-unipotent semigroups, *Nigerian N. N. Sci.* **5**, 245–8.

Watterson, G. A. and Guess, H. A. (1977). Is the most frequent allele the oldest? *Theoret. Pop. Biol.* **11**, 141–60.

Wilkinson, R. (1983). A description of *E*-unitary inverse semigroups. *Proc. R. Soc. Edinb.* Ser. A **95**, 239–42.

Yamada, M. (1967). Regular semigroups whose idempotents satisfy permutation identities. *Pac. J. Math.* **21**, 371–92.

Yamada, M. (1973). Orthodox semigroups whose idempotents satisfy a certain identity. *Semigroup Forum* **6**, 113–28.

Glossary of notation

Aut (A)	group of automorphisms of algebra A
$\mathscr{B}_X$	semigroup of binary relations on base set X
B_n	semigroup of binary relations of finite set of order n
$\mathscr{BS}(X)$	set of ordered triples (T, α, β), $T \in \mathscr{S}(X)$, $\alpha, \beta \in V(T)$
$c(\alpha)$	number of cyclic orbits of mapping α
$C(n)$	condition of semigroup presentation that no word is a product of fewer than n pieces
$\mathscr{D}$	Green's relation $\mathscr{L} \vee \mathscr{R}$
D_a	$\mathscr{D}$-class of a
$d(x)$	depth of vertex x
D_E	domain or biordered set E
dom α	domain of mapping α
Dom(U, S)	dominion of U in S
$d(u, v)$	distance between two vertices u and v
e^+, e^-	positive and negative edges associated with arc e
$E^\flat$	greatest congruence contained in equivalence relation E
$E(T)$	set of idempotents of subset T of S
$e(v)$	eccentricity of vertex v
F_X	free semigroup on base set X
FI_X	free inverse semigroup on base set X
$f(\alpha)$	number of fixed points of mapping α
G_X	free group on base set X
$\mathscr{G}_X$	full symmetric group on base set X
GB	condition that S is group-bound
$g(x)$	grasp of vertex x
$g(\alpha)$	gravity of mapping α
$(G; \mathscr{X}, \mathscr{Y})$	McAlister triple
$\mathscr{H}$	Green's relation $\mathscr{L} \cap \mathscr{R}$
H_a	$\mathscr{H}$-class of a
$h(x)$	height of vertex x
$\mathscr{I}_X$	symmetric inverse semigroup on base set X
it(α)	iterative range of mapping α
$\mathscr{J}$	Green's relation defined by equality of principal ideals
J_a	$\mathscr{J}$-class of a
$J(a)$	principal ideal generated by a

Glossary of notation | 251

$K(\Omega)$	kernel of orbit Ω of some mapping
$\ker \alpha$	kernel of mapping ($\ker \alpha = \alpha \circ \alpha^{-1}$)
$\mathscr{L}$	Green's relation defined by equality of left principal ideals
L_a	$\mathscr{L}$-class of a
$L(a)$	left principal ideal generated by a
$\mathscr{L}^o$	set of regular $\mathscr{L}$-classes of semigroup
$\mathscr{L}'$	set of $\succ\!\!\!-\!\!\!\prec$-classes of biordered set E
$l(x)$	level of vertex x
$LG(X;R)$	left graph of presentation $(X;R)$
$M^0[G; I, \Lambda; P]$	Rees semigroup with group G and sandwich matrix P
M_L, M_R, M_H, M_J	minimal condition on posets $S/\mathscr{L}$, $S/\mathscr{R}$, $S/\mathscr{H}$, and $S/\mathscr{J}$ respectively
M_L^*, M_R^*	minimal condition on $\mathscr{L}$-classes [$\mathscr{R}$-classes] within all $\mathscr{J}$-classes
$M(e,f)$	$\{g \in E : ge = g = fg\}$
$\mathbb{N}^0$	set of non-negative integers
$P(G, \mathscr{X}, \mathscr{Y})$	P-semigroup of McAlister triple $(G, \mathscr{X}, \mathscr{Y})$
$p\alpha^{-N}$	$\{x \in X : x\alpha^i = p \text{ for some } i > 0\}$
$PA(A)$	semigroup of all partial automorphisms on algebra A
$\mathscr{PT}(X)$	partial transformation semigroup on base set X
PT_n	partial transformation semigroup on finite set of order n
$\mathscr{R}$	Green's relation defined by equality of right principal ideals
R_a	$\mathscr{R}$-class of a
$R(a)$	right principal ideal generated by a
$\mathscr{R}^0$	set of regular $\mathscr{R}$-classes of S
$\mathscr{R}'$	set of $\leftrightarrow$ classes of a biordered set E
R^*	congruence generated by relation R
R^{-1}	inverse relation of relation R
R^s	least symmetric and reflexive relation containing relation R
R^c	equivalence relation generated by R on set of all R-words
$r(G)$	radius of graph G
$\operatorname{ran} \alpha$	range of mapping α
$r(x)$	reach of vertex x
$\operatorname{Reg}(S)$	set of regular elements of S
REP	representation extension property
$RG(X;R)$	right graph of presentation $(X;R)$
S	semigroup
S^1	semigroup with identity 1
S^0	semigroup with zero 0
S^*	dual of semigroup S
$S(a)$	set of strong inverses of a
SAP	strong amalgamation property
S/E	S factored by equivalence relation E
S/I	S factored by Rees congruence associated with ideal I

$S(e_1,\ldots,e_n)$	sandwich set of $e_1,\ldots,e_n$
$SpAP$	special amalgamation property
stran α	stable range of mapping α
$\mathrm{Sub}(A)$	lattice of all subalgebras of algebra A
$\mathscr{S}(X)$	transversal of isomorphism classes of word trees on X
$\mathscr{T}_X$	full transformation semigroup on base set X
T_n	full transformation semigroup on finite set of order n
$V(a)$	set of inverses of a
$V(S)$	$\{(a,b)\in S\times S: b\in V(a)\}$
$V(G)$	set of vertices of graph G
WAP	weak amalgamation property
X_n	$\{1,2,\ldots,n\}$
$Z(C)$	cycle of component C of a graph
$\Gamma(H)$	Schützenberger group of $\mathscr{H}$-class H
σ	minimum group congruence
η	minimum semilattice congruence
ι	identity relation
λ_a	left inner translation mapping of S by a
μ	maximum idempotent-separating congruence
ν	$\{(a,b): a=eb=bf \text{ for some } e,f\in E(S^1)\}$
$\bar{\pi}$ or $\phi(\pi)$	label of walk π
$\Pi(\alpha,\beta)$	unique path from α to β in word tree
$\rho^\natural$	natural homomorphism associated with congruence ρ
ρ_a	right inner translation mapping of S by a
ρ/σ	factor congruence of ρ on σ
τ	cycle number of a random permutation
Ω	orbit of a mapping
ω	universal relation
$\langle A\rangle$	subsemigroup generated by A
$\leq_l$	left partial order on S
$\leq_r$	right partial order on S
∂M	boundary of directed map M
$\|M\|$	number of regions of directed map M
$[p_1=q_1, p_2=q_2,\ldots]$	semigroup variety with equations as given
$[S_i; U]_{i\in I}$	semigroup amalgam with core U
$S*T$	free product of S and T
$S*_U T$	free product of S and T amalgamating U
$u\equiv v$	u is literally equal to v
$|X|$	cardinality of set X
$(X;R)$	presentation with generators X and relations R
$\langle X;R\rangle$	semigroup with presentation $(X;R)$
$[X;R]$	group with presentation $(X;R)$
$\succ\!\!-$	left arrow of a biordered set
$\rightarrow$	right arrow of a biordered set
$\prec$	pre-order symbol

Index

absolutely closed semigroup 143
amalgam 159, 165
  core of 159
  (weakly, strongly) embeddable 159
  special 159, 160
(strong, weak) amalgamation base 160, 162–3
  special 160
amalgamation pair 161, 162
amalgamation property
  special 159, 161
  strong 159, 161
  weak 159, 161
angle 77
  included edge(s) 77
  included region(s) 77
  interior 77
  sides of 77
  sink- 77, 183–4
  source- 77, 183–4
anti-automorphism 11
anti-homomorphism 11
anti-isomorphism 11
anti-representation 11
arc 69
(left, right) arrow 60
automorphism 2
  group acting on the left 98
  partial 5
  of word trees 83

Baer–Levi semigroup 14, 37, 157
band(s) 2, 157
  commutative 2
  of groups 43
  (left, right) normal 43
  medial 43
  normal 43
  rectangular 2, 64, 141
  of type-T semigroups 37
  0-rectangular 42
basic products 113
bimorphism 61
binary relation(s) 5
  semigroup of 5
biorder 60
biordered set(s) 60–6, 110–41
  axioms of 111–12

  of bands 127–30, 141
  of completely regular semigroups 137
  of eventually regular semigroups 142
  isomorphism between 112
  M-biordered 62, 67
  morphism 61, 112, 119
  rectangular 67
  regular 62, 66, 137, 141
  of regular semigroups 130–3
  solid 135, 137, 141
  subset 113, 120
(0-) bisimple semigroup 16, 22, 36
(left, right) boundary 74, 177
Brandt semigroup 42, 156, 165
bridge 71, 217

(left, right) cancellative semigroup 3, 12, 29, 79
central element 39
centre of a tree 69, 83
chain conditions 23
closed subset of the free group 95
commutative 2, 43, 154
  regular semigroup 40
  variety 155–6
completely regular
  element 4, 73
  semigroup 4, 37–41, 72, 73, 157
completely (0-) simple semigroup 33–7, 66, 126, 141
completely semisimple 41, 59, 91, 152–3
complete sequence 136
component
  cyclic 71
  even (odd) 209
  number 197–9
  simple 76
  two-sided 74
congruence(s) 6
  (left, right) congruence extension property 165–6
  generated by a relation 6
  idempotent-determined 108
  idempotent-consistent 50, 123
  idempotent-pure 108
  idempotent-separating 50–3, 57, 58, 61, 67, 104, 108
  idempotent-surjective 50

congruence(s) (*cont.*)
  identity 7
  intersection of 6
  left (right) 6
  least band 58
  least group 45, 53–4, 58, 98, 99
  least inverse 11, 59
  least semilattice 37, 55
  minimum contained in an equivalence 7
  Rees 7
  trace of 54
  universal 6
consistent pairing of trees 209–10
convex subset 95
core of a semigroup 55, 59, 124
cycle 68
left (right) cycle 78, 183–4
cyclic conjugate 179

$\mathcal{D}$ (Green's relation) 15
defect of a mapping 3, 230
derivation diagram 74
diagram 74
  group 183
    (source- sink-) reduced 185
  involutary 178
  R- 181–2
  semigroup 183
  over a semigroup 178
  short 170–1
  source-reduced (sink-reduced) 78
  T-minimal 169–70
digraph 69
  complete 69
  functional 70
  strong 69
  weak 70
  weakly connected 70
dipath 69
direct product 2
directed map 177
dominate 143
dominion 143, 160, 165
dual 2
  directional 80
  reflection 80
  semigroup 11, 80

edge(s) 68
  coinitial (coterminal) 77
  negative 69
  oriented 69
  positive 69
egg-box diagram 16
elementary R-transition 7

endomorphism 2, 153
epimorphism 2, 13, 143
epimorphically embedded 143
equational class 10
equivalence(s) 5, 7
  classes 5
  generated by a relation 6
  intersection of 6
Euler–Mascheroni constant 198
eventual regularity 49–52, 67
even-odd offspring 206

factor set 5
free
  group 92, 178
  inverse semigroup 10–12, 81–97, 108
  monogenic inverse semigroup 81, 92
  product with amalgamation 160–1
  product of semigroups 161
  semigroup 9–10, 15
full transformation semigroup 3, 12, 29, 30, 148–9, 154, 157, 166, 187–200, 204–36
full subset 59

generalized inverse semigroup 48, 57, 149–51
generator(s) 3, 10
  and relations 9–11
Gluskin's Lemma 212
graph 68
  acyclic 68
  centre of 69
  left (right) graph 77–9
  radius of 69
gravity of a mapping 223, 230
Green's relations 15–16
  in a regular $\mathcal{D}$-class 19
Green's Lemma 17, 51, 135
group 3, 12
  of automorphisms 5
  embeddable 14, 186
  left (right) 13, 14
  inverse 33
  maximal 18
  Schützenberger 20–1
  symmetric 4
  of units 5, 31
group-bound semigroup 4, 12, 21, 23–8, 36, 48

$\mathcal{H}$ (Green's relation) 15
$\mathcal{H}$-trivial 21
Hasse diagram 24
homomorphism 2
  with involution 15

Index

ideal (left, right) 3
  extension 7, 141
  generated by a set 3
  (0-) minimal 29, 37
  order 98
  (left, right) principal 3, 67, 204
  0-minimal left (right) 35–6
idempotent 1
  central 39
  primitive 33, 38, 42
idempotent-separating congruence 50–3, 57, 58, 61
  maximum 51–3, 57, 67
identity element 1
  right (left) 2
index 4, 73
inflation of a semigroup 15
iterative range 192–7
interface 169, 183
in-tree (out-tree) 70
  root-directed 71
inverse (of an element) 4
  Moore–Penrose 4
  of a relation 7
  strong 212
inverse semigroup 4, 151, 156–7, 165
  E-unitary 98–107
  free 11–12, 15, 81–97, 108
    monogenic 81, 92
  strict 107
  symmetric 4, 30, 45, 105, 154, 157
  0-simple 42
inverse subsemigroup(s) 104, 212, 222
  entire 106
involution 11
Isbell's Zigzag Theorem 144–58
  for commutative semigroups 165
isomorphism 2
  of biordered sets 61, 112
  of Rees semigroups 37
  of word trees 83

$\mathcal{J}$ (Green's relation) 15

kernel
  of a homomorphism 5
  of a mapping 5
  of an orbit of a mapping 72
  of a semigroup 28

$\mathcal{L}$ (Green's relation) 15
label(s)
  (left, right) boundary 74
  group 178
  semigroup 178
  of a walk 83

labelling function 178
  involutary 178
Lallement's Lemma 9, 15, 50
lattice
  of (left, right) congruences 7
  of equivalences 7
  of subalgebras 5
local submonoid(s) 65
  subsemigroup(s) 43

McAlister's P-Theorem 103
McAlister's Second Theorem 106, 108
McAlister triple 97
main permutation 72, 73
minimal conditions 23–8
minimal idempotent generating set 231–6
monogenic semigroup 3, 12
monoid 1
  bicyclic 31, 36, 37, 58
  bisimple 22
  embedding 22
  free 10
monomorphism 2, 13
  of word trees 83

natural mapping 5
nilpotent(s) 36, 236
null semigroup 2, 29

orbit(s) of a mapping 72
  acyclic 223
  cyclic 223, 236
  of general type 223
  singleton 223
orthodox semigroup 9, 32, 42, 58, 59

partial order(s) 44–9, 79
  compatibility with multiplication 49, 65–6
  on Green's factor sets 16
  left (right) partial order on a semigroup 45–6
  natural
    partial order on idempotents 2, 57
    partial order on an inverse semigroup 44–5, 48
    partial order on a regular semigroup 46
    partial order on a semigroup 46–9, 57
partial transformation(s) 4
  semigroup of 4, 157, 166, 212, 223
path 68
  (α, β)- 82
  length of 82
Penrose generalized inverse 4
period 4, 73

periodic semigroup 4, 12, 36
permutation identity 156, 157–8
piece 168
pre-inverse 136, 139
pre-order 16, 60, 218
presentation 10
  cycle-free 77, 78
  of the free inverse semigroup 10
  finite 10
  of a monoid 10
  positive 182
  reduced 176
  of a semigroup 10
Preston–Wagner Theorem 8
principal factor(s) 29, 41, 91
P-semigroup(s) 98–109
  equivalent 108–9

quasi-order 16
quasi-inverse 73

$\mathscr{R}$ (Green's relation) 15
rank 3, 188, 235, 236
receiver 70, 74, 184–5
rectangular band 2, 38, 43
rectangular group 42
regular
  $\mathscr{D}$-class 19
  element 4
  sandwich matrix 35
  semigroup 4
Rees congruence 7
Rees matrix semigroup 34, 41, 42
Rees–Suschkewitsch Theorem 35
region(s) 74
  inversely labelled 183, 184
  symmetrically labelled 169
relation
  compatibility relation on symmetric inverse semigroup 45, 107
  congruence generated by 7
  equivalence 6, 7
  identity 7
  inverse of a 7
  Nambooripad 46
  universal 6
relational product 5
representation(s) 3, 51
  of a biordered set 115
  extended (left, right) regular 3, 34
  (left, right) regular 3
  Preston–Wagner 9
  Scheiblich 92–7, 108
representation extension property 161, 162, 164, 166

right (left) identity 2
root of an in-tree 71
R-word 168

sandwich set(s) 55, 134–41
  eventually non-empty 142
  of an n-tuple of idempotents 136
  of a pair of idempotents 55, 56, 61, 112
  of a pair of elements 64, 66
sandwich matrix 34
  regular 35
saturated subset (of a congruence) 108
Schein left (right) canonical form 95, 108
semiband 56, 59, 121, 123, 135, 137, 236
semidirect product 104
semigroup 1
  abundant 31
  absolutely closed 143
  archimedean 43
  Baer–Levi 14, 27, 37, 157
  of binary relations 5
  (0-) bisimple 16, 22, 36
  Brandt 42, 156, 165
  (left, right) cancellative 3, 12, 29, 79
  closed 143, 162
  C(n) 168–76
  commutative 2, 43, 154
  commutative regular 40
  completely regular 4, 37–41, 72, 157
  completely semisimple 41, 59, 91, 152–3
  completely (0-) simple 33–7, 42, 66, 126, 141
  Croisot–Teissier 28
  dense 143
  diagram 73–80
  dual 11
  E-solid 59
  E-unitary 59
  eventually regular 49–52, 59, 67, 123
  finitely presented 10
  finite 0-simple 34
  free 9–10, 15
  free inverse 10–12, 15, 81–97
  free monogenic inverse 81, 92
  full transformation 3, 12, 29, 148–9, 154, 157, 166, 187–200, 204–36
  fundamental 53, 123
  generalized inverse 48, 57, 149–51
  group-bound 4, 12, 21, 23–38, 36, 48, 59
  idempotent-consistent 50, 123, 124
  inflation of a 15
  inverse, see inverse semigroup
  left (right) isolated 148
  left (right) inner 148
  left (right) reductive 14
  left (right) simple 3, 12, 148
  left (right) zero 2, 65, 166

Index | 257

locally finite 33
locally inverse 48, 57, 66
locally orthodox 67
locally regular 43
locally $\mathscr{L}$- ($\mathscr{R}$-) unipotent 65, 67
locally T–semigroup 43, 65
monogenic 3, 12
M-semigroup 62, 67
nilpotent generated 236
nowhere commutative 43
null 2, 157
of order-preserving mappings 187, 200–2, 203–4
orthodox 9, 32, 42, 58, 59, 67
partial transformation 4, 30, 154, 157, 166, 212, 223
periodic 4, 12
P-semigroup 98–109
regular 4
reductive 14
Rees 34, 41, 42
residually finite 91
saturated 143
semisimple 29, 43
simple 3, 16, 30, 31
of singular mappings 223, 230–1, 233–5
small overlap 168
symmetric inverse, see inverse semigroup
(left, right) T-nilpotent 158
left (right) unipotent 32, 67
weakly cancellative 42
0-simple 28, 36, 42
semilattice 2
congruence 37–8
inflation of a 15
local 43
lower 2
of completely simple semigroups 38
of free inverse semigroup 97
of groups 39–40, 157
of rectangular bands 38
singular mappings 223, 230–1, 233–5
sink 70, 74
solidity of M(e, f) 129
source 70, 74
stable range 71, 72, 192–7
strong semilattice
 of abelian groups 39
 of groups 39
 of semigroups 39
subdiagram 75
subgroup 4
subsemigroup 2
 dense 143
 full 54
 generated by a set 3
 self-conjugate 54
symmetric group 4

symmetric inverse semigroup 4, 45
symmetrized subset 179

tournament 69
trail 68
transition 7
 elementary R- 7
translation (s) (left, right) 3
 linked 20
transmitter 70, 74, 184–5
tree 68
 even (odd) 206
 centre of a 39, 83
 parent 206–9
 word 82
  birooted 88–91

union of groups 3, 33, 66, 157
(left, right) unit 31
(left, right) unitary subset 56–7
universal relation 6
(u, v)-diagram 74

variety 10
 commutative 155–6
 heterotypical 156
vertices 68
 central 69
 depth of 214
 distance between 69
 eccentricity of 69
 extremal 78
 grasp of 216
 height of 214
 hyperbolic 79
 level of 216
 reachable 70
 superfluous 168

walk 68
 (α, β)- 82
 closed 68
 directed 69
 left (right) 78
 length of a 68, 69
 null 68, 82
 open 68
 (proper) segment of 68
 spanning 82
 two-sided 69
Wallis product 201
word(s)
 cyclically reduced 179
 down- (up-) 176
 freely equal 179

word(s) (*cont.*)
  literally equal 178
  positive 178
  reduced 92
  shuffle- 176
  trivial 179
word problem
  for C(3) semigroups 175
  for free inverse semigroup 91
  one-relator 167, 186
word tree 82
  birooted 88–91

0-direct union 26, 42
zero-divisor 33
zero element 1
  left (right) 2
zigzag(s) 144, 162, 165
  equivalent 147
  left-inner (right-inner) 148
  length of 144
  spine of 144
  type I, type II(a), type 11(b) 163–4
  value of 144